An Introduction to
REAL
ANALYSIS

An Introduction to
REAL
ANALYSIS

Ravi P. Agarwal
Cristina Flaut
Donal O'Regan

CRC Press
Taylor & Francis Group
Boca Raton London New York

CRC Press is an imprint of the
Taylor & Francis Group, an **informa** business

A CHAPMAN & HALL BOOK

CRC Press
Taylor & Francis Group
6000 Broken Sound Parkway NW, Suite 300
Boca Raton, FL 33487-2742

© 2018 by Taylor & Francis Group, LLC
CRC Press is an imprint of Taylor & Francis Group, an Informa business

No claim to original U.S. Government works

Printed on acid-free paper
Version Date: 20171215

International Standard Book Number-13: 978-0-8153-9685-7 (Hardback)

Visit the Taylor & Francis Web site at
http://www.taylorandfrancis.com

and the CRC Press Web site at
http://www.crcpress.com

Dedicated to our mothers:

Godawari Agarwal[†]
Elena Paiu[†] and Maria Paiu
Eileen O'Regan[†]

Contents

Preface   xi

1 Logic and Proof Techniques   1

2 Sets and Functions   11

3 Real Numbers   19

4 Open and Closed Sets   29

5 Cardinality   37

6 Real-Valued Functions   45

7 Real Sequences   53

8 Real Sequences (Contd.)   63

9 Infinite Series   73

10 Infinite Series (Contd.)   83

11 Limits of Functions   89

12 Continuous Functions   99

13 Discontinuous Functions   111

14 Uniform and Absolute Continuities and Functions
   of Bounded Variation   119

15 Differentiable Functions   129

16 Higher Order Differentiable Functions   141

17 Convex Functions   151

18 Indeterminate Forms   159

19 Riemann Integration   167

20 Properties of the Riemann Integral   177

21 Improper Integrals   189

22 Riemann-Lebesgue Theorem   197

23 Riemann-Stieltjes Integral   203

24 Sequences of Functions   211

25 Sequences of Functions (Contd.)   217

26 Series of Functions   229

27 Power and Taylor Series   237

28 Power and Taylor Series (Contd.) **243**

29 Metric Spaces **251**

30 Metric Spaces (Contd.) **257**

Bibliography **267**

Index **269**

Preface

Real analysis (traditionally, mathematical analysis, or the theory of functions of real variables) is a branch of mathematics which deals with real numbers and real-valued functions of a real variable. While engineers learn and use calculus (which is mainly based on manipulations), real analysis is considered a rigorous/precise version of calculus. Real analysis also involves several topics which are of interest to only mathematicians.

Although several exceptional books on real analysis have been written, the present rigorous and transparent introductory text can be used directly in class for students of mathematics. In fact, we provide a compact, but thorough, introduction to the subject in *An Introduction to Real Analysis*. This book is intended for senior undergraduate and for beginning graduate one-semester courses. Gifted high school students can also benefit from the style of the serious mathematics discussed in this book.

The subject matter has been organized in the form of theorems and the proofs, and the presentation is rather unconventional. It comprises of 30 class-tested chapters that the authors have given mainly to math major students at various institutions over a period of almost 40 years. It is our belief that the content in a particular chapter, together with the problems therein, provides fairly adequate coverage of the topic under study.

A brief description of the topics covered in this book follows. We begin **Chapter 1** with the definition of mathematical statements, and introduce some logical connectives. We also discuss some commonly used methods to prove mathematical results. In **Chapter 2**, we introduce elementary set theory which is woven into the fabric of modern mathematics. In fact, the language of set theory is used to precisely define nearly all mathematical objects. In **Chapter 3**, we provide the structure of real numbers by means of certain axioms which in turn are used to prove several interesting results. In particular, we show that between any two real numbers there are an infinite number of rational and irrational numbers. The main purpose of **Chapter 4** is to define open and closed subsets of $\mathcal{R}$, and prove the Bolzano-Weierstrass and Heine-Borel theorems which are extremely important results in analysis. In **Chapter 5**, we discuss sizes of finite as well as infinite sets. We show how differently the elements of infinite sets are saturated, which, in particular, show that there are many different sizes of infinite sets. In **Chapter 6**, we define the concept of a function, and introduce several special functions which are of common use, and then define functions which have certain properties. In **Chapters 7 and 8**, we introduce the convergence and divergence of real sequences, prove several properties of convergent sequences, provide Cauchy's if and only if criterion for the convergence of a given sequence, show that it is always possible to extract a convergent sequence from any bounded sequence, thoroughly examine the convergence of monotone sequences, and prove the famous Cesaro-Stolz theorem. In **Chapters 9 and 10**, we provide several easily verifiable tests (comparison test, limit comparison test, ratio test, root test, Kummer's test, Raabe-Duhamel's test, and the logarithmic ratio test), which guarantee the convergence or divergence of a given series. We also provide some specific tests for the convergence of series in terms of arbitrary signs. Some important results on the rearrangements of terms of such series are also discussed.

In **Chapter 11**, the concept of a limit to real functions is introduced. This is one of the most important concepts for the development of analysis. In fact, without properly understanding the limit of a function at a point, one cannot appreciate/understand the later chapters. Among several other results, here we show that one-sided limits always exist for monotone functions. **Chapter 12** begins with several equivalent definitions of continuity of a function at a point which are closely connected with the concept of limits. Then, we show that continuous functions over closed and bounded intervals possess several remarkable properties. In **Chapter 13**, we classify different types of discontinuities (removable, first kind, second kind, mixed, and infinite) and prove some interesting results. In real-world applications, discontinuous functions are at least as important as continuous functions. In **Chapter 14**, we introduce uniform and absolute continuities. Uniform continuity is stronger than continuity, whereas absolute continuity is stronger than uniform continuity. Here we will also discuss functions of bounded variation and prove the famous Jordan decomposition theorem. In **Chapter 15**, we address the differentiability of functions, which is also a stronger concept than continuity. For differentiable functions we prove some major results, such as Rolle's theorem, the mean-value theorem, Cauchy's mean-value theorem, and the Darboux theorem. These results have a wide range of applications. In **Chapter 16**, we introduce higher order differentiable functions, establish Taylor's theorem and use it to provide sufficient conditions for the existence of an extremum. **Chapter 17** deals with the convexity of a function at a point and in an interval, and prove some fundamental properties of convex functions. Convexity of functions plays an important role in a wide variety of applications, especially in probability theory and optimization theory. In **Chapter 18**, we discuss L'Hôpital's rule (originally due to John Bernoulli) which enables us to determine limits of functions that are not only in an indeterminate form $0/0$, but also ∞/∞, and perhaps even $\infty - \infty$, $0 \cdot \infty$, 0^0, ∞^0 and 1^∞.

In **Chapters 19 and 20**, we provide a rigorous treatment of the Riemann integration on bounded intervals. The main purpose of Chapter 19 is to provide necessary and/or sufficient conditions for a function to be Riemann integrable. We show that unbounded functions are not integrable, and also not every bounded function is integrable. We also provide some easily verifiable conditions which ensure the existence of the Riemann integration of functions. In Chapter 20, we study various (mainly algebraic) properties of Riemann integrable functions. Here we prove the fundamental theorem of calculus, the mean-value theorem of integral calculus, Weierstrass's mean-value theorem, and Bonnet's mean-value theorem. We also establish formulas for the chain rule, change of variables, and integration by parts. If we extend the interval or the function is unbounded at a point(s), then the integral is called improper. In **Chapter 21**, we define improper integrals as the limit of Riemann integrals. Here we also discuss absolutely integrable and conditionally integrable functions on an infinite interval. This chapter concludes with Riemann's integral test for series and Cauchy's principal value for improper integrals. In **Chapter 22**, we prove the famous Riemann-Lebesgue theorem which shows that a function f is Riemann integrable on $[a, b]$ if and only if it is almost everywhere continuous, that is, if and only if the set of discontinuities of f on $[a, b]$ has a measure of zero. In **Chapter 23**, we study the Riemann-Stieltjes integral, which is a generalization of the Riemann integral. Instead of dealing with just one function f (integrand), it deals with two functions f (integrand) and g (integrator). The importance of the Riemann-Stieltjes integral over the Riemann integral becomes obvious when the function g is discontinuous. Here we also prove a result which relates the Riemann integral with the Riemann-Stieltjes integral.

In **Chapters 24 and 25**, we study the convergence of sequences of functions. We introduce pointwise and uniform convergences, and through several examples show that pointwise convergence lacks several important properties, provide some necessary and sufficient criteria for uniform convergence, and prove that the deficiencies of pointwise convergence are regained by uniform convergence. In **Chapter 26**, we define pointwise, uniform, and absolute convergence of a series of functions in terms of the sequence of its partial sums. We prove two important results known as term-by-term integration and term-by-term differentiation. For the uniform convergence of a series, we provide easily verifiable tests due to Weierstrass, Abel, and Dirichlet. In **Chapters 27 and 28**, we discuss a special type of series of functions known as the power series, which can be considered as a polynomial of infinite degree. Because of their simplicity in differentiation and integration, power series play a dominate role in approximation theory, especially when finding the solutions of ordinary differential equations. We provide criteria for the convergence and divergence of a power series, and introduce the terms *radius of convergence* and *interval of convergence*. Next, for a given function we find a power series whose sum is exactly the same as the given function. For this, we define analytic functions and develop Taylor's series and its particular case, Maclaurin's series.

In **Chapters 29 and 30**, we extend some topological properties inherited by the real line to a general setting. We begin with the definition of a metric space, and show that several results established in previous chapters carry over to an arbitrary metric space. As applications, we state the famous Weierstrass approximation theorem, and prove one of the most widely used results known as Banach's contraction mapping principle. We conclude the book with an application of this result to the solution of an initial value problem.

In this book, there are 122 examples which explain each concept and demonstrate the importance of every result. Two types of the 289 problems are also included, those that illustrate the general theory and others designed to complete our text material. The problems form an integral part of the book, and every reader is urged to attempt most, if not all of them. For the convenience of the reader, we have provided answers or hints to all the problems.

In writing a book of this nature, no originality can be claimed, only a humble attempt has been made to present the subject as simply, clearly, and accurately as possible. The illustrative examples are usually very simple, keeping the average student in mind.

It is earnestly hoped that *An Introduction to Real Analysis* will serve the inquisitive reader as a starting point in this rich, vast, and ever-expanding field of knowledge.

We would like to express our appreciation to Professors Bashir Ahmad (Saudi Arabia), Ferhan M. Atici (USA), Mircea Balaj (Romania), Leonid Berezansky (Israel), Alberto Cabada (Spain), Claudio Cuevas (Brazil), Manuel De la Sen (Spain), Alexander I. Domoshnitsky (Israel), Marléne Frigon (Canada), Erdal Karapinar (Turkey), Anthony T. Lau (Canada), Carlos Lizama (Chile), Giuseppe Marino (Italy), Gradimir V. Milovanovic (Serbia), M. Mursaleen (India), Juan Jose Nieto (Spain), Sotiris Ntouyas (Greece), Hemant Pathak (India), Adrian Petrusel (Romania), Bessem Samet (Saudi Arabia), Jozsef Sandor (Romania), Sin E. Takahasi (Japan), Patricia Wong (Singapore), and Alexander Zaslavski (Israel) for their suggestions and criticisms. We also thank Aastha Sharma at CRC Press/Taylor & Francis, Delhi Office, for her support and cooperation.

Ravi P. Agarwal
Cristina Flaut
Donal O'Regan

Chapter 1

Logic and Proof Techniques

We begin this chapter with the definition of mathematical statements, and introduce some logical connectives that we will use frequently in this book. We will also discuss some commonly used methods to prove mathematical results.

By a mathematical *statement* or *proposition*, we mean an unambiguous composition of words that is true or false. For example, two plus two is four is a true statement, and two plus three is seven is a false statement. However, $x - y = y - x$ is not a proposition, because the symbols are not defined. If $x - y = y - x$ for all x, y real numbers, then this is a false proposition; if $x - y = y - x$ for some real numbers, then this is a true proposition. Help me please, and your place or mine—are also not statements. A single letter is always used to denote a statement. For example, the letter p may be used for the statement eleven is an even number. Thus, $p : 11$ is an even number. A statement is said to have *truth value T* or *F* according as the statement is true or false. For example, the truth value of $p : 1 + 2 + \cdots + 10 = 55$ is T, whereas for $p : 1^2 + 2^2 + 3^2 = 15$ is F. The knowledge of truth value of a statement enables us to replace it by some other "equivalent" statement. From given statements, new statements can be produced by using the following standard logical connectives:

1. *Negation*, $\sim$: If p is a statement, then its negation $\sim p$ is the statement not p. The truth value of $\sim p$ is F or T according as the truth value of p is T or F. Thus, if p : seven is even number, then $\sim p$: seven is not an even number, or seven is an odd number.

2. *Implication*, $\Rightarrow$: If from a statement p another statement q follows, we say p implies q and write $p \Rightarrow q$. The truth value of $p \Rightarrow q$ is F only when p has truth value T and q has the truth value F. For example, $x = 7 \Rightarrow x^2 = 49$. If n is an even integer, then $n + 1$ is an odd integer.

3. *Conjuction*, $\wedge$: The statement p and q is denoted as $p \wedge q$ and is called the conjunction of the statements p and q. The truth value of $p \wedge q$ is T only when both p and q are true. For example, the statement 4 is positive and -7 is negative can be viewed as a conjuction of two statements $p : 4$ is positive and $q : -7$ is negative.

1

4. *Disjunction,* $\vee$: The statement p or q is denoted as $p \vee q$ and is called the disjunction of the statements p and q. The truth value of $p \vee q$ is F only when both p and q are false. For example, if p : Scott is a member of the financial committee, and q : Scott is a member of the executive committee, then $p \vee q$: Scott is a member of the financial committee or of the executive committee.

Two statements p and q are said to be *equivalent* if one implies the other, i.e., $(p \Rightarrow q) \wedge (q \Rightarrow p)$, and we denote this as $p \Leftrightarrow q$. Propositions which involve the phrases such as if and only if (iff), is equivalent to, or the necessary and sufficient condition, are of the type $p \Leftrightarrow q$. For example, ABC is an equilateral triangle $\Leftrightarrow AB = BC = CA$.

The following table gives the truth values of different compositions of statements. Tables of this type are called *truth tables.*

p	q	$\sim p$	$\sim q$	$p \Rightarrow q$	$p \wedge q$	$p \vee q$	$p \Leftrightarrow q$
T	T	F	F	T	T	T	T
T	F	F	T	F	F	T	F
F	T	T	F	T	F	T	F
F	F	T	T	T	F	F	T

Truth tables can be used to decide whether or not two compound statements are logically equivalent. For example, it can easily be seen that $p \Rightarrow q$ is equivalent to $\sim q \Rightarrow \sim p$. The implication $\sim q \Rightarrow \sim p$ is called *contrapositive* of $p \Rightarrow q$.

Some statements involve a variable such as x, or n. These statements are usually written as $p(x)$, or $p(n)$. For example, $p(x) : x^2 - 3x + 2 = 0$ is a statement for every specific real value of x. In fact, the truth value of $p(1)$ and $p(2)$ is T, whereas for each real $x \neq 1, 2$ the truth value of $p(x)$ is F. As an another example, $p(n) : 1 + 2 + \cdots + n = n(n+1)/2$ is a statement for every positive integer (natural number) n, and its truth value is T. Often, for the phrases like for all$\cdots$, for each$\cdots$, for every$\cdots$, we use the *universal quantifier* $\forall$. Thus, the statement for every real x, $x^2 - 3x + 2 = 0$, which is false, in symbols can be written as $\forall$ real x, $p(x)$. Similarly, the statement for every natural number n, $1 + 2 + \cdots + n = n(n+1)/2$ can be written as $\forall$ natural number n, $p(n)$. For the phrases such as there exists$\cdots$, there is at least one$\cdots$, we use the *existential quantifier* $\exists$. The symbol $\ni$ is often used in mathematics for the phrase such that. Thus, the statement there exists a real x such that $x^2 - 3x + 2 = 0$, which is true, in symbols can be written as $\exists$ real $x \ni p(x)$.

Basic statements which do not contradict themselves to the best of human knowledge are called *axioms*, or *postulates*. Axioms are accepted and used without a doubt. Obvious or self-evident axioms are preferred. A mathematical theory is based on axioms. For example, the *axiom of Archimedes* states that for any given real number x there exists an integer n such that n is greater than x.

A *theorem* consists of some propositions $H_1, H_2, \cdots, H_n$ called *hypotheses*, and a proposition C called its *conclusion*. A theorem is true provided $H_1 \wedge H_2 \wedge \cdots \wedge H_n \Rightarrow C$. A *formal proof* of a theorem consists of a sequence of propositions, ending with the conclusion C, that are regarded *valid*. To be valid a proposition may be one of the hypotheses, may be derived or inferred from the earlier propositions. A formal proof with a valid sequence or propositions is called a *valid proof*. Even if one of the propositions is invalid, then the argument is called a *fallacy*. A *lemma* is a simple theorem which is used to prove the main theorem. Often, complicated proofs are easier to understand when they are established using a series of lemmas. A *corollary* is a proposition that can be established directly from a theorem that has been proved.

We will now discuss some commonly used methods to prove mathematical results.

(1). *Direct proof.* The most natural proof is the direct proof in which the hypotheses $H_1, H_2, \cdots, H_n$ are shown to imply the conclusion C, i.e., $H_1 \wedge H_2 \wedge \cdots \wedge H_n \Rightarrow C$. For example, consider the *Pythagorean Theorem.* If a and b are the lengths of the two legs of a right triangle, and c is the length of the hypothenuse, then $a^2 + b^2 = c^2$. The following direct proof was discovered by U.S. Representative James A. Garfield (1831–1881), 5 years before he became the 20th president of the United States. Figure 1.1 shows three triangles forming half of a square with sides of length $a + b$. The angles A, B, and D satisfy the relations

$$A + B = 90° \quad \text{and} \quad A + B + D = 180°.$$

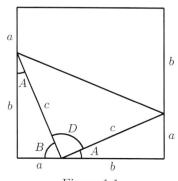

Figure 1.1

Thus, $D = 90°$, and hence all the three triangles are right triangles. The area of the half square is

$$\frac{1}{2}(a+b)^2 = \frac{1}{2}(a^2 + 2ab + b^2),$$

while the equivalent total area of the three triangles is

$$\frac{1}{2}ab + \frac{1}{2}c^2 + \frac{1}{2}ab.$$

Equating these two expressions, we get

$$a^2 + 2ab + b^2 = ab + c^2 + ab \quad \text{or} \quad a^2 + b^2 = c^2.$$

Sometimes it is convenient to convert the proposition $H_1 \vee H_2 \vee \cdots \vee H_n \Rightarrow C$ to its equivalent form $(H_1 \Rightarrow C) \wedge (H_2 \Rightarrow C) \wedge \cdots \wedge (H_n \Rightarrow C)$. Thus, a result can be proved by considering several *cases*. For example, we shall show that for every natural number n, $n^3 + n$ is even. For this we consider the following two cases:

(i).   Suppose n is even. Then, $n = 2k$ for some natural number k, and so $n^3 + n = 8k^3 + 2k = 2(4k^3 + k)$, which is even.

(ii).   Suppose n is odd. Then, $n = 2k + 1$ for some natural number k, and so $n^3 + n = (8k^3 + 12k^2 + 6k + 1) + (2k + 1) = 2(4k^3 + 6k^2 + 4k + 1)$, which is even.

However, the following direct proof is more elegant. Clearly, $n^3 + n = n(n^2 + 1)$. If n is even, then $n(n^2 + 1)$ is even. If n is odd, then n^2 is odd, and hence $n^2 + 1$ is even, and so $n(n^2 + 1)$ is even.

(2). *Indirect proof.* In this type of proof an equivalent statement is used to arrive at the result. Indirect proofs are of two different types. In the first type we use the *contrapositive statement* $\sim C \Rightarrow \sim (H_1 \wedge H_2 \wedge \cdots \wedge H_n)$, whereas the second type is a proof by *contradiction*, i.e., $\sim C \wedge H_1 \wedge H_2 \wedge \cdots \wedge H_n \Rightarrow$ a contradiction.

As an example of the first type of proof, we shall show that if $n + m \geq 53$, then $n \geq 27$ or $m \geq 27$. For this the contrapositive statement is $\text{not}(n \geq 27$ or $m \geq 27)$ implies $\text{not}(n+m \geq 53)$. Since $\text{not}(n \geq 27$ or $m \geq 27)$ is equivalent to $\text{not}(n \geq 27)$ and $\text{not}(m \geq 27)$, i.e., $(n \leq 26)$ and $(m \leq 26)$, the contrapositive statement is the same as if $n \leq 26$ and $m \leq 26$, then $n + m \leq 52$. But this statement immediately follows from the general property of inequalities: for all real numbers a, b, c, d if $a \leq c$ and $b \leq d$, then $a + b \leq c + d$.

To illustrate the method of proof by contradiction first we shall show that if n^2 is even for some integer n, then n itself is even. Suppose for contradiction that n is odd. Then, $n = 2m + 1$ for some integer m. But, then $n^2 = 4m^2 +$

$4m + 1$ is odd, which contradicts that n^2 is even. Now we shall show that $\sqrt{2}$ is irrational. Suppose for contradiction that $\sqrt{2}$ is rational. Then, $\sqrt{2} = p/q$ where p and q are natural numbers. By reducing the fraction if necessary we can assume that p and q have no common factors. In particular, p and q are not both even. Since $2 = p^2/q^2$ we have $p^2 = 2q^2$ and so p^2 is even. This implies that p is even. Hence, $p = 2k$, where k is some integer. But then, $(2k)^2 = 2q^2$ and therefore $q^2 = 2k^2$. Thus, q^2 and hence q is also even. This implies that both p and q are both even, which contradicts our earlier statement. Hence, $\sqrt{2}$ is irrational.

(3). *Mathematical induction.* To establish the truth of an infinite list of propositions the *principle of mathematical induction* is frequently used. This method can be described as follows: All the propositions $p(m), p(m+1), \cdots$, where $m \geq 0$ is an integer, are true provided (a) $p(m)$ is true, and (b) $p(n)$ implies $p(n+1)$ for all $n \geq m$. The first step (a) is called the *basis of induction*, whereas the second step (b) is called the *inductive step*. To illustrate this method, we will show that for all integers $n \geq 3$ the following inequality holds:

$$n > \left(1 + \frac{1}{n}\right)^n. \tag{1.1}$$

Let $p(n) : n > (1 + 1/n)^n$, $n \geq 3$. Clearly, $p(3)$ is true. Now suppose that $p(n)$ is true for a given natural number $n \geq 3$. Then, from (1.1), we have

$$n + 1 > \left(1 + \frac{1}{n}\right)^n + 1 > \left(1 + \frac{1}{n+1}\right)^n + 1 \tag{1.2}$$

and

$$n + 1 > n > \left(1 + \frac{1}{n+1}\right)^n,$$

which implies

$$1 > \frac{1}{n+1}\left(1 + \frac{1}{n+1}\right)^n. \tag{1.3}$$

Combining (1.2) and (1.3), we get

$$n + 1 > \left(1 + \frac{1}{n+1}\right)^n + \frac{1}{n+1}\left(1 + \frac{1}{n+1}\right)^n = \left(1 + \frac{1}{n+1}\right)^{n+1}.$$

Hence, $p(n+1)$ holds whenever $p(n)$ holds. By the principle of mathematical induction, we conclude that $p(n)$ is true for all $n \geq 3$.

In mathematical analysis we also mention *constructive proofs* and *nonconstructive proofs*. A constructive proof either specifies the solution or indicates how it can be determined by some algorithm. As an example of constructive proofs, we shall show that there is a positive integer that can be written as the sum of cubes of positive integers in two different ways. Indeed after considerable computation we find that $1729 = 10^3 + 9^3 = 12^3 + 1^3$. A nonconstructive proof establishes the existence of a solution, but gives no indication how to find it. As an example of nonconstructive proofs, we shall show that there exist irrational numbers a and b such that a^b is rational. Consider the irrational numbers $a = b = \sqrt{2}$. If the number $a^b = \sqrt{2}^{\sqrt{2}}$ is rational, we are done. If $\sqrt{2}^{\sqrt{2}}$ is irrational, we consider the numbers $a = \sqrt{2}^{\sqrt{2}}$ and $b = \sqrt{2}$ so that

$$a^b = \left(\sqrt{2}^{\sqrt{2}}\right)^{\sqrt{2}} = \sqrt{2}^{\sqrt{2}\sqrt{2}} = \sqrt{2}^2 = 2$$

is rational. Note that in this proof we could not find irrational numbers a and b such that a^b is rational.

An unproved statement based on guess or opinion, preferably based on some experience or other source of wisdom is called a *conjecture*. Once a conjecture is proved it becomes a theorem. Among all the known conjectures the most famous is the Goldbach conjecture: Every even integer greater than 2 is the sum of two prime numbers. For example, $30 = 7+23 = 11+19 = 13+17$. This conjecture was made in the year 1742. Although it has been verified for all even numbers up to 10^{14}, for every even integer it remains to be proved. Now consider the function $f(n) = n^2 - 79n + 1601$ for all natural numbers n. If we compute $f(n)$ for various values of n, we always seem to get a prime number, e.g., $f(1) = 1523$, $f(2) = 1447$, $f(39) = f(40) = 41$. Thus, based on this experience we might make the conjecture that for each natural number n, $p(n) : n^2 - 79n + 1601$ is a prime number. However, this statement cannot be proved because it happens to be false. In fact, while it is true for $n = 1, 2, \cdots, 79$ for $n = 80$, $p(80) = 80^2 - 79 \times 80 + 1601 = 1681 = 41^2$. An example of this type which shows that the statement is false is called a *counterexample*. Thus, to disprove a statement only one counterexample is sufficient.

Problems

1.1. Use mathematical induction to prove that for all natural numbers n:

(i). $n^5 - n$ is a multiple of 5.

(ii). $n^7 - n$ is a multiple of 7.

1.2. For all natural numbers n is $3^{2n} - 8n + 1$ divisible by 64?

1.3. Prove the following by induction.

(i). $2^n > n^2$ for all integers $n \geq 5$.

(ii). $2^n \geq n^3$ for all integers $n \geq 10$.

(iii). $(4/3)^n > n$ for all integers $n \geq 7$.

(iv). $\frac{1}{1\cdot2} + \frac{1}{2\cdot3} + \cdots + \frac{1}{n(n+1)} = \frac{n}{n+1}$ for all natural numbers n.

(v). $1^3 + 2^3 + \cdots + n^3 = (1 + 2 + \cdots + n)^2$ for all natural numbers n.

(vi). $1 + r + r^2 + \cdots + r^n = (1 - r^{n+1})/(1 - r)$, $r \neq 1$ for all natural numbers n.

1.4. The *Fibonacci numbers* F_n are recursively defined as follows:

$$F_1 = F_2 = 1, \quad F_n = F_{n-1} + F_{n-2} \quad \text{for} \quad n \geq 3.$$

Use induction to prove that for all natural numbers n:

(i). $F_{m+n} = F_{m-1}F_n + F_m F_{n+1}$ for all $m \geq 2$.

(ii). $F_1 + F_2 + \cdots + F_n = F_{n+2} - 1$.

(iii). $F_1 + 2F_2 + \cdots + nF_n = (n+1)F_{n+2} - F_{n+4} + 2$.

1.5. Use mathematical induction to show that for all natural numbers n:

$$S_n^{(1)} = 1 + 2 + \cdots + n = \frac{n(n+1)}{2}.$$

$$S_n^{(2)} = S_1^{(1)} + S_2^{(1)} + \cdots + S_n^{(1)} = \frac{n(n+1)(n+2)}{2 \cdot 3}.$$

$$S_n^{(3)} = S_1^{(2)} + S_2^{(2)} + \cdots + S_n^{(2)} = \frac{n(n+1)(n+2)(n+3)}{2 \cdot 3 \cdot 4}.$$

$$\cdots$$

$$S_n^{(k)} = S_1^{(k-1)} + S_2^{(k-1)} + \cdots + S_n^{(k-1)} = \frac{n(n+1)\cdots(n+k)}{2 \cdot 3 \cdots (k+1)}.$$

1.6. For all real numbers x, y and natural number n, prove the *binomial theorem*

$$(x+y)^n = \sum_{r=0}^{n} \binom{n}{r} x^{n-r} y^r \quad \text{where} \quad \binom{n}{r} = \frac{n!}{r!(n-r)!}.$$

1.7. Let $a_1, a_2, \cdots, a_n$ be positive numbers strictly less than 1. Show that for integers $n \geq 2$,

$$(1 - a_1)(1 - a_2) \cdots (1 - a_n) > 1 - (a_1 + a_2 + \cdots + a_n).$$

1.8. If $x \geq -1$ and n is a natural number, prove *Bernoulli's inequality*

$$(1 + x)^n \geq 1 + nx.$$

When does the strict inequality hold? Deduce for $n > 1$ that

(i). $\left(1 - \frac{1}{n^2}\right)^n > 1 - \frac{1}{n}$.

(ii). $\left(\frac{n+1}{n}\right)^n > \left(\frac{n}{n-1}\right)^{n-1}$.

1.9. Prove by induction

(i). $(1+x)^n \geq 1 + nx + \frac{n(n-1)}{2}x^2$ if $x \geq 0$.

(ii). $(1+x)^n \geq 1 - nx + \frac{n(n-1)}{2}x^2$ if $0 \leq x < 1$.

1.10. For the given positive numbers $a_1, a_2, \cdots, a_n$ the *arithmetic mean* (A.M.), and the *geometric mean* (G.M.), respectively, are defined as A.M.$= (a_1 + a_2 + \cdots + a_n)/n$ and G.M.$= \sqrt[n]{a_1 a_2 \cdots a_n}$. Show that G.M.$\leq$ A.M., i.e.,

$$\sqrt[n]{a_1 a_2 \cdots a_n} \leq \frac{a_1 + a_2 + \cdots + a_n}{n}.$$

Answers or Hints

1.1. (i). Let $p(n)$ be the statement "$n^5 - n$ is a multiple of 5." Clearly, $p(1)$ is true since $1^5 - 1 = 0 = 5 \times 0$. Assume $p(n)$ is true for a given natural number n, i.e., $n^5 - n = 5k$ for some integer k. Then, $(n+1)^5 - (n+1) = (n^5 - n) + 5(n^4 + 2n^3 + 2n^2 + n) = 5(k + n^4 + 2n^3 + 2n^2 + n)$. Hence, $p(n)$ is true for all $n \geq 1$.

(ii). Similar to (i).

1.2. Let $p(n) = 3^{2n} - 8n + 1$, then $p(n+1) - p(n) = 64(9^{n-1} + 9^{n-2} + \cdots + 9 + 1)$. Thus, $p(n+1)$ is divisible by 64 implies $p(n)$ is divisible by 64. However, $p(2) = 66$.

1.3. (i). Let $p(n) : 2^n > n^2$. Then, $p(5)$ is true. Suppose $p(k)$ is true for some $k \geq 5$. Then, $2^{k+1} = 2^k \cdot 2 > 2k^2 > (k+1)^2$ since $k \geq 5$. Thus, $p(k+1)$ is true.

(ii). Let $p(n) : 2^n > n^3$. Then, $p(10)$ is true. Suppose $p(k)$ is true for some $k \geq 10$. Then, $2^{k+1} = 2^k + 2^k > k^3 + k^3 \geq k^3 + k^2 \left(3 + \frac{3}{k} + \frac{1}{k^2}\right)$ because $3 + \frac{3}{k} + \frac{1}{k^2} < k$. This implies $2^{k+1} > k^3 + 3k^2 + 3k + 1 = (k+1)^3$.

(iii). Similar to (i).

(iv). Let $p(n) : \frac{1}{1 \cdot 2} + \cdots + \frac{1}{n(n+1)} = \frac{n}{n+1}$. Then, $p(1)$ is true. Suppose $p(k)$ holds, then $\frac{1}{1 \cdot 2} + \cdots + \frac{1}{k(k+1)} + \frac{1}{(k+1)(k+2)} = \frac{k}{k+1} + \frac{1}{(k+1)(k+2)} = \frac{k+1}{k+2}$.

(v). We shall use $1 + 2 + \cdots + n = n(n+1)/2$. Let $p(n) : 1^3 + \cdots + n^3 = (1 + \cdots + n)^2$. Clearly, $p(1)$ is true. Suppose $p(k)$ is true, then $1^3 + \cdots + k^3 + (k+1)^3 = (1 + \cdots + k)^2 + (k+1)^3 = (1 + \cdots + k)^2 + (k+1)^2 + k(k+1)(k+1) = (1 + \cdots + k)^2 + (k+1)^2 + 2(k+1)(1 + \cdots + k) = (1 + \cdots + (k+1))^2$.

(vi). Let $p(n) : 1 + r + \cdots + r^n = (1 - r^{n+1})/(1 - r)$. Clearly, $p(1)$ is true. Suppose $p(k)$ is true, then $1 + r + \cdots + r^k + r^{k+1} = \frac{1 - r^{k+1}}{1 - r} + r^{k+1} = \frac{1 - r^{k+2}}{1 - r}$.

1.4. (i). For a fixed integer $m \geq 2$, we shall use induction on n. For $n = 1$, $F_{m+1} = F_{m-1} + F_m$. Asuume that $F_{m+n} = F_{m-1}F_n + F_m F_{n+1}$ for $1 \leq n \leq k$. Then, $F_{m+k+1} = F_{m+k} + F_{m+(k-1)} = (F_{m-1}F_k + F_m F_{k+1}) + (F_{m-1}F_{k-1} + F_m F_k) = F_{m-1}(F_k + F_{k-1}) + F_m(F_{k+1} + F_k) = F_{m-1}F_{k+1} + F_m F_{k+2}$.

(ii). $F_1 = F_3 - 1$. Assume $F_1 + \cdots + F_k = F_{k+2} - 1$, then $F_1 + \cdots + F_k + F_{k+1} = F_{k+1} + F_{k+2} - 1 = F_{k+3} - 1$.

(iii). $F_1 = 2F_3 - F_5 + 2 = 2 \times 2 - 5 + 2 = 1$. Assume $F_1 + 2F_2 + \cdots + kF_k = (k+1)F_{k+2} - F_{k+4} + 2$, then $F_1 + 2F_2 + \cdots + kF_k + (k+1)F_{k+1} = (k+1)\{F_{k+1} + F_{k+2}\} - F_{k+4} + 2 = (k+1)F_{k+3} - (F_{k+5} - F_{k+3}) + 2 = (k+2)F_{k+3} - F_{k+5} + 2$.

1.5. Use induction.

1.6. Use $\binom{n}{r} + \binom{n}{r-1} = \binom{n+1}{r}$.

1.7. Let $p(n) : (1 - a_1) \cdots (1 - a_n) > 1 - (a_1 + \cdots + a_n)$. When $n = 2$,

$(1-a_1)(1-a_2) = 1-(a_1+a_2)+a_1a_2 > 1-(a_1+a_2)$. Hence, $p(2)$ is true. Suppose $p(k)$ is true for some $k \geq 2$. Then,

$$(1-a_1)\cdots(1-a_k)(1-a_{k+1}) > [1-(a_1+\cdots+a_k)](1-a_{k+1})$$
$$= 1-(a_1+\cdots+a_{k+1})+a_{k+1}(a_1+\cdots+a_k)$$
$$> 1-(a_1+\cdots+a_{k+1}).$$

1.8. The inequality is obvious for $x = -1$. For $x > -1$, let $p(n)$ be $(1+x)^n \geq 1+nx$. Clearly, $p(1)$ is true. Suppose $p(k)$ is true for some $k \geq 1$, then $(1+x)^{k+1} \geq (1+kx)(1+x) \geq 1+(k+1)x$. If $n > 1$ and $x \neq 0$ the strict inequality, i.e., $(1+x)^n > 1+nx$ holds.

(i). Take $x = -1/n^2$, $n > 1$.
(ii). From (i), $\left(1+\frac{1}{n}\right)^n \left(1-\frac{1}{n}\right)^n > \left(1-\frac{1}{n}\right)$, $n > 1$.

1.9. Use induction.

1.10. Let $p(n): \sqrt[n]{a_1a_2\cdots a_n} \leq (a_1+a_2+\cdots+a_n)/n$. Clearly, $p(1)$ is true. Suppose $p(k)$ holds, to show $p(k+1)$, if necessary, we arrange the numbers $a_1, a_2, \cdots, a_{k+1}$ in increasing order, and let $A = \frac{a_1+\cdots+a_{k+1}}{k+1}$. Clearly, $(a_{k+1} - A)(A - a_1) \geq 0$. Let $R = a_1 + a_{k+1} - A \geq 0$. Consider, $a_1, a_2, \cdots, a_{k-1}, R$, so that $kA = a_1 + a_2 + \ldots + a_k + a_{k+1} - A = a_2 + \ldots + a_k + R$. Hence, $A^{k+1} = A^k A \geq a_2 \cdots a_k RA$. Finally, since $RA - a_1a_{k+1} = (a_1+a_{k+1}-A)A - a_1a_{k+1} = (A-a_1)(a_{k+1}-A) \geq 0$, it follows that $RA \geq a_1a_{k+1}$. Thus, we have $A^{k+1} \geq a_1 \cdots a_{k+1}$.

Chapter 2

Sets and Functions

In this chapter we will introduce elementary set theory. Although, this theory as an independent branch of mathematics was founded by the German mathematician George Cantor (1845–1918) in the second half of the 19th century, it turned out to be the unifying foundation to almost all branches of mathematics.

A *set* is a collection of objects distinct and distinguishably defined. Often, *aggregate*, *class*, *family* or *collection* are also used for a set. The objects in a set are called *elements*, *points* or *members* of the set. We denote sets by capital letters like A, B, C, etc., and use lowercase letters a, b, c, etc., to represent elements. If A is a set and a is an element of it, then it is customary to write this fact in notation as $a \in A$. If a does not belong to the set A, then we write it as $a \notin A$. A set is defined by its elements, and hence, either it is expressed by its elements or by assigning a certain property by which all the members are known. It is often useful to denote a set by putting braces around its elements. The elements within the braces can be written in any order. For example, $A = \{a, b, c, d\}$ denotes the set consisting four elements a, b, c, and d. Clearly, $b \in A$ whereas $z \notin A$. In case a set S is defined by assigning a certain property $p(s)$ regarding its elements s, we write $S = \{s : p(s)\}$. The following sets must be familiar to the reader:

$$\mathcal{N} = \{x : x = 1, 2, \cdots\}, \text{ the set of natural numbers,}$$
$$\mathcal{Z} = \{x : x = \cdots, -2, -1, 0, 1, 2, \cdots\}, \text{ the set of all integers,}$$
$$\mathcal{Q} = \{x : x = p/q, \ p, q \in \mathcal{Z} \text{ and } q \neq 0\}, \text{ the set of rational numbers,}$$
$$\mathcal{R} = \{x : x \text{ is a real number}\}, \text{ the set of real numbers.}$$

A set B is said to be a *subset* of a set A if every member of B is also a member of A. In such a case we write $B \subseteq A$ and say that B is contained in A. If $B \subseteq A$ and $A \subseteq B$, then we write $A = B$ and say that two sets A and B are *equal*. A *proper subset* of A is a subset B such that $A \neq B$ and we designate this by $B \subset A$. Clearly,

$$\mathcal{N} \subset \mathcal{Z} \subset \mathcal{Q} \subset \mathcal{R}.$$

11

For a, $b \in \mathcal{R}$ with $a < b$, the special subsets of $\mathcal{R}$, called *intervals* are defined as

$$[a, b] = \{x \in \mathcal{R} : a \leq x \leq b\}, \qquad (a, b) = \{x \in \mathcal{R} : a < x < b\}$$
$$[a, b) = \{x \in \mathcal{R} : a \leq x < b\}, \qquad (a, b] = \{x \in \mathcal{R} : a < x \leq b\}.$$

The set $[a, b]$ is called a *closed interval*, the set (a, b) is called an *open interval*, and the sets $[a, b)$ and $(a, b]$ are called *half-open* or *half-closed* intervals. We will also use the term *interval* for some unbounded subsets of $\mathcal{R}$:

$$[a, \infty) = \{x \in \mathcal{R} : a \leq x\}, \qquad (a, \infty) = \{x \in \mathcal{R} : a < x\}$$
$$(-\infty, b] = \{x \in \mathcal{R} : x \leq b\}, \qquad (-\infty, b) = \{x \in \mathcal{R} : x < b\}.$$

Using this notation, we can write $\mathcal{R} = (-\infty, \infty)$. Although, $\mathcal{R}$ has no endpoint, by convention it is accepted as both open and closed. The intervals $[a, \infty)$ and $(-\infty, b]$ have only one endpoint and therefore are accepted as closed.

The *union* of two sets A and B denoted as $A \cup B$ is the set of all elements which belong to A or B (or both), i.e.,

$$A \cup B = \{x : x \in A \ \text{ or } \ x \in B\}.$$

The *intersection* of two sets A and B denoted as $A \cap B$ is the set of all elements which belong to A and B, i.e.,

$$A \cap B = \{x : x \in A \ \text{ and } \ x \in B\}.$$

The *complement* of a set B with respect to a set A denoted as $A \backslash B$ is the set of elements which belong to A but not to B, i.e.,

$$A \backslash B = \{x : x \in A \ \text{ and } \ x \notin B\}.$$

A set which has no elements is called an *empty set* and we denote it by $\emptyset$, e.g., $\{x \in \mathcal{R} : x^2 + 1 = 0\}$. It is clear that the empty set is unique, and for any given set A, $\emptyset \subseteq A$. Two sets A and B are said to be *disjoint* if they have no common element, i.e., $A \cap B = \emptyset$. The largest set which contains all elements likely to be considered during some specified mathematical treatment is called the *universal set*, and we denote it by U. Often, we write $U \backslash A$ as A^c, and call it the complement of the set A.

The following simple laws for sets can be proved rather easily.

1. Idempotent laws:
(i). $A \cup A = A$, (ii). $A \cap A = A$.

2. Commutative laws:
(i). $A \cup B = B \cup A$, (ii). $A \cap B = B \cap A$.

3. Associative laws:

(i). $(A \cup B) \cup C = A \cup (B \cup C)$, (ii). $(A \cap B) \cap C = A \cap (B \cap C)$.

4. Distributive laws:

(i). $A \cup (B \cap C) = (A \cup B) \cap (A \cup C)$, (ii). $A \cap (B \cup C) = (A \cap B) \cup (A \cap C)$.

5. Identity laws:

(i). $A \cup \emptyset = A$, $A \cup U = U$, (ii). $A \cap \emptyset = \emptyset$, $A \cap U = A$.

6. Complement laws:

(i). $A \cup A^c = U$, $(A^c)^c = A$, (ii). $A \cap A^c = \emptyset$, $U^c = \emptyset$.

7. De Morgan laws:

(i). $(A \cup B)^c = A^c \cap B^c$, (ii). $(A \cap B)^c = A^c \cup B^c$.

Now let A and B be sets, and suppose that with each element a of A there is associated a subset of B which we denote by C_a. The set whose elements are the sets C_a is denoted by $\{C_a\}$ and is called a *family* of sets. The union and intersection of these sets C_a is defined by

$$S = \cup \{C_a : a \in A\} = \{x : x \in C_a \text{ for some } a \in A\}$$

and

$$P = \cap \{C_a : a \in A\} = \{x : x \in C_a \text{ for all } a \in A\}.$$

If $A = \{1, 2, \cdots, n\}$, then we write

$$S = \cup_{m=1}^{n} C_m \quad \text{and} \quad P = \cap_{m=1}^{n} C_m,$$

and if $A = \mathcal{N}$, the usual notation is

$$S = \cup_{m=1}^{\infty} C_m \quad \text{and} \quad P = \cap_{m=1}^{\infty} C_m.$$

De Morgan laws hold for arbitrary unions and intersections also, i.e.,

$$(\cup \{C_a : a \in A\})^c = \cap \{C_a : a \in A\}^c \tag{2.1}$$

and

$$(\cap \{C_a : a \in A\})^c = \cup \{C_a : a \in A\}^c. \tag{2.2}$$

The *Cartesian product* or *cross product* of two sets A and B denoted by $A \times B$ is defined as the set of all *ordered pairs* (a, b) with $a \in A$ and $b \in B$, i.e.,

$$A \times B = \{(a, b) : a \in A \text{ and } b \in B\}.$$

Two points (a, b), $(c, d) \in A \times B$ are said to be *equal* if $a = c$ and $b = d$. Clearly, the xy-plane can be considered as the Cartesian product $\mathcal{R} \times \mathcal{R}$. Similarly, the set of all ordered n elements $(x_1, x_2, \cdots, x_n)$ with $x_1 \in A_1, x_2 \in A_2, \cdots, x_n \in A_n$ is called an *ordered n-tuple* and is written as $A_1 \times A_2 \times \cdots \times A_n$.

Let A and B be sets. A *relation* on $A \times B$ is any subset R of $A \times B$. The *domain* of R is the collection of $a \in A$ such that $(a, b) \in R$. When $(a, b) \in R$ we usually write aRb. If $B = A$, then the relation $R \subseteq A \times A$ is called a *relation on A.* A relation R on a set A is called an *equivalence relation* if for all $a, b, c \in A$ the following properties hold:

1. aRa, reflexive property.
2. If aRb, then bRa symmetric property.
3. If aRb and bRc, then aRc transitive property.

Let A and B be sets. A *function* between A and B is a nonempty relation $f \subseteq A \times B$ such that if $(a, b) \in f$ and $(a, b') \in f$, then $b = b'$. Interchangeably for function the words *mapping, map, transformation,* or *correspondence* are also sometimes used. Sets of all first, and second elements of f are respectively called *domain* and *range* of f, i.e.,

$$\begin{aligned} \text{domain } f &= \text{D}(f) &= \{a \in A : \exists\, b \in B \text{ such that } (a, b) \in f\} \\ \text{range } f &= \text{R}(f) &= \{b \in B : \exists\, a \in A \text{ such that } (a, b) \in f\}. \end{aligned}$$

If the domain of f is equal to the set A, then we say f is a function from A to B and write it as $f : A \to B$. If $(x, y) \in f$, we say that f maps x onto y or that y is the image of x under f, and write it as $y = f(x)$. The set $\{(x, f(x)) : x \in A\}$ is called the *graph* of f. From the definition of equality of ordered pairs it follows that two functions f and g are equal iff they have the same domain and values, i.e., $\text{D}(f) = \text{D}(g)$ and $f(x) = g(x)$ for all $x \in \text{D}(f)$.

A function $f : A \to B$ is called *surjective* or *onto* if $B = \text{R}(f)$. For example, the function $y = f(x) = x^2 : [-1, 1] \to [0, 1]$ is surjective, whereas $y = f(x) = x^2 : [-1, 1] \to [-1, 1]$ is not surjective.

A function $f : A \to B$ is called *injective* or *one-to-one* if for all x, $x' \in A$, $f(x) = f(x')$ implies that $x = x'$, i.e., if no two distinct elements in the domain of f are assigned to the same element in the range. For example, the function $y = f(x) = x^2 : [0, 1] \to [0, 1]$ is injective, whereas $y = f(x) = x^2 : [-1, 1] \to [0, 1]$ is not injective because $f(-1) = f(1) = 1$.

A function $f : A \to B$ is called *bijective* or *one-to-one correspondence* if it is both surjective and injective. For example, the function $y = f(x) = x^2 : [0, 1] \to [0, 1]$ is bijective.

Suppose f and g are two functions with respective domains A_1 and A_2. If $A_1 \subset A_2$ and $f(x) = g(x)$, $x \in A_1$ we say that g is an *extension* of f to A_2, or that f is the *restriction* of g to A_1. For example, the function $y = g(x) = x^2$, $x \in [-1, 1]$ is an extension of $y = f(x) = x^2$, $x \in [0, 1]$.

If $f : A \to B$ and $g : B \to C$, then the *composite function* $g \circ f : A \to C$ is defined by $g \circ f(x) = g(f(x))$, $x \in A$, i.e., the image of x under $g \circ f$ is defined to be the image of $f(x)$ under g. Thus, in terms of ordered pairs, we have

$$g \circ f = \{(a, c) \in A \times C : \exists\, b \in B \text{ such that } (a, b) \in f \text{ and } (b, c) \in g\}.$$

For example, if f, $g : \mathcal{R} \to \mathcal{R}$ are defined by $f(x) = x^3 + 2$, and $g(x) = 5x + 1$, then $(g \circ f)(x) = g(f(x)) = g(x^3 + 2) = 5(x^3 + 2) + 1 = 5x^3 + 11$. For this example, the composite function $f \circ g$ also exists and $(f \circ g)(x) = f(g(x)) = f(5x + 1) = (5x + 1)^3 + 2 = 125x^3 + 75x^2 + 15x + 3$. Thus, even if both the functions $g \circ f$ and $f \circ g$ are defined, these may not be equal.

If $C \subseteq A$, then $f(C)$ is defined as $\{f(x) : x \in C\}$. The set $f(C)$ is called the *image* of C under f. If $S \subseteq B$, then $f^{-1}(S)$ is defined as $\{x \in A : f(x) \in S\}$, the set of all points in the domain of f whose images are in S. If $S = \{y\}$, we will write $f^{-1}(y)$ instead of $f^{-1}(\{y\})$. The set $f^{-1}(S)$ is called the *inverse image* of S under f. Note that the symbol f^{-1} has not been defined so far. For example, for the function $y = f(x) = x^2$, $x \in \mathcal{R}$ it is clear that $f([0,2]) = [0,4]$, $f^{-1}(9) = \{-3,3\}$, $f^{-1}(-11) = \emptyset$, $f^{-1}([0,9]) = [-3,3]$.

The following results provide some relationships between images and inverse images of subsets of A and B.

Theorem 2.1. Let $f : A \to B$. If C and D are subsets of A, then the following hold:

(1). $C \subseteq f^{-1}[f(C)]$.
(2). $C \subseteq D \Rightarrow f(C) \subseteq f(D)$.
(3). $f(C \cup D) = f(C) \cup f(D)$.
(4). $f(C \cap D) \subseteq f(C) \cap f(D)$.

Proof. (1). $x \in C \Rightarrow f(x) \in f(C) \Rightarrow x \in f^{-1}[f(C)]$, and hence $C \subseteq f^{-1}[f(C)]$.

(2). $x \in C \Rightarrow f(x) \in f(C)$ and $x \in C \subseteq D \Rightarrow x \in D \Rightarrow f(x) \in f(D)$. Thus, $C \subseteq D \Rightarrow f(C) \subseteq f(D)$.

(3). From (2), $C \subseteq C \cup D \Rightarrow f(C) \subseteq f(C \cup D)$ and $D \subseteq C \cup D \Rightarrow f(D) \subseteq f(C \cup D)$, and hence, $f(C) \cup f(D) \subseteq f(C \cup D)$. Next, $y \in f(C \cup D) \Rightarrow \exists x \in C \cup D$ such that $y = f(x) \Rightarrow x \in C \lor x \in D$ such that $y = f(x) \Rightarrow y \in f(C) \lor f(D) \Rightarrow y \in f(C) \cup f(D)$, i.e., $f(C \cup D) \subseteq f(C) \cup f(D)$. Combining the two cases, we get $f(C \cup D) = f(C) \cup f(D)$.

(4). From (2), $C \cap D \subseteq C \Rightarrow f(C \cap D) \subseteq f(C)$ and $C \cap D \subseteq D \Rightarrow f(C \cap D) \subseteq f(D)$, and hence, $f(C \cap D) \subseteq f(C) \cap f(D)$.

Theorem 2.2. Let $f : A \to B$. If S and T are subsets of B, then the following hold:

(1). $f[f^{-1}(S)] \subseteq S$.
(2). $S \subseteq T \Rightarrow f^{-1}(S) \subseteq f^{-1}(T)$.
(3). $f^{-1}(S \cup T) = f^{-1}(S) \cup f^{-1}(T)$.
(4). $f^{-1}(S \cap T) = f^{-1}(S) \cap f^{-1}(T)$.
(5). $f^{-1}(B \backslash S) = A \backslash f^{-1}(S)$.

Proof. (1). $y \in f[f^{-1}(S)] \Rightarrow \exists\, x \in f^{-1}(S)$ such that $y = f(x) \Rightarrow y \in S$.

(2). $x \in f^{-1}(S) \Rightarrow f(x) \in S \subseteq T \Rightarrow f(x) \in T \Rightarrow x \in f^{-1}(T)$. Thus, $f^{-1}(S) \subseteq f^{-1}(T)$ if $S \subseteq T$.

(3). $x \in f^{-1}(S \cup T) \Leftrightarrow f(x) \in S \cup T \Leftrightarrow f(x) \in S \vee f(x) \in T \Leftrightarrow x \in f^{-1}(S) \vee x \in f^{-1}(T) \Leftrightarrow x \in f^{-1}(S) \cup f^{-1}(T)$. Thus, $f^{-1}(S \cup T) = f^{-1}(S) \cup f^{-1}(T)$.

(4). $x \in f^{-1}(S \cap T) \Leftrightarrow f(x) \in S \cap T \Leftrightarrow f(x) \in S \wedge f(x) \in T \Leftrightarrow x \in f^{-1}(S) \wedge x \in f^{-1}(T) \Leftrightarrow x \in f^{-1}(S) \cap f^{-1}(T)$. Thus, $f^{-1}(S \cap T) = f^{-1}(S) \cap f^{-1}(T)$.

(5). $x \in f^{-1}(B \backslash S) \Leftrightarrow f(x) \in B \wedge f(x) \notin S \Leftrightarrow x \in f^{-1}(B) \wedge x \notin f^{-1}(S) \Leftrightarrow x \in A \backslash f^{-1}(S)$.

In (1) and (4) of Theorem 2.1 and (1) of Theorem 2.2, $\subseteq$ cannot be replaced by $=$. In fact, if we consider $A = \mathcal{R}$, $B = [-1, 1]$, $f = \sin$, $C = [0, \pi/2]$ and $D = [\pi/2, 5\pi/2]$, then $C \cap D = \{\pi/2\}$, $f(C \cap D) = \{1\}$ and $f(C) \cap f(D) = [0, 1] \cap [-1, 1] = [0, 1]$, and hence $f(C \cap D) \subset f(C) \cap f(D)$. In the following result we add certain restrictions so that equality holds.

Theorem 2.3. Let $f : A \to B$. If C and D are subsets of A, and S is a subset of B, then the following hold:

(1). If f is injective, then $C = f^{-1}[f(C)]$.

(2). If f is surjective, then $f[f^{-1}(S)] = S$.

(3). f is injective iff $f(C \cap D) = f(C) \cap f(D)$.

Proof. See Problem 2.4.

Problems

2.1. Prove De Morgan laws (2.1) and (2.2).

2.2. Show that

(i). If A_i, $i \in \mathcal{N}$ are sets forming a decreasing sequence, i.e., $A_1 \supseteq A_2 \supseteq \cdots$ and $\cap_{i=1}^{\infty} A_i = \emptyset$, then $A_1 = \cup_{i=1}^{\infty}(A_i \backslash A_{i+1})$. Note that the sets $A_i \backslash A_{i+1}$ are disjoint.

(ii). If A_i are disjoint sets and $B_n = \cup_{i=n+1}^{\infty} A_i$, then $\cap_{n=1}^{\infty} B_n = \emptyset$. Note that B_n form a decreasing sequence.

(iii). If $A = \cup_{i=1}^{\infty} A_i$, then $A = \cup_{i=1}^{\infty}\left(A_i \backslash \cup_{j=1}^{i-1} A_j\right)$; here $\cup_{j=1}^{0} A_j = \emptyset$. Note that sets $A_i \backslash \cup_{j=1}^{i-1} A_j$ are disjoint.

2.3. Find examples of relations with the following properties.

(i).   Reflexive, but not symmetric and not transitive.

(ii).   Symmetric, but not reflexive and not transitive.

(iii).   Transitive, but not reflexive and not symmetric in $\mathcal{R}$.

(iv).   Reflexive and symmetric, but not transitive in $\mathcal{R}$.

(v).   Reflexive and transitive, but not symmetric.

(vi).   Symmetric and transitive, but not reflexive.

2.4.   Prove Theorem 2.3.

2.5.   Let $f : A \rightarrow B$ and $g : B \rightarrow C$. Show that

(i).   If f and g are surjective, then $g \circ f$ is surjective.

(ii).   If f and g are injective, then $g \circ f$ is injective.

(iii).   If f and g are bijective, then $g \circ f$ is bijective.

Answers or Hints

2.1.   We shall prove (2.1). Suppose $x \in (\cup\{C_a : a \in A\})^c$. Then, $x \notin \cup\{C_a : a \in A\}$ and so $x \notin C_a$ for all $a \in A$. Thus, $x \in C_a^c$ for all $a \in A$, and hence $x \in \cap\{C_a : a \in A\}^c$. Therefore, $(\cup\{C_a : a \in A\})^c \subseteq \cap\{C_a : a \in A\}^c$. Now suppose $x \in \cap\{C_a : a \in A\}^c$. Then, $x \in C_a^c$ for all $a \in A$. Thus, $x \notin C_a$ for all $a \in A$, and hence $x \notin \cup\{C_a : a \in A\}$. Therefore, $x \in (\cup\{C_a : a \in A\})^c$. This shows that $\cap\{C_a : a \in A\}^c \subseteq (\cup\{C_a : a \in A\})^c$.

2.2.   (i). It is clear that the right-hand side is contained in A_1. Conversely, if $x \in A_1$ there exists a greatest number i, say, s for which $x \in A_s$. Then, $x \in A_s \backslash A_{s+1}$, which implies that x belongs to the right-hand side.

(ii). If $x \in B_n$ for some n, then $x \in A_s$ for some $s > n$. But then, $x \notin A_i$ for any $i > s$ and therefore, $x \notin B_n$ if $n \geq s$. Thus, there are no elements belonging to all B_n.

(iii). If $x \in A$, then there exists the least number i, say, s for which $x \in A_s$. If $s = 1$ we have $x \in A_1$. If $s > 1$, then $x \in A_s \backslash \cup_{j=1}^{s-1} A_j$. In both cases x belongs to the right-hand side. The converse is obvious.

2.3.   (i). Consider $A = \{a, b, c\}$ and $R \subseteq A \times A$, $R = \{(a,a), (b,b), (c,c), (a,b), (b,c)\}$.

(ii). Consider $A = \{a, b, c\}$ and $R \subseteq A \times A$, $R = \{(a,b), (b,a), (a,c), (c,a)\}$.

(iii). Define $a < b$ iff there is an element $c \in \mathcal{R}$, $c > 0$ such that $b = a + c$.

(iv). xRy iff $|x - y| \leq 2$.

(v). Consider $A = \{a, b, c\}$ and $R \subseteq A \times A$, $R = \{(a,a), (b,b), (c,c), (a,b), (b,c), (a,c)\}$.

(vi). xRy iff $(x - y)^2 > 0$.

2.4. We shall prove part (3). If $f(C \cap D) = f(C) \cap f(D)$ for all C and D subsets of A, then f is injective. For this, consider $x, y \in A$ such that $x \neq y$. Let $C = \{x\}$, $D = \{y\}$. From $f(C \cap D) = f(C) \cap f(D)$, we have $f(C \cap D) = f(\emptyset) = \emptyset = f(C) \cap f(D)$, therefore $f(C) \neq f(D)$. This means that $f(x) \neq f(y)$, and hence f is injective. Conversely, from Theorem 2.1(4) we already know that $f(C \cap D) \subseteq f(C) \cap f(D)$, and hence it suffices to show that $f(C) \cap f(D) \subseteq f(C \cap D)$. Let $y \in f(C) \cap f(D)$. Then, $y \in f(C)$ and $y \in f(D)$. Thus, there exist points $x_1 \in C$ and $x_2 \in D$ such that $f(x_1) = y$ and $f(x_2) = y$. Now since f is injective and $f(x_1) = y = f(x_2)$ we must have $x_1 = x_2$, i.e., $x_1 \in C \cap D$. But then, $y = f(x_1) \in f(C \cap D)$.

2.5. We shall prove part (i). Since g is surjective $R(g) = C$, and hence for any $c \in C$ there exists $b \in B$ such that $g(b) = c$. Now since f is surjective, there exists $a \in A$ such that $f(a) = b$. But then, $(g \circ f)(a) = g(f(a)) = g(b) = c$, so $g \circ f$ is surjective.

Chapter 3

Real Numbers

In this chapter we will provide the structure of real numbers by means of certain axioms which in turn are used to prove several interesting results. In particular, we will show that between any two real numbers there is an infinite number of rational and irrational numbers.

We begin with the following axioms formulated by the Italian mathematician G. Peano (1858–1932) for the natural numbers $1, 2, \cdots$:

1. 1 is a natural number.
2. For each natural number n there is a unique successor $n + 1$.
3. 1 is not the successor of any natural number.
4. Two natural numbers are equal if their successors are equal.
5. Any set of natural numbers which contains 1 and the successor of every natural number p whenever it contains p is the set $\mathcal{N}$ of natural numbers.

These axioms completely define the set of natural numbers $\mathcal{N}$. The axiom 5 is the principle of mathematical induction discussed in Chapter 1. In order to make subtraction possible, i.e., to solve equations of the form $x + n = m$, where $n, m \in \mathcal{N}$, we extend the set of natural numbers to the set of all integers $\mathcal{Z}$. Next to make division possible, i.e., to solve equations of the form $nx = m$, the set of rational numbers is defined. In Chapter 1, we have seen that $\sqrt{2}$ is not a rational number, and hence the set $\mathcal{Q}$ has certain gaps. To fill these gaps irrational numbers are introduced. The totality of rational and irrational numbers is the set of real numbers $\mathcal{R}$. In what follows the set $\mathcal{R}$ acts as the universal set U. Thus, $\mathcal{R} \backslash \mathcal{Q}$ is the set of irrational numbers. Now we shall provide the structure of real numbers by means of certain axioms which in turn are used to deduce further results. The axioms for real numbers are classified as follows:

(a). *Extended axiom.* This axiom states that $\mathcal{R}$ has at least two distinct elements.

(b). *Field axioms.* Real numbers are combined by two fundamental operations, namely, addition $(+)$ and multiplication $(\times$ or $\cdot)$. The addition operation satisfies the following axioms:

(A1). *The closure law.* $x, y \in \mathcal{R} \Rightarrow x + y \in \mathcal{R}$.

(A2). *The associative law.* x, y, $z \in \mathcal{R} \Rightarrow (x + y) + z = x + (y + z)$.

(A3). *The commutative law.* x, $y \in \mathcal{R} \Rightarrow x + y = y + x$.

(A4). *Identity for addition.* There is a real number 0 (zero) such that $x + 0 = x$, $\forall\, x \in \mathcal{R}$.

(A5). *Inverse for addition.* For each $x \in \mathcal{R}$ there is a real number $-x$ such that $x + (-x) = 0$.

The multiplication operation satisfies the following axioms:

(M1). *The closure law.* x, $y \in \mathcal{R} \Rightarrow x \cdot y \in \mathcal{R}$.

(M2). *The associative law.* x, y, $z \in \mathcal{R} \Rightarrow (x \cdot y) \cdot z = x \cdot (y \cdot z)$.

(M3). *The commutative law.* x, $y \in \mathcal{R} \Rightarrow x \cdot y = y \cdot x$.

(M4). *Identity for multiplication.* There is a real number 1 (one) such that $x \cdot 1 = x$, $\forall\, x \in \mathcal{R}$.

(M5). *Inverse for multiplication.* For each $x \in \mathcal{R}$ except $x = 0$ there is a real number $1/x$ such that $x \cdot 1/x = 1$.

(D). *The distributive law.* x, y, $z \in \mathcal{R} \Rightarrow x \cdot (y + z) = (x \cdot y) + (x \cdot z)$.

In view of the above axioms (A1) to (A5), (M1) to (M5), and (D) the set of real numbers $\mathcal{R}$ is called a *field*. It is clear that the set of rational numbers $\mathcal{Q}$ is also a field.

(c). *Order axioms.* Real numbers possess an ordering $<$ called *less than*. This ordering satisfies the following axioms:

(O1). *The trichotomy law.* For every pair x, $y \in \mathcal{R}$ exactly one of the following holds $x < y$, $x = y$, $y < x$.

(O2). *The transitive property.* If x, y, $z \in \mathcal{R}$ and $x < y$, $y < z$, then $x < z$.

(O3). *The additive property.* If x, y, $z \in \mathcal{R}$ and $x < y$, then $x + z < y + z$.

(O4). *The multiplicative property.* If x, y, $z \in \mathcal{R}$ and $x < y$, $0 < z$, then $xz < yz$.

In view of the above axioms (O1) to (O4) the field of real numbers $\mathcal{R}$ is called an *ordered field*. Clearly, the set of rational numbers $\mathcal{Q}$ is also an ordered field. The above axioms (O1) to (O4) can be expressed in terms of the ordering $>$ called *greater than*. A real number x is said to be *positive* or *negative* according as $a > 0$, or $a < 0$. Usually, the set of positive and negative numbers are respectively denoted as $\mathcal{R}^+$ and $\mathcal{R}^-$. Thus, it follows that $\mathcal{R} = \mathcal{R}^+ \cup \{0\} \cup \mathcal{R}^-$. Often we shall write $x \le y$ to mean that either $x < y$ or $x = y$. A similar meaning holds for $x \ge y$.

The set of real numbers $\mathcal{R}$ also has the *completeness axiom*, however, to state this we need the following boundedness definitions of subsets of $\mathcal{R}$. The subset $S \subset \mathcal{R}$ is said to be *bounded above* if $\exists M \in \mathcal{R}$ such that $x \in S \Rightarrow x \le$

M. The number M is called an *upper bound* (*u.b.*) of S. Similarly, the subset $S \subset \mathcal{R}$ is said to be *bounded below* if $\exists m \in \mathcal{R}$ such that $x \in S \Rightarrow x \geq m$. The number m is called a *lower bound* (*l.b.*) of S. The set S is said to be *bounded* if it is bounded above as well as below. Clearly, S is bounded iff $S \subseteq [m, M]$ for some interval $[m, M]$ of finite length. The set $\mathcal{N}$ is bounded below by 1 but not above, whereas the set $\{1/2, 2/3, 3/4, \cdots\}$ is bounded above by 1 and below by $1/2$. If the set S has at least one upper bound then there are infinitely many upper bounds greater than it. If S has no upper bound, then S is said to be *unbounded above*. It follows that the empty set $\emptyset$ is neither bounded below or above, nor unbounded. If there is a least number among the upper bounds of the set S, then this number is called the *least upper bound* (l.u.b.), or *supremum* of the set S, and often it is denoted as sup S. Similarly, if the set S has at least one lower bound then there are infinitely many lower bounds smaller than it. If S has no lower bound, then S is said to be *unbounded below*. If there is a greatest number among the lower bounds of the set S, then this number is called the *greatest lower bound* (g.l.b.), or *infimum* of the set S, and we usually denote it as inf S. The sup S and/or inf S of the set S may not be members of S, e.g., the set $(1, 2)$ (open interval) does not contain either its inf or its sup which are 1 and 2, respectively. From these definitions it follows that supremum and infimum of sets, if they exist, are unique. We also note that if u is the supremum of S then for every $\epsilon > 0 \; \exists$ at least one element $x \in S$ such that $u \geq x > u - \epsilon$. Similarly, if ℓ is the infimum of S then for every $\epsilon > 0 \; \exists$ at least one element $x \in S$ such that $\ell \leq x < \ell + \epsilon$.

A number σ is said to be the *greatest* (*maximal*, or *largest*) member of a set S if $\sigma \in S \wedge x \in S \Rightarrow x \leq \sigma$. If such a number exists then it is unique and is also the supremum of S. The set S may or may not have a greatest number, e.g., the set $\{x : 1 < x \leq 3\}$ has 3 as the greatest number, whereas the set $\{x : 1 \leq x < 3\}$ has no greatest number. Similarly, a number ρ is said to be the *least* (*minimal*, or *smallest*) member of a set S if $\rho \in S \wedge x \in S \Rightarrow x \geq \rho$. If such a number exists then it is unique and is also the infimum of S. The maximal and minimal numbers of S we write as max S and min S, respectively.

For the existence of supremum and infimum of nonempty sets bounded above and below, respectively, is the following completeness axiom:

(d). *Completeness axiom.* If S is any nonempty subset of $\mathcal{R}$ that is bounded above, then S has a supremum in $\mathcal{R}$. If the set S is bounded below, then by considering the set $\tilde{S} = \{x : -x \in S\}$ we can restate the completeness axiom as follows: If S is any nonempty subset of $\mathcal{R}$ that is bounded below, then S has an infimum in $\mathcal{R}$. The completeness axiom is also known as the *continuity axiom* in $\mathcal{R}$.

In view of the axioms (a) to (d) the set of real numbers $\mathcal{R}$ is called a *complete ordered field*.

The proofs of many important results such as the existence of maxima and minima, the intermediate-value theorem, Rolle's theorem, the mean-value theorem, and so on, depend strongly on the completeness axiom. This axiom also has several other implications, some of which we shall now state and prove. We begin by showing that this axiom enables us to prove the principle of mathematical induction. For this, we state this principle in an alternative form in the following theorem.

Theorem 3.1. If the proposition $p(m_0)$ is true, and $p(m)$ is true whenever $p(n)$ is true for all n such that $m_0 \leq n < m$, then $p(n)$ is true for all $n \geq m_0$.

Proof. Let the set $S = \{r : p(r)$ is false$\} \neq \emptyset$. Then, by the completeness axiom S has a least member, say, ℓ, so that $p(\ell)$ is false. Thus, $m < \ell \Rightarrow m \notin S$, i.e., $p(m)$ is true for all $m_0 \leq m < \ell$. But then by the hypothesis $p(\ell)$ is true. This contradiction implies that S must be empty. Hence, $p(n)$ is true for all $n \geq m_0$.

Theorem 3.2. The set of rational numbers $\mathcal{Q}$ is not complete.

Proof. Consider the set $S = \{x : x \in \mathcal{Q}^+, \ x^2 < 3\} \subset \mathcal{Q}$, where $\mathcal{Q}^+$ is the set of positive rational numbers. Since $1 \in S$, $S \neq \emptyset$ and $2^2 > 3 \Rightarrow S$ is bounded above. We shall show that S has no supremum in $\mathcal{Q}$. If $\lambda \in \mathcal{Q}$ is supremum of S, $\lambda \geq 1$ and so $\lambda \in \mathcal{Q}^+$. Then, by (O1) one and only one of (i) $\lambda^2 > 3$, (ii) $\lambda^2 = 3$, and (iii) $\lambda^2 < 3$ holds.

(i). If $\lambda^2 > 3$ and $\lambda \in \mathcal{Q}^+$, then $\eta = (2\lambda + 3)/(\lambda + 2) \in \mathcal{Q}^+$ and

$$\lambda - \frac{2\lambda + 3}{\lambda + 2} = \frac{\lambda^2 - 3}{\lambda + 2} > 0 \Rightarrow \lambda > \frac{2\lambda + 3}{\lambda + 2}.$$

On the other hand,

$$3 - \left(\frac{2\lambda + 3}{\lambda + 2}\right)^2 = \frac{3 - \lambda^2}{(\lambda + 2)^2} < 0 \Rightarrow \left(\frac{2\lambda + 3}{\lambda + 2}\right)^2 > 3.$$

Hence, there exists an upper bound η for S smaller than λ, and so $\lambda \neq \sup S$ if $\lambda^2 > 3$.

(ii). If $\lambda^2 = 3$ and $\lambda \in \mathcal{Q}^+$, let $\lambda = p/q$ where $p, q \in \mathcal{N}$ and have no common factors. Then, $p^2 = 3q^2$. Since the square of every integer has integer factors in pairs, 3 is a factor of p^2 and so of p, together with being a factor of q. Thus, 3 is a factor of both p and q. This contradicion shows that $\lambda \notin \mathcal{Q}^+$, and so $\lambda \notin \mathcal{Q}$.

(iii). If $\lambda^2 < 3$ and $\lambda \in \mathcal{Q}^+$, then $\eta = (2\lambda + 3)/(\lambda + 2) \in \mathcal{Q}^+$ and $\eta > \lambda$, $\eta^2 < 3$, i.e., $\eta \in S$. Thus, λ is not even an upper bound for S. Hence, $\lambda \neq \sup S$ if $\lambda^2 < 3$.

From the above three cases it follows that $\sup S \notin \mathcal{Q}$. Thus, $\mathcal{Q}$ is not complete.

From Theorem 3.2 it is clear that $\lambda^2 = 3$ and $\lambda \notin \mathcal{Q}$. Thus, $\lambda \in \mathcal{R}$ is necessarily irrational. This establishes the existence of irrational numbers.

Theorem 3.3 (Archimedean Property). If x, $y \in \mathcal{R}$ with $x > 0$, then there exists an $n \in \mathcal{N}$ such that $nx > y$.

Proof. If $y \leq 0$ the result is obvious. For $y > 0$ let the result be false, so that $nx \leq y$, $\forall n \in \mathcal{N}$. Thus, the set $S = \{nx : n \in \mathcal{N}\}$ is nonempty and bounded above. Hence, from the completeness axiom, $u = \sup S$ exists. Then, $nx \leq u$, $\forall n \in \mathcal{N} \Rightarrow (n+1)x \leq u$, $\forall n \in \mathcal{N} \Rightarrow nx \leq u - x$, $\forall n \in \mathcal{N}$. Thus, $u - x (< u)$ is also an upper bound of S. This contradicts that $u = \sup S$. Therefore, the assumption $nx \leq y$, $\forall n \in \mathcal{N}$ is false, and now the result follows from (O1).

Theorem 3.4 (Dedekind's Property). Let A and B be two nonempty subsets of $\mathcal{R}$ such that

(i). $A \cup B = \mathcal{R}$,

(ii). $x \in A \wedge y \in B \Rightarrow x < y$.

Then, either A has the greatest member, or B has the least member.

Proof. In view of (ii), the nonempty set A is bounded above. If A has the greatest member, we are done. If A does not have the greatest member, then since B is the set of upper bounds of A, the completeness axiom implies that B has the least member.

Theorem 3.5. Dedekind's property is equivalent to the completeness axiom in $\mathcal{R}$.

Proof. In Theorem 3.4 it is shown that the completeness axiom in $\mathcal{R}$ implies Dedekind's property. Thus, we need to show that Dedekind's property implies the completeness axiom. Let S be a nonempty set bounded above, and let the sets A and B be defined by

$$A = \{x : x \text{ is not an u.b. of } S\} \quad \text{and} \quad B = \{y : y \text{ is an u.b. of } S\}.$$

Clearly, both A and B are nonempty, disjoint, and (i) $A \cup B = \mathcal{R}$, (ii) $x \in A \wedge y \in B \Rightarrow x < y$. Thus, by Dedekind's property either A has the greatest member, or B has the least member. Let λ be the greatest member of A. Then, $\lambda \in A \Rightarrow \lambda \notin B \Rightarrow \exists$ an $z \in S$ such that $\lambda < z$. Now since $\lambda < (z + \lambda)/2 \in B$, $(z+\lambda)/2$ is an u.b. of S. On the other hand, $(z+\lambda)/2 < z \in S \Rightarrow (z+\lambda)/2$ is not an u.b. of S. Therefore, our assumption on λ is false, and thus the set B has the least member. Hence, the set of upper bounds of a nonempty set S bounded above has a least member. This is precisely our completeness axiom in $\mathcal{R}$.

Theorem 3.6. Between any two real numbers there are infinitely many rational numbers.

Proof. Let x, $y \in \mathcal{R}$ be such that $x < y$. Then, by Theorem 3.3 $\exists\, n \in \mathcal{N}$ such that $n(y-x) > 1$, i.e., $ny > nx + 1$. Further in view of Problem 3.11(i), $\exists\, m \in \mathcal{Z}$ such that $m > nx \geq m - 1$. Thus, $ny > nx + 1 \geq m > nx$. This implies $y > (m/n) > x$, where $n \in \mathcal{N}$ and $m \in \mathcal{Z}$. Thus, the rational number m/n lies between x and y. The arguments can now be repeated with the distinct numbers x and m/n, and the process can be continued. Thus, between x and y there are infinitely many rational numbers.

Theorem 3.7. Between any two real numbers there are infinitely many irrational numbers.

Proof. Let x, $y \in \mathcal{R}$ be such that $x < y$. Choose λ any positive irrational number, say, $\sqrt{2}$. Then, by Theorem 3.3, $\exists\, n \in \mathcal{N}$ such that $n(y-x) > \lambda$. Thus, it follows that

$$y > x + \frac{\lambda}{n} > x + \frac{\lambda}{2n} > x.$$

Now since the difference of the numbers $x + (\lambda/n)$ and $x + (\lambda/2n)$ is the irrational number $(\lambda/2n)$, both of these numbers cannot be rational. Thus, an irrational number lies between x and y. The arguments can now be repeated and the process can be continued. Thus, between x and y there are infinitely many irrational numbers.

We call a set $S \subseteq \mathcal{R}$ *dense* in $\mathcal{R}$ provided $S \cap I \neq \emptyset$ for every interval I. From Theorems 3.6 and 3.7 it follows that the set of rational numbers $\mathcal{Q}$ and the set of irrational numbers $\mathcal{R}\backslash\mathcal{Q}$ are dense in $\mathcal{R}$. However, the set $\mathcal{Z}$ and any bounded interval are not dense in $\mathcal{R}$.

A set $S \subseteq \mathcal{R}$ is called *nowhere dense* if for each interval I there is an interval $I_1 \subseteq I$ such that $I_1 \cap S = \emptyset$. It is clear that the set $\mathcal{Z}$ is nowhere dense. If S is an interval or contains an interval, then S is not nowhere dense. Although they contain no intervals, the sets $\mathcal{Q}$ and $\mathcal{R}\backslash\mathcal{Q}$ are also not nowhere dense.

It follows that if S is nowhere dense, then $\mathcal{R}\backslash S$ is dense in $\mathcal{R}$. However, its converse is not true. For example, the set $\mathcal{R}\backslash\mathcal{Q}$ is dense in $\mathcal{R}$, but the set $\mathcal{Q}$ is not nowhere dense.

A line on which a positive direction has been specified is called a *directed straight line*, e.g., in the Euclidean plane x-axis. The real numbers can be represented on a directed straight line as points. In fact we have the following axiom:

(e). *Dedekind-Cantor axiom of continuity of a straight line.* For every real number there corresponds a unique point of a directed straight line, and conversely for every point on this line there corresponds a unique real number.

Thus, there is one-to-one correspondence between real numbers and the points on a directed straight line. This allows us to call points for real numbers, and the real line for the directed straight line.

Problems

3.1. Let $x, y, a \in \mathcal{R}$. Show that

(i). $x < y + \epsilon, \ \forall \, \epsilon > 0$ iff $x \le y$.

(ii). $x > y - \epsilon, \ \forall \, \epsilon > 0$ iff $x \ge y$.

(iii). $-\epsilon < a < \epsilon, \ \forall \, \epsilon > 0$ iff $a = 0$.

3.2. Let $a, \ b \in \mathcal{R}^+ \cup \{0\}$, and $p, \ q \in \mathcal{N}$. Show that

$$\frac{a^{p+q} + b^{p+q}}{2} \ge \left(\frac{a^p + b^p}{2}\right)\left(\frac{a^q + b^q}{2}\right).$$

3.3. Find the supremum and infimum (if they exist) of the following sets and state when they belong to the sets.

(i). $\{(1/n) + (-1)^n : n \in \mathcal{N}\}$.

(ii). $\{x : 0 \le x \le \sqrt{2} \ \text{and} \ x \in \mathcal{Q}\}$.

(iii). $\{x : x^2 + x - 1 < 0\}$.

(iv). $\{x \in \mathcal{R} : 3x + 5 < 4x - 7\}$.

(v). $\{x \in \mathcal{Q} : -1 \le x \le \sqrt{3}\}$.

(vi). $\{2 + 2^{-n} : n \in \mathcal{N}\}$.

3.4. Suppose A is a bounded nonempty subset of $\mathcal{R}$ and B is a nonempty subset of A. Prove or disprove each of the following assertions.

(i). $\inf A \le \inf B \le \sup B \le \sup A$.

(ii). If $\inf A = \sup A$, then A contains exactly one element.

(iii). If $\inf A = \inf B$ and $\sup A = \sup B$, then $A = B$.

(iv). If $\inf A = \inf B$ and $\sup A = \sup B$, then the set $C = \{x : x \in A \ \text{and} \ x \notin B\}$ contains at most two elements.

3.5. Let A and B be two bounded nonempty subsets of $\mathcal{R}$. Define $A + B = \{a + b : a \in A, \ b \in B\}$, $\lambda A = \{\lambda a : a \in A\}$, where $\lambda \in \mathcal{R}$. Prove that

(i). $\sup(A + B) = \sup A + \sup B$.

(ii). $\sup \lambda A = \begin{cases} \lambda \sup A & \text{if} \ \lambda \ge 0 \\ \lambda \inf A & \text{if} \ \lambda < 0. \end{cases}$

(iii). $\sup(A \cup B) = \max\{\sup A, \ \sup B\}$.

(iv). $\inf(A \cup B) = \min\{\inf A, \ \inf B\}$.

3.6. Let A and B be nonempty bounded subsets of $\mathcal{R}^+$. Define $AB = \{xy : x \in A, \ y \in B\}$. Assume that $\sup A$, $\sup B$, and $\sup AB$ exist and are positive. Show that for each $y \in B$, $\sup A \le (\sup AB)/y$. Deduce that $(\sup A)(\sup B) \le \sup AB$.

3.7. Let A be a nonempty bounded set of nonnegative numbers and $B = \{x : x^2 \in A, \ x \geq 0\}$. Prove that B is bounded and $\sup B = \sqrt{\sup A}$.

3.8. Show that the set $S = \{x \in \mathcal{R} : \ x < 5\}$ has no maximum element.

3.9. State whether $\sup S$, $\inf S$, $\max S$, and $\min S$ exist, and if so determine their values, where

(i). $\ S = \{x : \ 1 + 2x \leq 1/(1 - 2x), \ x \neq 1/2\}$.

(ii). $\ S = \{x : \sqrt{2 + x} \geq x, \ x \geq -2\}$.

3.10. A subset A of real numbers is said to be inductive if (a) 1 is in A, and (b) $k + 1$ is in A whenever k is in A. Show that

(i).   The set of positive real numbers is inductive.

(ii).   The set of positive real numbers unequal to $1/2$ is inductive.

(iii). The set of positive real numbers unequal to 5 is not inductive.

(iv). Ω is inductive, where $\Omega = \cap\{A : A \text{ is inductive}\}$.

3.11. Use the Archimedean property to prove the following.

(i).   For any $x \in \mathcal{R} \ \exists$ a unique $n \in \mathcal{Z}$ such that $n + 1 > x \geq n$.

(ii).   If $x \in \mathcal{R}^+ \ \exists$ a unique $n \in \mathcal{N}$ such that $n(n+1)/2 > x \geq n(n-1)/2$.

3.12. Show that the square root of any rational number, which is not a perfect square, is irrational.

3.13. If $a < b$ are real numbers, then show that there exists both a rational number and an irrational number between a and b.

3.14. If $a \in \mathcal{R} \backslash \mathcal{Q}$ and $0 \neq b \in \mathcal{Q}$, prove that $a + b$ and ab are irrational.

Answers or Hints

3.1. We shall prove only part (i). Suppose to the contrary that $x < y + \epsilon$, $\forall \ \epsilon > 0$ but $x > y$. Set $\epsilon_0 = x - y > 0$ and observe that $y + \epsilon_0 = x$. Hence, by (O1), $y + \epsilon_0$ cannot be greater than x. But this contradicts the hypothesis for $\epsilon = \epsilon_0$. Thus, $x \leq y$. Conversely, suppose $x \leq y$ and $\epsilon > 0$ is given. Either $x < y$ or $x = y$. If $x < y$ then by (O2) and (O3), $x + 0 < y + 0 < y + \epsilon \Rightarrow \ x < y + \epsilon$. If $x = y$ then by (O3), $x < y + \epsilon$. Thus, $x < y + \epsilon$, $\forall \ \epsilon > 0$ in either case.

3.2. $a^{p+q} + b^{p+q} \geq a^p b^q + a^q b^p$ is the same as $(a^p - b^p)(a^q - b^q) \geq 0$.

3.3. (i). $\sup = 3/2$ (in the set), $\inf = -1$.

(ii). $\sup = \sqrt{2}$, $\inf = 0$ (in the set).

(iii). $\sup = (-1 + \sqrt{5})/2$, $\inf = (-1 - \sqrt{5})/2$.

(iv). $x > 12$, no sup; inf $= 12$ (not in the set).

(v). sup $= \sqrt{3}$ (not in the set); inf $= -1$ (in the set).

(vi). sup $= 5/2$ (in the set); inf $= 2$ (not in the set).

3.4. (i). Let $x \in B$, then inf $B \le x \le$ sup B. Since inf $A \le x$ for all $x \in A$ and $B \subseteq A$, inf A is a lower bound of B, so inf $A \le$ inf B. The proof for sup $B \le$ sup A is similar.

(ii). Let x and y be any two distinct elements of A such that $x < y$. Then, inf $A \le x < y \le$ sup A. But, this contradicts inf $A =$ sup A.

(iii) and (iv) are false. For example, consider $A = [0,1]$ and $B = \{0,1\}$.

3.5. (i). For $a \in A$ and $b \in B$, we have $a + b \le$ sup $A +$ sup B. Thus, sup $A +$ sup B is an upper bound of $A + B$. Let $\epsilon > 0$. Then, there exists $a_1 \in A$ and $b_1 \in B$ such that $a_1 >$ sup $A - \epsilon$, $b_1 >$ sup $B - \epsilon$. Hence, $a_1 + b_1 >$ sup $A +$ sup $B - 2\epsilon$. Therefore, $\sup(A + B) =$ sup $A +$ sup B.

(ii). If $\lambda \ge 0$, then $\lambda a \le \lambda$ sup A. Let $\epsilon > 0$. Then, there exists $a_1 \in A$ such that $a_1 >$ sup $A - \epsilon$. Hence, $\lambda a_1 > \lambda$ sup $A - \lambda\epsilon$. Thus, $\sup(\lambda A) = \lambda$ sup A. If $\lambda < 0$, then $\lambda a \le \lambda$ inf A. Let $\epsilon > 0$. Then, there exists $a_2 \in A$ such that $a_2 <$ inf $A + \epsilon$. Hence, $\lambda a_2 > \lambda$ inf $A + \lambda\epsilon = \lambda$ inf $A - \epsilon_1$, where $\epsilon_1 = -\lambda\epsilon > 0$. Thus, $\sup(\lambda A) = \lambda$ inf A.

(iii). If $a \in A \cup B$, then $a \in A$ or $a \in B$. Hence, $a \le$ sup A or $a \le$ sup B. In either case, $a \le \max\{\text{sup } A, \text{sup } B\}$, i.e., $\sup(A \cup B) \le \max\{\text{sup } A, \text{sup } B\}$. Now since, $A \subset A \cup B$, $\sup(A \cup B) \ge$ sup A, and similarly $\sup(A \cup B) \ge$ sup B. Hence, $\sup(A \cup B) = \max\{\text{sup } A, \text{sup } B\}$.

(iv). The proof is similar to that of (iii).

3.6. Clearly, $xy \le$ sup AB for each $x \in A, y \in B$. Hence, sup AB is an upper bound of AB. Therefore, $x \le (\text{sup } AB)/y$. This implies that $(\text{sup } AB)/y$ is an upper bound of A for each $y \in B$. Thus, sup $A \le (\text{sup } AB)/y$. Now since sup $A > 0$, we have $y \le$ sup $AB/$ sup A, i.e., sup $AB/$ sup A is an upper bound of B. Hence, sup $B \le$ sup $AB/$ sup A, i.e., $(\text{sup } A)(\text{sup } B) \le$ sup AB.

3.7. Let $k > 0$ be such that $0 \le a \le k$ for all $a \in A$. Clearly, sup $A \ge 0$. Now $x \in B$ implies $x^2 \in A$ and $x \ge 0$, and hence $0 \le x^2 \le k$, or $0 \le x \le \sqrt{k}$. Thus, B is bounded. We also have sup $B \ge 0$. Next, $x \in B$ implies $x^2 \le$ sup A and $x \ge 0$ and hence $0 \le x \le \sqrt{\text{sup } A}$. This gives sup $B \le \sqrt{\text{sup } A}$. Finally, $x \in A$ implies $\sqrt{x} \in B$ and hence $\sqrt{x} \le$ sup B, or $x \le (\text{sup } B)^2$, which implies sup $A \le (\text{sup } B)^2$, hence $\sqrt{\text{sup } A} \le$ sup B. Therefore, sup $B = \sqrt{\text{sup } A}$.

3.8. Suppose max $S = k$. Then, $x \le k$ for all $x \in S$ and $k \in S$, i.e., $k < 5$. Let $\ell = (k+5)/2 < (5+5)/2 = 5$. But, $\ell = (k+5)/2 > (k+k)/2 = k$. We have $\ell \in S$ and $\ell > k$. This contradicts the fact that max $S = k$.

3.9. (i). If $x > 1/2$, $(1 + 2x)(1 - 2x) \ge 1$, i.e., $1 - 4x^2 \ge 1$ which is not possible. If $x < 1/2$, $1 - 4x^2 \le 1$ which holds for $x \in (-\infty, 1/2)$. Thus, sup $S = 1/2$, no max, no inf, no min.

(ii). For $-2 \leq x < 0$ the inequality $\sqrt{2+x} \geq x$ is obvious. For $x > 0$, $\sqrt{2+x} \geq x$ implies $2+x \geq x^2$, or $(2-x)(x+1) \geq 0$ which holds provided $0 \leq x \leq 2$. Thus, $S = [-2, 2]$. Clearly, $\sup S = \max S = 2$, $\inf S = \min S = -2$.

3.10. (i). Obvious.

(ii). The set contains 1. Suppose the set contains k but not $k+1$. Then, $k+1 = 1/2$, i.e., $k = -1/2$, which is impossible.

(iii). The set contains 4 but not $4+1$.

(iv). $1 \in \Omega$ because 1 is in every inductive set. Suppose $k \in \Omega$. Then, k is in every inductive set. Thus, $k+1$ is in every inductive set, i.e., $k+1 \in \Omega$.

3.11. (i). By the completeness axiom the set $\{p : p < x \wedge p \in \mathcal{Z}\}$ being bounded above, has a supremum, say, n, $n \in \mathcal{Z}$. Thus, $x \geq n \wedge n+1 > x$, i.e., $n+1 > x \geq n$, $n \in \mathcal{Z}$. Since n is supremum the uniqueness of n follows.

(ii). From part (i), for the real number $\sqrt{(2x+1/4)} + 1/2 \, \exists$ a unique $n \in \mathcal{N}$ such that $n+1 > \sqrt{(2x+1/4)} + 1/2 \geq n$.

3.12. Let m/n be a positive fraction in its lowest terms, and suppose, if possible, that $p^2/q^2 = m/n$, where p/q is also in its lowest terms. Then, $np^2 = mq^2$. Since p and q have no common factor, every factor of q^2 must divide n. Therefore, $n = \lambda q^2$, where λ is an integer. It follows that $m = \lambda p^2$. But m and n also have no common factor. Therefore, λ must be 1, and $m = p^2$, $n = q^2$. Thus, the square root of m/n cannot be rational, unless m and n are both perfect squares.

In particular, if $n = 1$ we deduce that the integer m cannot be the square root of a rational number unless it is a perfect square. Alternatively, if $p^2/q^2 = m$ and λ, $\lambda + 1$ are two integers between which p/q lies, then $p^2 - mq^2 = 0$ and $\lambda q < p < (\lambda+1)q$. Now consider the identity $(mq - \lambda p)^2 - m(p - \lambda q)^2 = (\lambda^2 - m)(p^2 - mq^2) = 0$. From this it follows that $(mq - \lambda p)/(p - \lambda q)$ is another fraction whose square is m. But the denominator of this fraction is less than q, which contradicts the assumption that m is the square of the fraction p/q which is in its lowest terms.

3.13. Let n be any integer greater than $2\sqrt{2}/(b-a)$ so that $1/n < \sqrt{2}/n < (b-a)/2$. Let m be an integer such that $m \leq na < m+1$. It then follows that

$a < \frac{m}{n} + \frac{1}{n} < \frac{m}{n} + \frac{\sqrt{2}}{n} < a + \frac{b-a}{2} < b$.

Thus, $(m+1)/n$ and $(m+\sqrt{2})/n$ are the required rational and irrational numbers.

3.14. Let $b = \frac{\alpha}{\beta}$. If $a + b = \frac{\gamma}{\delta}$, we have $a = \frac{\gamma\beta - \alpha\delta}{\beta\delta} \in \mathcal{Q}$, which is false. If $ab = \frac{\tau}{\sigma}$, we have $a = \frac{\tau\beta}{\sigma\alpha} \in \mathcal{Q}$, which is also false.

Chapter 4

Open and Closed Sets

In Chapter 2 we introduced open and closed intervals. Here we will define open and closed subsets of $\mathcal{R}$, and prove the Bolzano-Weierstrass and Heine-Borel theorems which are extremely important results in analysis.

We begin this chapter with the concept of a neighborhood. Let $x \in \mathcal{R}$. A set $N = N(x, \epsilon)$ is called a *neighborhood* (in short *nbd*) of x if there exists an $\epsilon > 0$ such that $(x - \epsilon, x + \epsilon) \subset N$. Equivalently, N is an nbd of x if there exists an open interval $I(x)$ such that $x \in I(x) \subset N$. The open interval $(x - \epsilon, x + \epsilon)$ is often referred to as ϵ-nbd of x, and is denoted as $I_\epsilon(x)$. The set $I_\epsilon^*(x) = (x - \epsilon, x) \cup (x, x + \epsilon)$ is called a *deleted nbd* of the point x. A set T is called an nbd of the set S if T is an nbd of each point of S. Clearly, $\mathcal{R}$ is an nbd of each of its points, interval (a, b) is an nbd of itself, and $\emptyset$ is an nbd of each of its points because there is no point of which it is not an nbd, however, every finite set of points is not an nbd of any of the points of $\mathcal{R}$.

A point x is called an *interior point* of a set S if there exists an open interval $I(x)$ such that $x \in I(x) \subset S$. Thus, a set is an nbd of a point iff the point is an interior point of the set. It follows that

1. If x is an interior point of each of the sets S and T, then x is also an interior point of $S \cap T$.

2. If x is an interior point of the set S and $S \subset T$, then x is also an interior point of the set T.

The set of all interior points of a set S is called the *interior* of S and is denoted by S^i. Clearly, $S^i \subseteq S$, and $\mathcal{N}^i = \mathcal{Z}^i = \mathcal{Q}^i = \emptyset$, $\mathcal{R}^i = \mathcal{R}$.

A point x is called an *exterior point* of a set S if there exists an nbd N such that $N \cap S = \emptyset$. The set of all exterior points of S is called the *exterior* of S and is denoted as S^e.

A point x is called a *boundary point* of the set S if it is neither an interior point nor an exterior point of S. The set of all boundary points of S is called the *boundary* of S and is denoted as S^b. It is clear that $\mathcal{R} = S^i \cup S^e \cup S^b$.

A set S is called *open* iff it is an nbd of each of its points, i.e., $S = S^i$.

Thus, $\mathcal{R}$, (a, b) and $(a, b) \cup (c, d)$, $b \neq c$, are open sets, but $[a, b)$ is not an open set. The set $\emptyset$ is open because there is no point in it of which it is not an nbd. However, a nonempty open set must contain infinitely many points. It also follows that every open set is a union of open intervals, i.e., $S = \cup_{x \in S} I(x)$. The following results can be proved rather easily.

(O1). S^i of the set S is the largest open set contained in S.

(O2). $S \subset T \Rightarrow S^i \subseteq T^i$.

(O3). The union of an arbitrary collection of open sets is open.

(O4). The intersection of any finite number of open sets is open.

The result (O4) is false for the intersection of an arbitrary collection of open sets. For example, $\cap_{n \in \mathcal{N}} (a - 1/n, a + 1/n) = \{a\}$ is not an open set.

A point x is called a *limit point* (or an *accumulation point*) of the set S iff for every nbd N of x, $N \cap S$ is an infinite set. Alternatively, a point x is a limit point of the set S iff each interval (a, b) containing x contains at least one point of S other than x. Sets $\mathcal{N}$, $\mathcal{Z}$, $\emptyset$ have no limit points, whereas every $x \in \mathcal{R}$ is a limit point of the set $\mathcal{Q}$. The limit points of (a, b), $(a, b]$, $[a, b)$, or $[a, b]$ are the points of $[a, b]$.

Let S be a nonempty set bounded above. If $\sigma = \sup S \notin S$, then σ is a limit point of S. Indeed, for $\epsilon > 0$, $\exists y \in S$ such that $\sigma - \epsilon < y < \sigma \Rightarrow$ nbd $(\sigma - \epsilon, \sigma + \epsilon)$ of σ contains a point y of S other than σ. This implies that σ is a limit point of S. Similarly, if S is a nonempty set bounded below and if its infimum does not belong to S, then it is a limit point of S. It also follows that if x is not a limit point of S, then it is not a limit point of any of its subsets; however, if x is a limit point of S, then it is a limit point of every set which contains S.

The set of all limit points of S is called the *derived set* of S and is denoted by S'. It is clear that $\mathcal{N}' = \emptyset$, $\mathcal{Q}' = \mathcal{R}$, $\mathcal{R}' = \mathcal{R}$ and $\emptyset' = \emptyset$. The set $S \cup S'$ is called the *closure* of S and is represented by $\overline{S}$, or cl S. It follows that $S \subseteq \overline{S}$, $\overline{\mathcal{N}} = \mathcal{N}$, $\overline{\mathcal{Q}} = \mathcal{R}$, $\overline{\mathcal{R}} = \mathcal{R}$ and $\overline{\emptyset} = \emptyset$.

A set S is called *closed* iff each of its limit points belongs to S, i.e., $S' \subseteq S$. Equivalently, S is closed iff its complement $\mathcal{R} \backslash S$ is open. Thus, $[a, b]$ is closed, the set $[a, b] \cup [c, d]$, $b \neq c$ is closed, every finite set S is closed because $S' = \emptyset \subseteq S$, since $\mathcal{R}' = \mathcal{R} \subseteq \mathcal{R}$ and $\emptyset' = \emptyset \subseteq \emptyset$ both $\mathcal{R}$ and $\emptyset$ are closed; however, the set $\mathcal{Q}$ is not closed because $\mathcal{Q}' = \mathcal{R} \not\subseteq \mathcal{Q}$, whereas $[a, b)$ is neither open nor closed. Thus, if a set is not open (closed), then it need not be closed (open). The following results can be shown easily.

(C1). A set S is closed iff $S = \overline{S}$.

(C2). For any set S, S' is closed.

(C3). The intersection of an arbitrary collection of closed sets is closed.

(C4). The union of any finite number of closed sets is closed.

The result (C4) is false for the union of an arbitrary collection of closed sets. For example, $\cup_{n \in \mathcal{N}} [a + 1/n, b - 1/n] = (a, b)$ is not a closed set.

Clearly, a finite set cannot have limit points. We have also remarked that the infinite sets $\mathcal{N}$ and $\mathcal{Z}$ which are unbounded have no limit points. However, every bounded infinite set has at least one limit point. This classical result which is of fundamental importance in analysis we shall prove now.

Theorem 4.1 (Bolzano-Weierstrass Theorem). Every bounded infinite set S of real numbers has at least one limit point. This limit point may not belong to S.

Proof. Since S is bounded and infinite, there exists an interval $[m, M]$ such that $S \subseteq [m, M]$ and $[m, M] \cap S$ is an infinite set. Consider the set

$$T = \{x : (-\infty, x) \cap S \text{ is finite}\}.$$

Clearly, $m \in T$, and hence $T \neq \emptyset$. Now we claim that M is an upper bound of T. If not, there exists $x_0 \in T$ such that $x_0 > M$. But this implies that $[m, M] \cap S \subset (-\infty, x_0) \cap S$ is infinite and so $x_0 \notin T$. Thus, by the completeness axiom T has the supremum, say, $\tau = \sup T$. If $\tau \in (a, b)$, $a < \tau \Rightarrow a$ is not an upper bound of T and so there exists an $\eta \in T$ such that $a < \eta \leq \tau \Rightarrow (-\infty, a] \cap S$ is finite, because $(-\infty, \eta) \cap S$ is finite. But $\eta < b \Rightarrow b \notin T \Rightarrow (-\infty, b) \cap S$ is infinite. Thus, $(a, b) \cap S = [(-\infty, b) \cap S] \backslash (-\infty, a] \cap S$ is infinite. But this means that τ is a limit point of the set S. Finally, we note that with a slight modification in the arguments the set T can be taken as $\{x : (x, \infty) \cap S \text{ is finite}\}$.

We shall now prove the following interesting results.

Theorem 4.2. Let S be a bounded infinite set. Then,

(1). S' is bounded.

(2). S' has smallest (minimum) and greatest (maximum) members.

Proof. (1). Since S is bounded and infinite from Theorem 4.1 it follows that $S' \neq \emptyset$, and there exists an interval $[m, M]$ such that $S \subseteq [m, M]$. Let $x \notin [m, M]$, then either $x < m$ or $x > M$. If $x < m$, nbd $(x - \epsilon, m)$ of x contains no points of $[m, M]$ and so of $S \Rightarrow x \notin S'$. If $x > M$, nbd $(M, x + \epsilon)$ of x contains no points of $[m, M]$ and so of $S \Rightarrow x \notin S'$. Thus, $x \notin [m, M] \Rightarrow x \notin S'$. Hence, $S' \subseteq [m, M]$, i.e., S' is bounded and the bounds of S are the bounds of S'.

(2). Since S' is nonempty and bounded, it has infimum. Let $\lambda = \inf S'$. Then, for any $\epsilon > 0 \, \exists$ a point $\xi \in S'$ such that $\lambda \leq \xi < \lambda + \epsilon$ and so $\xi \in (\lambda - \epsilon, \lambda + \epsilon)$. This implies that $(\lambda - \epsilon, \lambda + \epsilon)$ is an nbd of $\xi \in S'$, and so contains infinitely many points of S. Since $(\lambda - \epsilon, \lambda + \epsilon)$ contains λ and infinitely many points of S for every $\epsilon > 0$, λ is a limit point of S, i.e., $\lambda \in S'$. Hence, S' contains the infimum λ. The proof of $\sup S' \in S'$ is similar.

Theorem 4.3. The closure of the set S is the smallest closed set containing S, and so $(\overline{\overline{S}}) = \overline{S}$.

Proof. Let A be a closed set such that $S \subseteq A$. Then, by Problem 4.2(i), $\overline{S} \subseteq \overline{A} = A$. Since $S \subseteq \overline{S}$, the smallest closed set containing S is $\overline{S}$. Now since $\overline{S}$ is closed it is the smallest closed set containing itself. Therefore, $(\overline{\overline{S}}) = \overline{S}$.

The smallest and greatest numbers, say, ρ and σ of the derived set S' of an infinite bounded set S, which in view of Theorem 4.2 exist, are denoted by $\underline{\lim}S$ and $\overline{\lim}S$ and are called *infima* (or *lower*) and *suprema* (or *upper*) *limits* of S. Clearly, $\rho \le \sigma$ and for any $\epsilon > 0$, $(\rho - \epsilon, \sigma + \epsilon)$ contains infinitely many members of S and at the most only a finite number of members of S can lie outside the interval $(\rho - \epsilon, \sigma + \epsilon)$.

The collection $\mathcal{O} = \{O_i\}$ of open sets O_i is called an *open cover* of the set S if $S \subseteq \cup_i O_i$. If $\mathcal{O}$ and $\mathcal{P}$ are open covers of S and $\mathcal{P} \subseteq \mathcal{O}$, then $\mathcal{P}$ is called a *subcover* of S. The set S is said to be *compact* iff every open cover of S contains a finite subcover.

Theorem 4.4 (Heine-Borel Theorem). The set S is compact iff it is closed and bounded.

Proof. Suppose that S is compact. For each $n \in \mathcal{N}$, let $I_n = (-n, n)$. Clearly each I_n is open, and $S \subseteq \cup_{n=1}^{\infty} I_n$. Thus, $\{I_n\}$ is an open cover of S, and since S is compact, there exist finitely many integers $n_1, n_2, \cdots, n_m$ such that $S \subseteq \cup_{k=1}^{m} I_{n_k} = I_M$, where $M = \max\{n_1, n_2, \cdots, n_m\}$. Thus, it follows that $|x| < M$ for all $x \in S$, and hence S is bounded.

Now suppose that S is not closed. Then, there will be a point $p \in (\text{cl } S) \backslash S$. For each $n \in \mathcal{N}$, let $U_n = \mathcal{R} \backslash \text{cl } N(p, 1/n)$. Clearly, each U_n is an open set, and

$$\cup_{n=1}^{\infty} U_n = \mathcal{R} \backslash \cap_{n=1}^{\infty} \text{cl } N(p, 1/n) = \mathcal{R} \backslash \{p\} \supseteq S.$$

Thus, $\{U_n\}$ is an open cover of S, and since S is compact, there exists finitely many integers $n_1 < n_2 < \cdots < n_m$ such that $S \subseteq \cup_{k=1}^{m} U_{n_k}$. However, since $U_k \subseteq U_n$ if $k \le n$, it follows that $S \subseteq U_{n_m}$. But then, $S \cap N(p, 1/n_m) = \emptyset$, contradicting our choice of $p \in (\text{cl } S) \backslash S$.

Conversely, suppose that S is closed and bounded. Let $\mathcal{O}$ be an open cover of S. For each $x \in \mathcal{R}$, let $S_x = S \cap (-\infty, x]$ and $B = \{x : S_x$ is covered by a finite subcover of $\mathcal{O}\}$. Since S is closed and bounded, Problem 4.5 implies that S has a minimum, say, ρ. Then, $S_\rho = \{\rho\}$, and this is certainly covered by a finite subcover of $\mathcal{O}$. Thus, $\rho \in B$, and so B is nonempty. If we can show that B is not bounded above, then it will contain a number p greater than $\sup S$. But then, $S_p = S$, and since $p \in B$, we can conclude that S is compact. To this end, we suppose that B is bounded above, and let $\sigma = \sup B$. We shall show that $\sigma \in S$ and $\sigma \notin S$ both lead to contradictions. If $\sigma \in S$, then since $\mathcal{O}$ is an open cover of S, there exists O_0 in $\mathcal{O}$ such that $\sigma \in O_0$. Since O_0 is open, there exists an interval $[x_1, x_2]$ in O_0 such that $x_1 < \sigma < x_2$. Since $x_1 < \sigma$ and $\sigma = \sup B$, there exists $O_1, O_2, \cdots, O_m$ in $\mathcal{O}$ that cover S_{x_1}. But then, $O_0, O_1, O_2, \cdots, O_m$ cover S_{x_2}, so that $x_2 \in B$. This contradicts $\sigma = \sup B$. On the other hand, if $\sigma \notin S$, then since S is closed there exists an $\epsilon > 0$ such that $N(\sigma, \epsilon) \cap S = \emptyset$. But then, $S_{\sigma-\epsilon} = S_{\sigma+\epsilon}$. Since $\sigma - \epsilon \in B$, we have $\sigma + \epsilon \in B$, which again contradicts $\sigma = \sup B$. Thus, B is not bounded above, and this implies that S is compact.

In the Heine-Borel theorem both the properties of S, that it is bounded and closed are necessary. For this first consider the bounded but not closed set $A = (0, 1]$. Let $\mathcal{O} = \{(1/n, 3) : n \in \mathcal{N}\}$. Clearly, $\mathcal{O}$ is an open cover of A, but no finite subset of $\mathcal{O}$ covers A. Now consider the closed but not bounded set $B = [0, \infty)$. Let $\mathcal{O} = \{(n - 3, n) : n \in \mathcal{N}\}$. Clearly, $\mathcal{O}$ is an open cover of B, but no finite subset of $\mathcal{O}$ covers B.

Problems

4.1. Let S, $T \subseteq \mathcal{R}$. Prove that

(i). $S \subseteq T \Rightarrow S' \subseteq T'$.

(ii). $(S \cup T)' = S' \cup T'$.

(iii). $(S \cap T)' \subseteq S' \cap T'$, and give an example to show that $\subseteq$ cannot be replaced by $=$.

4.2. Let S, $T \subseteq \mathcal{R}$. Prove that

(i). $S \subseteq T \Rightarrow \overline{S} \subseteq \overline{T}$.

(ii). $\overline{S \cup T} = \overline{S} \cup \overline{T}$.

(iii). $\overline{\overline{S}} = \overline{S}$.

4.3. Show that if the set S is bounded above or below, then so is S' with the bounds of S and has greatest or smallest member accordingly, provided $S' \neq \emptyset$.

4.4. Show that infima and suprema of the set S are also infima and suprema of $\overline{S}$ and are contained in $\overline{S}$ according as S is bounded below or above.

4.5. Let S be a nonempty closed bounded subset of $\mathcal{R}$. Show that S has a maximum and a minimum.

4.6. Show that every open set can be represented as a countable union of pairwise disjoint open intervals.

4.7. Prove that

(i). The intersection of an arbitrary collection of compact sets is compact.

(ii). The union of any finite number of compact sets is compact.

4.8. Show that a closed set either contains an interval or else is nowhere dense.

Answers or Hints

4.1. We shall prove part (ii). $S \subseteq S \cup T \Rightarrow S' \subseteq (S \cup T)'$ and $T \subseteq S \cup T \Rightarrow T' \subseteq (S \cup T)'$. Thus, $S' \cup T' \subseteq (S \cup T)'$. If $x \in (S \cup T)'$, $\forall$ nbd N of x, $N \cap (S \cup T)$ is infinite $\Rightarrow N \cap S \vee N \cap T$ is infinite $\Rightarrow x \in S' \vee x \in T' \Rightarrow x \in S' \cup T' \Rightarrow (S \cup T)' \subseteq S' \cup T'$.

4.2. We shall prove part (iii). $\overline{\overline{S}} = (S \cup S') \cup (S \cup S')' = (S \cup S') \cup (S' \cup S'') = (S \cup S') \cup S' = S \cup S' = \overline{S}$, where we have used the fact that $S'' \subseteq S'$.

4.3. See the proof of Theorem 4.2.

4.4. See the proof of Theorem 4.2.

4.5. Since S is bounded above $\sigma = \sup S$ exists. Thus, given $\epsilon > 0$, $\sigma - \epsilon$ is not an upper bound for S. If $\sigma \notin S$, there exists $x \in S$ such that $\sigma - \epsilon < x < \sigma$. But this implies that σ is a limit point of S. However, since S is closed $\sigma \in S$, and hence $\sigma = \max S$.

4.6. Let A be an open subset of $\mathcal{R}$. Let $x \in A$ and let I_x be the largest open interval in $\mathcal{R}$ such that $x \in I_x \subseteq A$. If $x, y \in A$ are such that $I_x \cap I_y \neq \emptyset$, then $I_x \cup I_y$ is still an open interval, so we have $I_x \cup I_y = I_x = I_y$. Thus, for $x, y \in A$ we have that I_x and I_y are disjoint or equal. Therefore, A is a union of pairwise disjoint open intervals in $\mathcal{R}$, i.e., $A = \bigcup \{I_x, x \in U\}$, where $U \subseteq A$. Finally, we prove that this union is countable. Let A_q be the set of rational points in A and let I be the collection $\{I_x, x \in U\}$. Define the map $\sigma : A_q \to I$ as follows: for each $q \in A_q$ let $\sigma(q)$ be that unique interval in I which contains q. Note that $\sigma : A_q \to I$ is a bijection. Thus, I is countable.

4.7. (i). We know that an arbitrary intersection of closed sets is closed. The arbitrary intersection of bounded sets is bounded, since the intersection is contained in every set of the intersection.

(ii). The union of any finite number of bounded sets is bounded and the union of any finite number of closed sets is closed.

4.8. Suppose the set S is closed and not nowhere dense. Then, there is some interval I_0 such that for each interval $I \subseteq I_0$, $I \cap S \neq \emptyset$. It suffices to show that $I_0 \subseteq S$. Let $x_0 \in I_0$. Then, every nbd of x_0 contains within it at least one point of S. But this implies that either $x_0 \in S$ or x_0 is a limit point of S. However, since S is closed $x_0 \in S$.

Chapter 5

Cardinality

The sizes of finite sets can always be compared by counting the number of elements; however, this approach cannot be extended to infinite sets. A more effective approach which is applicable to finite as well as infinite sets to compare their sizes is to begin with a one-to-one correspondence (bijective mapping) between the elements of the given sets. This approach is particularly useful to realize how differently the elements of infinite sets are saturated. In particular, this will show us that there are many different sizes of infinite sets. We begin this chapter with the following definitions.

Two sets A and B are said to be *equivalent*, or have the same *cardinal number*, provided there is a bijective mapping $f : A \to B$. The equivalence of these sets is written as $A \sim B$. It immediately follows that $\sim$ is an equivalence relation. Clearly, between two finite sets a bijective mapping can be set iff they have the same number of elements. The cardinal number of $\emptyset$ is always taken as 0, whereas for the set $\{1, 2, \cdots, n\}$ it is n. If a cardinal number is not finite, it is called *transfinite*.

Set A is said to be *countable* (*denumerable*) if A is equivalent to the set $\mathcal{N}$. The cardinal number of a denumerable set is denoted by $\aleph_0$ (the Hebrew letter called aleph zero). An *uncountable* set is an infinite set which is not countable. Countable infinite sets are considered of smallest size. The set $\mathcal{Z}$ is countable. For this, it suffices to note that $f : \mathcal{N} \to \mathcal{Z}$ defined by

$$f(n) = \begin{cases} n/2 & \text{if } n \text{ is even} \\ -(n-1)/2 & \text{if } n \text{ is odd,} \end{cases}$$

or

$$f(n) = \begin{cases} -n/2 & \text{if } n \text{ is even} \\ (n-1)/2 & \text{if } n \text{ is odd.} \end{cases}$$

is a bijective mapping. Similarly, it can be shown that the set of even (odd) integers is equivalent to $\mathcal{N}$. Thus, an infinite set can be equivalent to a proper subset of itself. As an another example, from the bijective mapping $f(x) = \tan x$ it follows that the interval $(-\pi/2, \pi/2) \sim \mathcal{R}$.

The following properties of countable and uncountable sets are fundamental.

37

(P1). Any infinite set has a countable subset.

(P2). Every subset of a countable set is at most countable.

(P3). The union of a countably many countable sets is also a countable set.

(P4). The Cartesian product of two countable sets A and B is countable.

(P5). If $A = B \cup C$, where B is an arbitrary infinite set and C is at most countable (finite or countable), then $A \sim B$.

(P6). If A is an uncountable set and B is a finite or countable subset of A then $A \backslash B \sim A$.

(P7). If $A \subset B$ and A is uncountable, then B is uncountable.

For illustration we shall prove (P4). We arrange the elements of the sets A and B as $A = \{a_1, a_2, \cdots\}$ and $B = \{b_1, b_2, \cdots\}$ so that $A \times B$ can be written as

$$\begin{array}{cccc} (a_1, b_1) & (a_1, b_2) & (a_1, b_3) & \cdots \\ (a_2, b_1) & (a_2, b_2) & (a_2, b_3) & \cdots \\ (a_3, b_1) & (a_3, b_2) & (a_3, b_3) & \cdots \\ \cdots & \cdots & \cdots & \cdots \end{array}$$

Now we define a bijective mapping $f : A \times B \to \mathcal{N}$ as follows: $f(a_1, b_1) = 1$, $f(a_2, b_1) = 2$, $f(a_1, b_2) = 3$, $f(a_1, b_3) = 4$, $f(a_2, b_2) = 5$, $f(a_3, b_1) = 6$, $f(a_4, b_1) = 7$, $f(a_3, b_2) = 8$, $f(a_2, b_3) = 9$, $f(a_1, b_4) = 10, \cdots$ (draw the diagonal zigzag path).

Now we shall prove the following important results.

Theorem 5.1. The set of all rational numbers $\mathcal{Q}$ is countable.

Proof. Clearly, $\mathcal{Q} = \cup_{n=1}^{\infty} A_n$, where A_n is the set of rational numbers with denominator n, i.e., $A_n = \{0/n, -1/n, 1/n, -2/n, 2/n, \cdots\}$. Now each A_n is equivalent to $\mathcal{Z}$ and is thus countable. Hence, the set $\mathcal{Q}$ is the countable union of countable sets, and hence from (P3) must be countable.

From Theorem 5.1 and (P2) it follows that the set of all rational numbers contained in any given interval is countable. However, these rational numbers cannot be arranged as an increasing sequence. This follows from the fact that there is no smallest rational number among the numbers exceeding a given rational number.

Theorem 5.2. Let the elements of a set A be specified by a finite number of parameters each of which can independently take on any value belonging to a countable set. Then, the set A is countable.

Proof. We write the elements of the set A as $a(\lambda_1, \lambda_2, \cdots, \lambda_n)$ where $\lambda_1, \lambda_2, \cdots, \lambda_n$ are parameters. We assume, without loss of generality, that these parameters are natural numbers. For each $a(\lambda_1, \lambda_2, \cdots, \lambda_n) \in A$ we set

$N(a) = \lambda_1 + \lambda_2 + \cdots + \lambda_n$. It is clear that $N(a)$ is a natural number and $N(a) \geq n$. Now given any $m \geq n$ let A_m denote the set of all elements of A for which $N(a) = m$. It is clear that every set A_m is finite and $A = \cup_{m=n}^{\infty} A_m$. Therefore, from (P3), set A is countable.

From Theorem 5.2 it immediately follows that the set $\mathcal{P}_n$ of all algebraic polynomials of a fixed degree n with rational coefficients

$$P_n(x) = p_0 x^n + p_1 x^{n-1} + \cdots + p_{n-1} x + p_n, \quad p_0 \neq 0, \quad p_i \in \mathcal{Q}, \ 0 \leq i \leq n$$

is countable. Thus, from (P3) we can conclude that the set of all algebraic polynomials $\mathcal{P} = \cup_{n=0}^{\infty} \mathcal{P}_n$ with rational coefficients is countable.

A real number is called *algebraic* if it is a zero of an algebraic polynomial with integer coefficients. For example, the number $\sqrt{7}$ is algebraic because it is a zero of the polynomial $x^2 - 7$. Since each algebraic polynomial can have only a finite number of distinct real zeros the countability of the set $\mathcal{P}$ immediately implies that the set of all algebraic numbers is countable. Real numbers which are not algebraic numbers are called *transcendental numbers*, e.g., the numbers e, π are transcendental.

Theorem 5.3. The set of all real numbers contained in the interval $(0, 1)$ is uncountable.

Proof. Suppose to the contrary that there is a bijective mapping f from $\mathcal{N}$ onto the interval $(0, 1)$. We write these numbers as $f(1), f(2), \cdots$. Now each number $f(n)$ has an infinite decimal expansion, so we can write these numbers as

$$\begin{aligned} f(1) &= 0.a_{11}a_{12}a_{13}\cdots \\ f(2) &= 0.a_{21}a_{22}a_{23}\cdots \\ f(3) &= 0.a_{31}a_{32}a_{33}\cdots \end{aligned}$$

$$\vdots$$

where each $a_{ij} \in \{0, 1, \cdots, 9\}$. Here some numbers such as $1/4 = 0.25000\cdots = 0.24999\cdots$ have more than one representation, but this will not create any problem. Now we construct a number $y = 0.b_1 b_2 b_3 \cdots$ where

$$b_n = \begin{cases} 4 & \text{if} \quad a_{nn} \neq 4 \\ 3 & \text{if} \quad a_{nn} = 4, \end{cases} \quad n \in \mathcal{N}.$$

This number y is obviously in the interval $(0, 1)$. But, y is not one of the numbers $f(n)$, because it differs from $f(n)$ at the nth decimal place. (Since none of the digits in y are 0 or 9, it is also not one of the numbers with two representations.) This contradiction completes the proof.

From Theorem 5.3 and (P7), it is clear that the set of all real numbers contained in the interval $[0, 1]$ is uncountable. Now from the transformation $y = a + (b - a)x$, which maps the interval $[0, 1]$ to $[a, b]$, it follows that any interval $[a, b]$ is uncountable. Next from (P6), it is immediate that half-open as well as open intervals are uncountable. Now from (P7), or from the equivalence $(-\pi/2, \pi/2) \sim \mathcal{R}$ it follows that $\mathcal{R}$ is uncountable. Finally, from (P6) it is clear that the set of transcendental numbers in any interval is uncountable.

A set A is said to have the *power of continuum* (cardinality), denoted as c, if it is equivalent to the set of real numbers contained in the interval $[0, 1]$. It is clear that any interval closed, open, half-open, finite or infinite has the power c. Also, the power of the set of transcendental numbers in any interval is c.

Now we shall prove the following interesting result.

Theorem 5.4. A finite or countable union of disjoint sets each of power c is itself of power c.

Proof. Let $A = \cup_n A_n$, where the union of disjoint sets is finite or countable, and each of the sets A_n has the power c. For each n we consider the interval $[n-1, n)$. Since every interval has the power of the continuum, it follows that $A_n \sim [n-1, n)$. Therefore, $A \sim \cup_n [n-1, n)$, and hence $A \sim [0, m)$ or $A \sim [0, \infty)$ according as the union is finite or countable. Thus, in either case A is equivalent to an interval, and hence its power is c.

Next we state the extensions of Theorems 5.2 and 5.4.

Theorem 5.5. Let the elements of a set A be specified by a finite number of parameters each of which can independently take on any value belonging to a set of power c. Then, the set A is also of power c.

Theorem 5.6. If $A = \cup_x A_x$, where x runs through a set of power c and each of the sets A_x is of power c, then the set A is of power c.

If in Theorem 5.5 we identify the coordinates x and y of the xy-plane as parameters, then it immediately follows that the set of all points in the square $0 \le x, y \le 1$, as well as the set of all points in the xy-plane is of power c. This means that it is possible to set up a bijective mapping between the points of the square $0 \le x, y \le 1$ and the interval $[0, 1]$.

Now to compare the sizes of infinite sets we note that for any two given sets A and B with cardinal numbers a and b (which are just symbols), only three possibilities can arise:

1. A is equivalent to some subset of B, and B is equivalent to some subset of A. In this case $A \sim B$ (Bernstein's theorem), and we have $a = b$.

2. A is equivalent to some subset of B, but B is not equivalent to any subset of A. In this case $a < b$.

3. B is equivalent to some subset of A, but A is not equivalent to any subset of B. In this case $a > b$.

This ordering of the cardinal numbers immediately allows us to conclude that $\aleph_0 < c$. The question whether there exists uncountable sets of cardinality less than c is known as the *continuum hypothesis*. In 1900, Hilbert included this question in his famous 23 unsolved problems. Later developments in set theory showed that the continuum hypothesis itself has some logical difficulties and requires a more rigorous setting. The following properties of the cardinal numbers $\aleph_0$, c can be proved: (a) $\aleph_0 \cdot \aleph_0 = \aleph_0$. (b) $c + \aleph_0 = c$. (c) $c + c = c$. (d) $c \cdot c = c$. (e) $2^{\aleph_0} = c$.

Finally, in this chapter we shall show that there are uncountable sets of cardinality greater than c. For this, we recall that the collection of all subsets of a given set A is called the *power set* of A, and is denoted by $\mathcal{P}(A)$. If a is the cardinality of A, then it follows that there are 2^a distinct subsets of A.

Theorem 5.7. For any set A, cardinality of $A <$ cardinality of $\mathcal{P}(A)$.

Proof. The function $f : A \to \mathcal{P}(A)$ defined by $f(x) = \{x\}$ is clearly injective, so cardinality of $A \le$ cardinality of $\mathcal{P}(A)$. To show that the cardinality of $A \ne$ cardinality of $\mathcal{P}(A)$, we will show that no function $g : A \to \mathcal{P}(A)$ can be surjective. Clearly, for each $x \in A$, $g(x)$ is a subset of A. Now for $x \in A$ either $x \in g(x)$, or $x \notin g(x)$. Let $B = \{x \in A : x \notin g(x)\}$. Since, $B \subseteq A$ it follows that $B \in \mathcal{P}(A)$. If g were surjective, then $B = g(y)$ for some $y \in A$. Now either $y \in B$ or $y \notin B$. But, we will show that both possibilities lead to contradictions. If $y \in B$, then $y \notin g(y)$ by the definition of B. But, $g(y) = B$, so $y \notin B$. On the other hand, if $y \notin B$, then $y \notin g(y)$, which implies that $y \in B$.

From Theorem 5.7, it follows that the cardinality of $\mathcal{N} = \aleph_0 <$ cardinality of $\mathcal{P}(\mathcal{N}) = 2^{\aleph_0} = c <$ cardinality of $\mathcal{P}(\mathcal{P}(\mathcal{N})) < \cdots$. Thus, we have an infinite sequence of transfinite cardinals each larger than the one preceding.

Problems

5.1. Suppose that $f : A \to B$ is a function from a nonempty finite set A to a nonempty finite set B, where cardinality $A =$ cardinality B. Show that f is injective iff f is surjective. Give a counterexample to show that this is false when A and B are both infinite sets.

5.2. Show that the set of intervals with rational numbers as endpoints is countable.

5.3. Let $\mathcal{O}$ be a collection of disjoint open subsets of $\mathcal{R}$. Prove that $\mathcal{O}$ is countable.

5.4. Give an example of a collection of disjoint closed subsets of $\mathcal{R}$ that is not countable.

5.5. Let S be a nonempty set. Show that the following conditions are equivalent.

   (i). S is countable.
   (ii). There exists an injective mapping $f : S \to \mathcal{N}$.
   (iii). There exists a surjective mapping $f : \mathcal{N} \to S$.

5.6. Let $I = [0,1]$. Remove the open middle third segment $(1/3, 2/3)$ and let A_1 be the set that remains, i.e., $A_1 = [0, 1/3] \cup [2/3, 1]$. Then, remove the open middle third segment from each of the two parts of A_1 and call the remaining set A_2. Thus, $A_2 = [0, 1/9] \cup [2/9, 1/3] \cup [2/3, 7/9] \cup [8/9, 1]$. Continue in this manner, i.e., given A_k, remove the open middle third segment from each of the closed segments whose union is A_k, and call the remaining set A_{k+1}. Clearly, $A_1 \supseteq A_2 \supseteq A_3 \supseteq \cdots$ and that for each $k \in \mathcal{N}$, A_k is the union of 2^k closed intervals each of length 3^{-k}. The set $C = \cap_{k=1}^{\infty} A_k$ is called the *Cantor set*.

   (i).   Prove that C is compact.
   (ii). Let $x = 0.a_1 a_2 a_3 \cdots$ be the base 3 (ternary) expansion of a number x in $[0, 1]$. Prove that $x \in C$ iff $a_n \in \{0, 2\}$ for all $n \in \mathcal{N}$.
   (iii). Prove that C is uncountable.
   (iv). Prove that C contains no intervals.
   (v).   Prove that $1/4 \in C$ but $1/4$ is not an endpoint of any of the intervals in any of the sets A_k, $k \in \mathcal{N}$.

Answers or Hints

5.1. Suppose f is injective. Let $A = \{a_1, \cdots, a_n\}$, cardinality $A = n$. Then, $f(A) = \{f(a_1), \cdots, f(a_n)\}$. Since f is injective $f(a_i) \neq f(a_j)$, $i \neq j$. Hence, cardinality $f(A) = n = $ cardinality B. Since $f(A) \subseteq B$, we must have $f(A) = B$, i.e., f is surjective. Conversely, if f is surjective, cardinality $f(A) = $ cardinality $B = n$. If $f(a_i) = f(a_j)$ for some $i \neq j$, then cardinality $\{f(a_1), \cdots, f(a_n)\} < n$, contradicting cardinality $f(A) = n$. Hence, f is injective. Now let $A = B = \mathcal{N}$. The function $f_1 : A \to B$, $f_1(n) = 2n$ is injective but not surjective. The function $f_2 : A \to B$, $f_2(2n) = n$, $f_2(2n-1) = n$ is surjective but not injective.

5.2. Let S be the collection of intervals I_α with rational numbers a_α, b_α ($a_\alpha < b_\alpha$) as endpoints (here $\alpha \in I$ and $I \subseteq \mathcal{R}$). Notice the map $f : S \to Q \times (Q \setminus \{0\})$ given by $f(I_\alpha) = \{a_\alpha, b_\alpha - a_\alpha\}$ is a bijection and $Q \times (Q \setminus \{0\})$ is countable.

5.3. Let S be the collection of intervals I_α with rational numbers as endpoints (here $\alpha \in I$ and $I \subseteq \mathcal{R}$) and $\mathcal{O}$ be the collection of disjoint open subsets $\mathcal{O}_\beta$ of $\mathcal{R}$ (here $\beta \in J$ and $J \subseteq \mathcal{R}$). Fix $\beta \in J$. Note it is easy to see (see Problem 4.6) that there is a collection of intervals I_α in S, say $\alpha \in \theta \subseteq I$, with $\mathcal{O}_\beta \subseteq \cup_{\alpha \in \theta} I_\alpha$ and $\mathcal{O}_\gamma \cap I_\alpha = \emptyset$ for $\gamma \in J \setminus \{\beta\}$ and $\alpha \in \theta$. Now apply Problem 5.2.

5.4. Note $\cup_{x \in \mathcal{R}} \{x\}$ is a collection of disjoint closed subsets of $\mathcal{R}$ which is not countable.

5.5. (i)$\Rightarrow$(ii) is immediate since a bijective map is injective. For (ii)$\Rightarrow$(i) let $f : S \to \mathcal{N}$ be an injective map. Note $f : S \to f(S)$ is a bijection and since $f(S)$ is a subset of $\mathcal{N}$ it is countable. Thus, S is countable.

(i)$\Rightarrow$(iii) is immediate. For (iii)$\Rightarrow$(ii), let $f : \mathcal{N} \to S$ be a surjective map and let $s \in S$, and $A_s = f^{-1}(s)$. Note that $A_s \subset \mathcal{N}$ and $A_s \cap A_r = \emptyset$ if $r, s \in S$ and $r \neq s$ (note if there exists $w \in A_s \cap A_r$ then $s = f(w) = r$) and $\mathcal{N} = \cup_{s \in S} A_s$. Let $a_s = \min A_s$. Finally, note that $g : S \to \mathcal{N}$ given by $g(s) = a_s$ is an injective map.

5.6. (i). It is clear that each A_k is closed and bounded, thus C which is the intersection of compact sets A_k must be compact.

(ii). We know that some points admit more than one representation in basis 3. For example $1/3$, which can be written as 0.1_3 or as $0.022222 \cdots_3$, $2/3$ can be written as 0.2_3 or as $0.12222 \cdots_3$. In the construction of the Cantor set, we remove the middle third, which contains the numbers with ternary numerals of the form $0.1aaaaa \cdots_3$ where $aaaaa \cdots_3$ is strictly between $00000 \cdots_3$ and $22222 \cdots_3$. Therefore, the remaining numbers are of the form $0.0xxxxx \cdots_3$, $1/3 = 0.1_3 = 0.022222 \cdots_3$, $2/3 = 0.122222 \cdots_3 = 0.2_3$, or numbers of the form $0.2xxxxx \cdots_3$. Since $x \notin (1/3, 2/3)$, $a_1 \neq 1$ and also since $x \notin (1/9, 2/9) \cup (7/9, 8/9)$, $a_2 \neq 1$, so by induction $a_n \neq 1$ for all n.

(iii). Suppose C is countable, so that we can find a bijective mapping $f : \mathcal{N} \to C$. Since C is countable, we have $C = \{x_1, x_2, \cdots, x_n, \cdots .\}$, where $x_i = 0.x_{i1}x_{i2} \cdots x_{in} \cdots$ with $x_{ij} \in \{0, 2\}$. We define $x = 0.x_1x_2 \cdots x_n \cdots$, where $x_i = 0$ if $x_{ii} = 2$ or $x_i = 2$ if $x_{ii} = 0$. We have $x \in C$ but $f(n) \neq x$ for all $n \in \mathcal{N}$, a contradiction.

(iv). At the k step we removed 2^{k-1} intervals of length $\frac{1}{3^k}$. Thus the length of the complement is $\sum_{k=1} \frac{2^{k-1}}{3^k} = \frac{1}{3} \sum_{k=1} \frac{2^{k-1}}{3^{k-1}} = \frac{1}{3} \sum_{k=1} \frac{2^k}{3^k} = 1$. Therefore, we removed a set of length 1 from $[0, 1]$ which has length 1.

(v). $\frac{1}{4} = 0.020202 \cdots \in C$. At each step of removal, each number $x = 0.x_1x_2 \cdots x_n \cdots$ with a ternary expansion containing 1 is removed and any number remaining must have the digit $x_1 = 0$ if $x \in [0, \frac{1}{3}]$ or 2 if $x \in [\frac{2}{3}, 1]$. Repeating this argument for each step of removal, at the kth step, any endpoint has one of this form: 2 at 3^{-k} position which repeats infinitely, or 0 at $3^{-(k-1)}$, that means we are done at this point. We remark that $\frac{1}{4}$ does not have this form.

Chapter 6

Real-Valued Functions

In mathematics the concept of a function is a basic notion, and the main goal of calculus is to analyze functions. In this chapter, we will first introduce several special functions which are of common use, and then define functions which have certain properties.

Let $A \subseteq \mathcal{R}$. If $f : A \to \mathcal{R}$ we call f a *real-valued function*. The set A is called the *domain* of f. If $x \in A$, then $f(x)$ is called the *value* of f at x. If $f : A \to \mathcal{R}$ and $g : A \to \mathcal{R}$, then the four arithmetical operations on the real numbers, namely, sum, difference, product, and quotient induce analogous operations on these functions, and are defined as follows:

$$
\begin{aligned}
(f + g)(x) &= f(x) + g(x) \\
(f - g)(x) &= f(x) - g(x) \\
(f \cdot g)(x) &= (fg)(x) = f(x)g(x) \\
\left(\tfrac{f}{g}\right)(x) &= \tfrac{f(x)}{g(x)} \quad \text{provided} \quad g(x) \neq 0,\ \forall\, x \in A.
\end{aligned}
$$

A function defined on a set A that maps each element of A into itself is called the *identity function* on A, and is denoted as i_A. Thus, $i_A(x) = x, \forall\, x \in A$. Now consider the function $f : A \to B \subseteq \mathcal{R}$. A function $f^{-1} : B \to A$ is said to be the *inverse* of f if $f^{-1} \circ f = i_A$ and $f \circ f^{-1} = i_B$, i.e., $f^{-1}(f(x)) = x, \forall\, x \in A$ and $f(f^{-1}(y)) = y, \forall\, y \in B$. A function is called *invertible* if it has an inverse. If f is one-to-one on the set A, then f^{-1} exists as a function defined on $f(A)$. Further, f^{-1} is one-to-one on $f(A)$ and the range of f^{-1} is the set A.

A function $f : A \to B$ is called a *constant function* if there is some $c \in B$ so that $f(x) = c$ for all $x \in A$. The value of a constant function does not change or vary as x varies over A.

The *signum function* is defined by

$$
(x) = \begin{cases} 1 & \text{if } x > 0 \\ 0 & \text{if } x = 0 \\ -1 & \text{if } x < 0. \end{cases}
$$

The domain of $f(x) = (x)$ is $\mathcal{R}$ and the range is the finite set $\{-1, 0, 1\}$.

Let S be a set and $A \subseteq S$. The function $\chi_A : S \to \{0, 1\}$ defined as

$$\chi_A(x) = \begin{cases} 1 & \text{if } x \in A \\ 0 & \text{if } x \in S \backslash A \end{cases}$$

is called the *characteristic function of A*. Thus the set A is characterized by the function χ_A, i.e., if C and D are subsets of S then $C = D$ iff $\chi_C = \chi_D$.

The *absolute-value function* is defined by

$$|x| = \begin{cases} x & \text{if } x \geq 0 \\ -x & \text{if } x < 0. \end{cases}$$

The domain of $f(x) = |x|$ is $\mathcal{R}$ and the range is $\mathcal{R}^+ \cup \{0\}$. For all $x, y \in \mathcal{R}$, the following properties of the absolute-value function are immediate:

1. $|x| = \max\{x, -x\}$, $\quad -|x| = \min\{x, -x\}$.
2. $|-x| = |x|$, $\quad |x| \geq x \geq -|x|$.
3. $|xy| = |x||y|$, $\quad \left|\frac{x}{y}\right| = \frac{|x|}{|y|}$, $\quad y \neq 0$.
4. $||x| - |y|| \leq |x + y| \leq |x| + |y|$ (the *triangle inequality*).

To show 4, it suffices to note that

$$(|x| + |y|)^2 = |x|^2 + |y|^2 + 2|x||y| \geq x^2 + y^2 + 2xy = (x + y)^2 = |x + y|^2$$

and

$$|x - y|^2 = (x - y)^2 = x^2 + y^2 - 2xy \geq |x|^2 + |y|^2 - 2|x||y| = ||x| - |y||^2.$$

The *distance* between any two real numbers x and y is defined by $|x - y|$. It follows that

$$|x - y| \begin{cases} = 0 & \text{iff } x = y \\ = |y - x| \\ \leq |x - z| + |z - y| & \text{for all } z \in \mathcal{R}. \end{cases}$$

If $f : A \to \mathcal{R}$ and $g : A \to \mathcal{R}$, then the functions $\max(f, g)$ and $\min(f, g)$ are defined as

$$\max(f, g)(x) = \max\{f(x), g(x)\}$$
$$\min(f, g)(x) = \min\{f(x), g(x)\}.$$

It follows that

$$\max(f, g) = \frac{|f - g| + f + g}{2}$$
$$\min(f, g) = \frac{-|f - g| + f + g}{2}.$$

For a given number $a \in \mathcal{R}$ the *positive* and *negative parts* are defined by

$$a^+ = \frac{|a| + a}{2} = \max\{a, 0\} = \begin{cases} a, & a \geq 0 \\ 0, & a < 0 \end{cases}$$

and

$$a^- = \frac{|a| - a}{2} = \max\{-a, 0\} = \begin{cases} 0, & a \geq 0 \\ -a, & a < 0. \end{cases}$$

It follows that $a^+ \geq 0$, $a^- \geq 0$, $a = a^+ - a^-$, and $|a| = a^+ + a^-$.

The *greatest-integer function* is defined by

$$[x] = \text{greatest integer which is less than or equal to } x.$$

Thus $[2.3] = 2$, $[-2.3] = -3$, and $[2] = 2$.

A function $f : A \to \mathcal{R}$ is said to be *bounded below* (*bounded above*) if there exists a real number m (M) such that $m \leq f(x)$ ($f(x) \leq M$). The numbers m and M are known as the *lower* and *upper* bounds of f on A. By the completeness axiom it is clear that f has infimum or supremum according as f is bounded below or above on A. These are written as $\inf_A f(x)$ and $\sup_A f(x)$. If $f(x)$ is not bounded below (above) we define $\inf_A f(x) = -\infty$, ($\sup_A f(x) = +\infty$). If a function is bounded below as well as above on A, then it is called a *bounded function* on A, otherwise it is called *unbounded* on A.

The following results can be proved rather easily.

(B1). If $f : A \to [a, b]$ and $\sup_A f(x) = \inf_A f(x)$, then f is a constant function.

(B2). If f is bounded on A, then

$$\sup_A f(x) = -\inf_A(-f(x)) \quad \text{and} \quad \inf_A f(x) = -\sup_A(-f(x)),$$
$$\sup_A |f(x)| = \max\{|\inf_A f(x)|, \ |\sup_A f(x)|\}.$$

(B3). If f and g are bounded on A and c is any real number, then the functions $f \pm g$, cf, $f \cdot g$ are bounded on A.

(B4). If f and g are bounded on A, then

$$\sup_A(f(x) + g(x)) \leq \sup_A f(x) + \sup_A g(x),$$
$$\inf_A(f(x) + g(x)) \geq \inf_A f(x) + \inf_A g(x).$$

In these relations $\leq$ cannot be replaced by $=$. For this it suffices to consider the functions $f(x) = x^2 + 1$ and $g(x) = 2 - x$ and $A = [0, 1]$.

A function $f : A \to \mathcal{R}$ is said to be *bounded at* a point $x_0 \in A$ provided there is an open interval J containing x_0 such that f is bounded on $J \cap A$. The following result as an application of the Heine-Borel theorem gives a condition for which boundedness at each point of a set implies boundedness on the set.

Theorem 6.1. If $f : A \to \mathcal{R}$ is bounded at each point $x_0 \in A$ and A is compact, then f is bounded on A.

Proof. Clearly, for each $x \in A$ there is an open interval $I_x = (x - \delta_x, x + \delta_x)$ such that f is bounded on $A \cap I_x$. The set of open intervals $\{I_x : x \in A\}$ is an open covering of A, and so by Theorem 4.4 there are finitely many of these open intervals, say, $I_{x_1}, I_{x_2}, \cdots, I_{x_n}$ such that $A \subseteq \cup_{k=1}^{n} I_{x_k}$. For each $k = 1, 2, \cdots, n$, f is bounded on $A \cap I_{x_k}$, and so there is an $M_k > 0$ such that $|f(x)| \leq M_k$ for all $x \in A \cap I_{x_k}$. Let $M = \max_{1 \leq k \leq n} M_k$. Now for any $x \in A$ there is a j with $1 \leq j \leq n$ such that $x \in I_{x_j}$. Hence, $x \in A \cap I_{x_j}$, and so $|f(x)| \leq M_j \leq M$. Thus, f is bounded on A.

A function $f : A \to B$ is said to be *monotone increasing* (nondecreasing) on A if $f(x_1) \leq f(x_2)$ for all $x_1, x_2 \in A$ such that $x_1 < x_2$. Similarly, f is said to be *monotone decreasing* (nonincreasing) on A if $f(x_1) \geq f(x_2)$ for all $x_1, x_2 \in A$ such that $x_1 < x_2$. A function which is either nondecreasing or nonincreasing is called a *monotone function*. If strict inequalities hold then f is respectively called *strictly increasing* and *strictly decreasing*. A function which is either strictly increasing or strictly decreasing is called a *strictly monotone function*. For example, the function $f(x) = a^x$, $x \in \mathcal{R}$ where $a > 0$ is one-to-one from $\mathcal{R}$ to $(0, \infty)$. It is strictly increasing or decreasing according as $a > 1$ or $a < 1$. For $a = 1$, $f(x) = 1$, $x \in \mathcal{R}$ and hence, it is a constant function. Similarly, the function $f(x) = |x|$ is strictly decreasing on $(-\infty, 0]$ and strictly increasing on $[0, \infty)$. The constant function as well as the function $f(x) = \sin x$, $x \in [0, 2\pi]$ is neither increasing nor decreasing. The following properties of monotone functions are fundamental.

(M1). If f is nondecreasing (nonincreasing) on A, then $-f$ is nonincreasing (nondecreasing) on A.

(M2). If $f : A \to B$ is monotone on A and $g : B \to C$ is monotone on B, then $g \circ f : A \to C$ is monotone on A.

(M3). If $f : A \to B$ is strictly monotone on A, then it is one-to-one function. Hence, the inverse function $f^{-1} : f(A) \to A$ exists, and is one-to-one.

Consider a set A with the property that if $x \in A$, then $-x \in A$. A function $f : A \to B$ is said to be an *even function* iff $f(x) = f(-x)$, $\forall x \in A$, and an *odd function* iff $f(x) = -f(-x)$, $\forall x \in A$. Clearly, the functions $\cos x$ and $\sin x$ are respectively even and odd on $\mathcal{R}$. If $f : \mathcal{R} \to \mathcal{R}$ then it can be decomposed into the sum of an even function and an odd function, i.e., $f = e + o$ where $e(x) = (f(x) + f(-x))/2$ and $o(x) = (f(x) - f(-x))/2$. If $f : \mathcal{R} \to \mathcal{R}$ is an odd function and $g : \mathcal{R} \to \mathcal{R}$ is an even function, then it follows that the functions $g \circ f$, $g \circ g$, $f \circ g$ are even. Most of the functions are neither even nor odd.

A function $f : \mathcal{R} \to \mathcal{R}$ is called *periodic* if there exists a number $\omega > 0$ such that $f(x + \omega) = f(x)$ for all $x \in \mathcal{R}$. If ω is the smallest such number, then it is called the *period* of f. Geometrically, this means that the graph of $f(x)$ repeats itself in successive intervals of length ω. For example, the functions $\sin x$ and $\cos x$ are periodic of period 2π.

Problems

6.1. Let $f : \mathcal{R} \to \mathcal{R}$ be defined by $f(x) = ax + b$, where a, b are real constants with $a \neq 0$. Find the values of a and b for which $f \circ f = i_{\mathcal{R}}$.

6.2. Sketch the graph of the following functions.

(i). $x - [x]$.
(ii). $[x] + \sqrt{x - [x]}$.
(iii). $[1/x]$, $x \neq 0$.
(iv). $[x]/x$, $x \neq 0$.
(v). $(x + 1)U(x + 1) - xU(x)$, where $U(x) = 0$ if $x < 0$ and 1 if $x \geq 0$.

6.3. Let $x, y \in \mathcal{R}$. Show that

(i). If $\epsilon > 0$, then $|x - y| < \epsilon \Leftrightarrow y - \epsilon < x < y + \epsilon$.
(ii). If $|x - y| < \epsilon$, $\forall \epsilon > 0$, then $x = y$.
(iii). If $0 < y < 1$ and $-1 < x < 1$, then $|x(1 - y)| < 1 + yx$.
(iv). If $2|x| \leq |y|$ and $y \neq 0$, then $|x|/|x - y| \leq 1$.

6.4. For any $x_1, x_2, \cdots, x_n \in \mathcal{R}$ prove the general triangle inequality
$$|x_1 + x_2 + \cdots + x_n| \leq |x_1| + |x_2| + \cdots + |x_n|.$$

6.5. Let $a_1, a_2, \cdots, a_n$ be real numbers. Show that for $n \geq 2$
$$|(1 + a_1)(1 + a_2)\cdots(1 + a_n) - 1| \leq (1 + |a_1|)(1 + |a_2|)\cdots(1 + |a_n|) - 1.$$

6.6. For any $x_k, y_k \in \mathcal{R}$, $k = 1, 2, \cdots, n$ prove

(i). The *Cauchy-Schwarz inequality*
$$\sum_{k=1}^{n} |x_k y_k| \leq \left(\sum_{k=1}^{n} x_k^2\right)^{1/2} \left(\sum_{k=1}^{n} y_k^2\right)^{1/2}.$$

(ii). The *Minkowski inequality*
$$\left[\sum_{k=1}^{n} (x_k + y_k)^2\right]^{1/2} \leq \left(\sum_{k=1}^{n} x_k^2\right)^{1/2} + \left(\sum_{k=1}^{n} y_k^2\right)^{1/2}.$$

6.7. Let $a_1, a_2, \cdots, a_n$ be positive real numbers. Prove that
$$\left(\sum_{k=1}^{n} a_k\right)\left(\sum_{k=1}^{n} \frac{1}{a_k}\right) \geq n^2.$$

Deduce that the *harmonic mean* $[(1/n)\sum_{k=1}^{n} 1/a_k]^{-1}$ is less than or equal to the arithmetic mean.

6.8.   Find all numbers x for which

(i). $|x+1| - |x-2| + |x-3| \le 2$.

(ii). $3 - x \ge 2/x$.

(iii). $|2x-1| < |x-1|$.

(iv). $||x+1| - |x-1|| < 1$.

(v). $\frac{x+4}{|x+1|} > -2$.

(vi). $\frac{x(x-2)}{|x-1|} < -2$.

6.9.   Let C and D be subsets of S. Show that

(i). $\chi_{C \cup D} = \max(\chi_C, \chi_D)$.

(ii). $\chi_{C \cap D} = \min(\chi_C, \chi_D) = \chi_C \chi_D$.

6.10.   Prove or disprove each of the following statements.

(i).   If f is bounded on A and g is unbounded on A, then $f + g$ is unbounded on A.

(ii).   If f and g are unbounded on A, then $f \cdot g$ is unbounded on A.

(iii).   If f and g are bounded below on A, then $f \cdot g$ is bounded below on A.

(iv).   If f is bounded on A and g is unbounded on A, then $f \cdot g$ is unbounded on A.

6.11.   Let m, M denote the infimum and supremum of a function $f : A \to \mathcal{R}$. Show that for all $x_1, x_2 \in A$, $\sup_A |f(x_1) - f(x_2)| = M - m$.

6.12.   Let $f, g : A \to \mathcal{R}$ be two bounded functions. Show that

(i). $|\sup_A f - \sup_A g| \le \sup_A |f - g|$ and $|\inf_A f - \inf_A g| \le \sup_A |f - g|$.

(ii).   If $|f(x) - f(y)| \le |g(x) - g(y)|$ for all $x, y \in A$, then $\sup_A f - \inf_A f \le \sup_A g - \inf_A g$.

Answers or Hints

6.1. $f \circ f(x) = x$, $x \in \mathcal{R}$ iff $a(ax + b) + b = x$, i.e., $a^2 - 1 = 0$ and $b(a + 1) = 0$, i.e., $a = 1, b = 0$ or $a = -1, b \in \mathcal{R}$.

6.3.   (ii).   Let $x \ne y$. Then $\epsilon = |x - y|/2 > 0$ would imply $|x - y| < (1/2)|x - y|$, a contradiction.

(iv). $|x - y| \ge |y| - |x| \ge \begin{cases} 2|x| - |x| = |x| \\ |y| - |y|/2 = |y|/2 > 0 \text{ since } y \ne 0. \end{cases}$

6.4.   Use induction.

6.5. Use the triangle inequality and mathematical induction.

6.6. (i). From Problem 1.10 with $n = 2$, we have

$$\frac{2|x_k||y_k|}{\left(\sum_{k=1}^{n} x_k^2\right)^{1/2} \left(\sum_{k=1}^{n} y_k^2\right)^{1/2}} \leq \frac{x_k^2}{\left(\sum_{k=1}^{n} x_k^2\right)} + \frac{y_k^2}{\left(\sum_{k=1}^{n} y_k^2\right)}.$$

Now sum both sides from $k = 1$ to n.

(ii). Note that $\sum_{k=1}^{n}(x_k + y_k)^2 = \sum_{k=1}^{n} x_k^2 + 2\sum_{k=1}^{n} x_k y_k + \sum_{k=1}^{n} y_k^2$. Now use part (i).

6.7. In Cauchy-Schwarz's inequality let $x_k = \sqrt{a_k}$ and $y_k = 1/\sqrt{a_k}$ to get $n^2 \leq \left(\sum_{k=1}^{n} a_k\right)\left(\sum_{k=1}^{n} \frac{1}{a_k}\right)$. Hence, $\left(\frac{1}{n}\sum_{k=1}^{n} \frac{1}{a_k}\right)^{-1} \leq \frac{1}{n}\sum_{k=1}^{n} a_k$.

6.8. (i). Let $f(x) = |x+1| - |x-2| + |x-3| - 2$. Then,

$$f(x) = \begin{cases} -(x+1) + (x-2) - (x-3) - 2 & \text{if } x \leq -1 \\ (x+1) + (x-2) - (x-3) - 2 & \text{if } -1 \leq x \leq 2 \\ (x+1) - (x-2) - (x-3) - 2 & \text{if } 2 \leq x \leq 3 \\ (x+1) - (x-2) + (x-3) - 2 & \text{if } x \geq 2 \end{cases}$$

$$= \begin{cases} -x - 2 & \text{if } x \leq -1 \\ x & \text{if } -1 \leq x \leq 2 \\ -x + 4 & \text{if } 2 \leq x \leq 3 \\ x - 2 & \text{if } x \geq 3. \end{cases}$$

Hence, $\{x : f(x) \leq 0\} = ((-\infty, -1] \cap [-2, \infty)) \cup ([-1, 2] \cap (-\infty, 0]) \cup ([2, 3] \cap [4, \infty)) \cup ([3, \infty) \cap (-\infty, 2]) = [-2, -1] \cup [-1, 0] \cup \emptyset \cup \emptyset = [-2, 0]$. Thus, $f(x) \leq 0$ when $-2 \leq x \leq 0$.

(ii). $x < 0$, or $1 \leq x \leq 2$.

(iii). $0 < x < 2/3$.

(iv). $-1/2 < x < 1/2$.

(v). $x \in (-\infty, -1) \cup (-1, \infty)$.

(vi). $x \in (2 - \sqrt{2}, 1) \cup (1, \sqrt{2})$.

6.9. We shall prove only part (i). If $x \in C \cup D$ then $\chi_{C \cup D}(x) = 1$. But either $x \in C$ or $x \in D$ (or both), and so either $\chi_C(x) = 1$ or $\chi_D(x) = 1$. Thus, $\max(\chi_C, \chi_D)(x) = 1$. Hence, $1 = \chi_{C \cup D}(x) = \max(\chi_C, \chi_D)(x)$, $x \in C \cup D$. If $x \notin C \cup D$ then $\chi_{C \cup D}(x) = 0$. But, $x \in (S \backslash C) \cap (S \backslash D)$ and hence $x \in S \backslash C$ and $x \in S \backslash D$ so that $\chi_C(x) = 0 = \chi_D(x)$. Thus, $\max(\chi_C, \chi_D)(x) = 0$. Hence, $0 = \chi_{C \cup D}(x) = \max(\chi_C, \chi_D)(x)$, $x \notin C \cup D$.

6.10. (i). True. Let $|f(x)| \leq K$, $x \in A$. For any $L > 0$, since g is unbounded on A, there exists a $x \in A$ such that $|g(x)| > L + K$. Hence, $|f(x) + g(x)| \geq |g(x)| - |f(x)| > L + K - K = L$.

(ii). False. $A = \mathcal{R}^+$, $f(x) = x$ and $g(x) = 1/x$.

(iii). False. $A = \mathcal{R}^+$, $f(x) = x$ and $g(x) = -1$.

(iv). False. $A = \mathcal{R}$, $f(x) = 0$ and $g(x) = x$.

6.11. Since $f(x_1) \le M$, $f(x_2) \ge m$, $\forall\, x_1, x_2 \in A$ we have $|f(x_1) - f(x_2)| \le M - m$. Also, for any $\epsilon > 0\, \exists\, y_1, y_2 \in A$ such that $f(y_1) > M - (\epsilon/2)$, $f(y_2) < m + (\epsilon/2)$. Thus, $f(y_1) - f(y_2) > M - m - \epsilon$, i.e., $|f(y_1) - f(y_2)| > M - m - \epsilon$, $\forall\, \epsilon > 0$. Thus, no number smaller than $M - m$ can be an upper bound of $|f(x_1) - f(x_2)|$, $\forall\, x_1, x_2 \in A$. Thus, $\sup_A |f(x_1) - f(x_2)| = M - m$.

6.12. (i). From $f = g + (f - g)$, $f - g \le |f - g|$, and B4 we have $\sup_A f \le \sup_A g + \sup_A(f-g) \le \sup_A |f-g| + \sup_A g$. Thus, $\sup_A f - \sup_A g \le \sup_A |f-g|$. Similarly, we obtain $\sup_A g - \sup_A f \le \sup_A |f-g|$. For the second inequality, we replace f with $-f$ and g with $-g$, and use B2.

(ii). Clearly, $f(x) - f(y) \le |g(x) - g(y)| \le \max\{f(x), g(x)\} - \min\{f(x), g(x)\} \le \sup_A g - \inf_A g$ and $\sup_A\{f(x) - f(y)\} \le \sup_A g - \inf_A g$. Thus, from Problem 3.5, $\sup_A\{f(x) - f(y)\} = \sup_A f(x) + \sup_A(-f(y)) = \sup_A f(x) - \inf_A f(x)$.

Chapter 7

Real Sequences

Sequences are frequently used in modeling discrete changes, and to approximate quantities which are not known exactly. In this chapter we will introduce the convergence and divergence of sequences, prove several properties of convergent sequences, and provide Cauchy's iff criterion for the convergence of a given sequence.

A function whose domain is the set of natural numbers $\mathcal{N}$ and the range is a subset of $\mathcal{R}$ is called a *real sequence*, or simply a *sequence*. Thus, $f : \mathcal{N} \to \mathcal{R}$ is a sequence. Usually, a sequence is denoted as $\{u_n\}$, i.e., by its images. We call u_n the nth term of the sequence. Sometimes the domain of a sequence is $\mathcal{N} \cup \{0\}$ or $\{n \in \mathcal{N}, \ n \geq m \in \mathcal{N}\}$. In this case we write $\{u_n\}_{n=0}^\infty$ or $\{u_n\}_{n=m}^\infty$, respectively. Clearly, the complete sequence can be defined by its nth term. For example, if $u_n = 1/n^2$, then the complete sequence $\{u_n\}$ is $\{1, \ 1/2^2, \ 1/3^2, \cdots\}$. We begin with our discussion of the limit of real sequences.

A sequence $\{u_n\}$ is said to *converge* to a real number u if for every $\epsilon > 0$ there exists a natural number N such that $|u_n - u| < \epsilon$ for all $n \geq N$, i.e., for all large n the terms of the sequence get close to u. This number u is called the *limit* of the sequence $\{u_n\}$, and interchangeably we will write it as $\lim_{n \to \infty} u_n = u$, $\lim u_n = u$, or $u_n \to u$. The value of the integer N depends on the choice of ϵ. If $\{u_n\}$ does not have a limit, we say that the sequence is *divergent*.

Thus, at this stage if we wish to show that a given sequence has a limit then we must first guess what the limit is!

Example 7.1. We shall show that $\lim_{n \to \infty}(1/n) = 0$. By the Archimedean property, given any $\epsilon > 0$ there exists a natural number N such that $1/N < \epsilon$. Now if $n \geq N$, then $|1/n - 0| = 1/n \leq 1/N < \epsilon$.

Example 7.2. We shall show that $\lim_{n \to \infty}(2n^2 + 2n)/(n^2 + 1) = 2$. Since

$$\left| \frac{2n^2 + 2n}{n^2 + 1} - 2 \right| = \left| \frac{2n - 2}{n^2 + 1} \right| < \frac{2n}{n^2} = \frac{2}{n}$$

for any given $\epsilon > 0$ if we choose N such that $N > 2/\epsilon$, then for $n \geq N$, we have $\left|(2n^2 + 2n)/(n^2 + 1) - 2\right| < 2/n \leq 2/N < \epsilon$.

Example 7.3. We shall show that $\{u_n\}$ diverges, where $u_n = (-1)^n$. Suppose $\lim_{n\to\infty} u_n = u$ exists. Let $\epsilon = 1/2$. Then, there exists an N such that for $n \geq N$, $|u_n - u| < 1/2$, i.e., $-1/2 < u_n - u < 1/2$. Since $2N$ and $2N + 1 > N$, we have $-1/2 < u_{2N} - u < 1/2$ and $-1/2 < u_{2N+1} - u < 1/2$, i.e., $-1/2 < 1 - u < 1/2$ and $-1/2 < -1 - u < 1/2$, i.e., $-1/2 < 1 - u < 1/2$ and $-1/2 < 1 + u < 1/2$ which implies $-1 < 2 < 1$, a contradiction.

Example 7.4. We shall show that $\{u_n\}$ diverges, where $u_n = n$. Suppose $\lim_{n\to\infty} u_n = u$ exists. Let $\epsilon = 1$. Then, there exists an N such that for $n \geq N$, $|u_n - u| < 1$, i.e., $-1 < u_n - u < 1$ which implies $u - 1 < n < u + 1$, a contradiction.

Often, the evaluation of limits of sequences directly by definition is not practical. However, the following general results facilitate the computation of limits.

Theorem 7.1. If $\lim_{n\to\infty} u_n$ exists, then it is unique.

Proof. Suppose that $\lim_{n\to\infty} u_n = u$ and $\lim_{n\to\infty} u_n = \tilde{u}$. Given any $\epsilon > 0$ there exists natural numbers N_1 and N_2 such that $|u_n - u| < \epsilon/2$, $n > N_1$ and $|u_n - \tilde{u}| < \epsilon/2$, $n > N_2$. Thus, if $n > N^* = \max\{N_1, N_2\}$, $|u_n - u| < \epsilon/2$ and $|u_n - \tilde{u}| < \epsilon/2$. But, then

$$|u - \tilde{u}| = |u - u_n + u_n - \tilde{u}| \leq |u_n - u| + |u_n - \tilde{u}| < \frac{\epsilon}{2} + \frac{\epsilon}{2} = \epsilon.$$

Since u and $\tilde{u}$ are fixed real numbers, and ϵ can be chosen arbitrarily small, the above inequality can hold only if $|u - \tilde{u}| = 0$, and hence $u = \tilde{u}$.

A sequence u is said to be *bounded below* or *above*, respectively, if there exist numbers m_1, m_2 such that $m_1 \leq u_n$ or $u_n \leq m_2, \forall n \in \mathcal{N}$. The numbers m_1, m_2 are called *lower* and *upper* bounds of u. If there exists numbers m_1 and m_2 such that $m_1 \leq u_n \leq m_2$, $n \in \mathcal{N}$ then the sequence u is said to be *bounded*. Clearly, a sequence is bounded iff there exists a number m such that $|u_n| \leq m$, $n \in \mathcal{N}$. Also notice that $\{u_n\}$ is bounded above iff $\{-u_n\}$ is bounded below. A sequence which is neither bounded below nor above is called *unbounded*. For example, the sequence $\{1 + (-1)^n\}$ is bounded with lower bound 0 and the upper bound 2, whereas the sequence $\{(-1)^n n\}$ is unbounded.

Theorem 7.2. If $\lim_{n\to\infty} u_n = u$ exists then the sequence $\{u_n\}$ is bounded.

Proof. Given $\epsilon = 1$ there is an $N \in \mathcal{N}$ such that $|u_n - u| < 1$ for all $n \geq N$. Thus, by the triangle inequality, $|u_n| < 1 + u$ for all $n \geq N$. If we choose $M = \max\{|u_1|, |u_2|, \cdots, |u_{N-1}|, 1 + u\}$, then clearly, $|u_n| \leq M$ for all $n \in \mathcal{N}$, and hence $\{u_n\}$ is bounded.

Thus, if a sequence is unbounded then it will necessarily diverge, see Example 7.4. We also remark that the converse of Theorem 7.2 is false, see Example 7.3; however, the following result holds.

Theorem 7.3. For every bounded sequence $\{u_n\}$ the range set $S = \{u_n : n \in \mathcal{N}\}$ has at least one limit point.

Proof. If S is finite, then for at least one $u \in S$, $u_n = u$ for infinitely many values of n, and so u is a limit point of S. If S is infinite, then since S is bounded, by Theorem 4.1 (Bolzano-Weierstrass) it has at least one limit point.

Theorem 7.4 (Squeeze Theorem). Suppose $\{x_n\}$, $\{u_n\}$ and $\{y_n\}$ are real sequences.

(1). If $x_n \to u$ and $y_n \to u$, and $x_n \leq u_n \leq y_n$, $n \geq N_0$ for some $N_0 \in \mathcal{N}$, then $u_n \to u$.

(2). If $x_n \to 0$ and $\{u_n\}$ is bounded, then $x_n u_n \to 0$.

Proof. (1). Let $\epsilon > 0$. There exists N_1, $N_2 \in \mathcal{N}$ such that $n \geq N_1$ implies $-\epsilon < x_n - u < \epsilon$ and $n \geq N_2$ implies $-\epsilon < y_n - u < \epsilon$. Set $N = \max\{N_1, N_2\}$. Then, for all $n \geq N$, $u - \epsilon < x_n \leq u_n \leq y_n < u + \epsilon$, i.e., $|u_n - u| < \epsilon$ for $n \geq N$. Thus, $u_n \to u$.

(2). Let $M > 0$ be such that $|u_n| \leq M$ for all $n \in \mathcal{N}$. Since $x_n \to 0$, for a given $\epsilon > 0$ we can choose an $N \in \mathcal{N}$ such that $n \geq N$ implies $|x_n| < \epsilon/M$. Then, for $n \geq N$ we have $|x_n u_n| < M(\epsilon/M) = \epsilon$, and this implies $x_n u_n \to 0$.

Example 7.5. From Theorem 7.4(2) it follows that the sequences $\{\sin(n^3 + n^2 + 7)/n^2\}$ and $\{[2^n - (-2)^n]/5^n\}$ tend to 0 as $n \to \infty$.

Theorem 7.5. Suppose $\{u_n\} \to u$ and $\{v_n\} \to v$ and c is a real number. Then,

(1). $\lim_{n\to\infty}(u_n \pm v_n) = u \pm v$.

(2). $\lim_{n\to\infty} cu_n = cu$.

(3). $\lim_{n\to\infty} u_n v_n = uv$.

(4). $\lim_{n\to\infty} u_n/v_n = u/v$ provided $v \neq 0$.

Proof. (1). Let $\epsilon > 0$. There exists N_1, $N_2 \in \mathcal{N}$ such that $n \geq N_1$ implies $|u_n - u| < \epsilon/2$ and $n \geq N_2$ implies $|v_n - v| < \epsilon/2$. Set $N = \max\{N_1, N_2\}$. Then, for all $n \geq N$,

$$|(u_n \pm v_n) - (u \pm v)| \leq |u_n - u| + |v_n - v| < \frac{\epsilon}{2} + \frac{\epsilon}{2} = \epsilon.$$

(2). If $c = 0$, the result is obvious. Suppose $c \neq 0$ and let $\epsilon > 0$. There exists an $N \in \mathcal{N}$ such that $n \geq N$ implies $|u_n - u| < \epsilon/|c|$. Hence, $|cu_n - cu| = |c||u_n - u| < |c|(\epsilon/|c|) = \epsilon$.

(3). In view of Theorem 7.2 the sequence $\{u_n\}$ is bounded, say, $|u_n| \leq M$, $n \in \mathcal{N}$ where $M > 0$. Let $\epsilon > 0$. There exists N_1, $N_2 \in \mathcal{N}$ such that $n \geq N_1$ implies $|u_n - u| < \epsilon/[2(1 + |v|)]$ and $n \geq N_2$ implies $|v_n - v| < \epsilon/(2M)$. The reason for choosing the bound for $|u_n - u|$ to be $\epsilon/[2(1 + |v|)]$ instead of $\epsilon/2|v|$ is that v may be zero. Set $N = \max\{N_1, N_2\}$. Then, for all $n \geq N$,

$$
\begin{aligned}
|u_n v_n - uv| &= |u_n v_n - u_n v + u_n v - uv| \\
&\leq |u_n||v_n - v| + |v||u_n - u| \\
&< M \tfrac{\epsilon}{2M} + |v| \tfrac{\epsilon}{2(1+|v|)} < \epsilon.
\end{aligned}
$$

(4). In view of (3) and the fact that $u_n/v_n = u_n \times 1/v_n$ it suffices to show that $\lim_{n \to \infty} 1/v_n = 1/v$. Let $\epsilon > 0$. There exists $N_1 \in \mathcal{N}$ such that $n \geq N_1$ implies $|v_n - v| < |v|/2$. Hence, if $n \geq N_1$ then $|v|/2 > |v| - |v_n|$, and so $|v_n| > |v|/2$. Also, there exists $N_2 \in \mathcal{N}$ such that $n \geq N_2$ implies $|v_n - v| < \epsilon|v|^2/2$. Set $N = \max\{N_1, N_2\}$. Then, for all $n \geq N$,

$$
\left| \frac{1}{v_n} - \frac{1}{v} \right| = \frac{|v_n - v|}{|v_n||v|} < \frac{(\epsilon|v|^2)/2}{(|v|/2)|v|} = \epsilon.
$$

Example 7.6. Consider the sequence $\{u_n\}$, where

$$
u_n = \left(\frac{3n^2 + 4}{n^2 + 5} \right) \left(\frac{3n^2 + 9n - 8}{4n^2 + 7} \right).
$$

Since

$$
u_n = \left(\frac{3 + 4/n^2}{1 + 5/n^2} \right) \left(\frac{3 + 9/n - 8/n^2}{4 + 7/n^2} \right)
$$

from Theorem 7.5 and the fact that $\lim_{n \to \infty} 1/n = \lim_{n \to \infty} 1/n^2 = 0$ it follows that $u_n \to (3/1) \times (3/4) = 9/4$.

Example 7.7. We shall show that for any given $a \in \mathcal{R}$ with $|a| < 1$, $\lim_{n \to \infty} a^n = 0$, and if $a > 0$, then $\lim_{n \to \infty} \sqrt[n]{a} = 1$. Let $x = (1/|a|) - 1$. Then, $|a| = 1/(1 + x)$, and from Bernoulli's inequality

$$
|a|^n = \frac{1}{(1 + x)^n} \leq \frac{1}{1 + nx} < \frac{1}{nx}.
$$

Hence, $||a|^n - 0| < 1/nx < \epsilon$ if $n > [1/x\epsilon]$. Thus, $|a|^n \to 0$, which implies $a^n \to 0$. If $a \geq 1$, let $\sqrt[n]{a} = 1 + h_n$, $h_n \geq 0$. Since $a = (1 + h_n)^n \geq 1 + nh_n > nh_n$ (again using Bernoulli's inequality), we have $0 \leq h_n < a/n$. Thus, $h_n \to 0$ as $n \to \infty$, and hence $\sqrt[n]{a} \to 1$. If $0 < a < 1$, let $a = 1/k$, $k > 1$. Then, $\sqrt[n]{a} = 1/(\sqrt[n]{k}) \to 1/1 = 1$.

Given two sequences $\{u_n\}$ and $\{v_k\}$ we say that $\{v_k\}$ is a *subsequence* of $\{u_n\}$ provided for each index k there is an index n_k such that $n_1 < n_2 < \cdots < n_k < \cdots$ and $v_k = u_{n_k}$ for all $k = 1, 2, \cdots$. For example, the sequence $\{k^2\}$ is a subsequence of $\{n\}$.

Theorem 7.6. If $\lim_{n \to \infty} u_n = u$ then every subsequence of $\{u_n\}$ is convergent with limit u.

Proof. Since $\lim_{n \to \infty} u_n = u$, for every $\epsilon > 0$ there exists an $N \in \mathcal{N}$ such that $|u_k - u| < \epsilon$, whenever $k \geq N$. If $\{u_{n_k}\}$ is a subsequence of $\{u_n\}$, then obviously $n_k \geq k$. Therefore, it is necessary that $|u_{n_k} - u| < \epsilon$, whenever $n_k \geq N$.

An immediate consequence of Theorem 7.6 is that if $\{u_n\}$ has one subsequence converging to u and a second subsequence converging to v and $u \neq v$, then $\{u_n\}$ must diverge. This fact easily shows that the sequence $\{(-1)^n\}$ considered in Example 7.3 diverges. Indeed, it has two subsequences $\{1, 1, \cdots\}$ and $\{-1, -1, \cdots\}$ which converge to different limits.

As we have indicated the definition of convergence of a sequence requires an advance knowledge of the limit which may not be feasible in certain cases. Usually, this difficulty is resolved by Cauchy's criterion of convergence. For this, we need the following basic definition.

A sequence $\{u_n\}$ is said to be a *Cauchy sequence* if for each $\epsilon > 0$ there exists a natural number N such that $|u_n - u_m| < \epsilon$ for all $m, n \geq N$, i.e., the terms of the sequence get close to each other.

Theorem 7.7 (Cauchy's Convergence Criterion). A sequence $\{u_n\}$ is convergent iff it is a Cauchy sequence.

Proof. Suppose $\lim u_n = u$. Then, given any $\epsilon > 0$ there is a natural number N such that $|u_n - u| < \epsilon/2$ whenever $n \geq N$. Thus, if $n, m \geq N$, we have

$$|u_n - u_m| = |u_n - u + u - u_m| \leq |u_n - u| + |u_m - u| < \frac{\epsilon}{2} + \frac{\epsilon}{2} = \epsilon$$

and hence $\{u_n\}$ is a Cauchy sequence.

Conversely, since $\{u_n\}$ is a Cauchy sequence, from Problem 7.9(i) it is bounded. But, then from Theorem 7.3 the range set $S = \{u_n : n \in \mathcal{N}\}$ has a limit point, say, u. We claim that $u_n \to u$. For this, given any $\epsilon > 0$ there exists a natural number N such that $|u_n - u_m| < \epsilon/2$ for all $m, n \geq N$. Further, since u is a limit point of S, the nbd $(u - \epsilon/2, u + \epsilon/2)$ contains infinitely many points of S. Thus, in particular, there exists an $m \geq N$ such that $|u_m - u| < \epsilon/2$. Hence, for any $n \geq N$, we have

$$|u_n - u| = |u_n - u_m + u_m - u| \leq |u_n - u_m| + |u_m - u| < \frac{\epsilon}{2} + \frac{\epsilon}{2} = \epsilon.$$

Therefore, $u_n \to u$.

It should be noted that a sequence which satisfies $u_{n+1} - u_n \to 0$ is not necessarily Cauchy. For this, consider the sequence $\{\ln n\}$. Clearly, it diverges, in fact it is not even bounded, but $\ln(n+1) - \ln n = \ln(n+1)/n \to \ln 1 = 0$.

A sequence $\{u_n\}$ is said to *diverge to* $+\infty$, and we write $u_n \to +\infty$ if for any real number $M > 0$ there is a natural number $N \in \mathcal{N}$ such that $n \geq N$ implies $u_n > M$. Similarly, $\{u_n\}$ is said to *diverge to* $-\infty$, and we write $u_n \to -\infty$ if for any real number $M > 0$ there is a natural number $N \in \mathcal{N}$ such that $n \geq N$ implies $u_n < -M$. If $\{u_n\}$ diverges but not to $+\infty$ or $-\infty$, we say that it *oscillates*. Thus, the sequence $\{n\}$ considered in Example 7.4 diverges to $+\infty$, whereas $\{(-1)^n\}$ discussed in Example 7.3, oscillates. Another example of an oscillatory sequence is $\{1, 2, 1, 3, 1, 4, 1, 5, \cdots\}$. If the sequences $\{u_n\}$ and $\{v_n\}$ are such that $u_n \leq v_n$ for all $n \in \mathcal{N}$, then $u_n \to +\infty$ implies $v_n \to +\infty$, and $v_n \to -\infty$ implies $u_n \to -\infty$.

Problems

7.1. Use the definition of limit to show that

(i). $\lim_{n\to\infty} \frac{n^2-n+5}{5n^2+2n-3} = \frac{1}{5}$.

(ii). $\lim_{n\to\infty}(\sqrt{3n+1} - \sqrt{3n}) = 0$.

(iii). $\lim_{n\to\infty} \frac{2^n}{n!} = 0$.

(iv). $\lim_{n\to\infty} \left[\frac{1}{1\cdot2} + \frac{1}{2\cdot3} + \cdots + \frac{1}{(n-1)n} \right] = 1$.

(v). $\lim_{n\to\infty} \frac{n+(-1)^n}{n^2+1} = 0$.

(vi). $\lim_{n\to\infty} \frac{\sqrt{n^2+n}}{n} = 1$.

7.2. Let $\{u_n\}$ and $\{v_n\}$ be two sequences of real numbers. Prove or disprove each of the following statements.

(i). If $\{u_n\}$ and $\{v_n\}$ are divergent, then $\{u_n + v_n\}$ is divergent.
(ii). If $\{u_n\}$ and $\{v_n\}$ are divergent, then $\{u_n v_n\}$ is divergent.
(iii). If $|u_n| \to 0$ as $n \to \infty$, then $u_n \to 0$ as $n \to \infty$.
(iv). If $\{|u_n|\}$ is convergent, then $\{u_n\}$ is convergent.
(v). If $\{u_n\}$ is convergent, then $\{u_n^2\}$ is convergent.
(vi). If $\{u_n^2\}$ is convergent, then $\{u_n\}$ is convergent.
(vii). If $\lim_{n\to\infty} u_n^2 = A^2$ and $\lim_{n\to\infty} u_n$ exists, then $\lim_{n\to\infty} u_n = A$.

7.3. Prove that if $\lim_{n\to\infty} u_n = u$, then $\lim_{n\to\infty} |u_n| = |u|$. Is the converse true if (i) $u = 0$, (ii) $u \neq 0$?

7.4. Suppose $\{u_n\} \to u$ and $\{v_n\} \to v$ and satisfy $u_n \leq v_n$, $n \geq N$ for some $N \in \mathcal{N}$. Show that $u \leq v$. Further, give an example to show that $u_n < v_n$ does not imply $u < v$.

7.5. Find the limit of each of the following sequences as $n \to \infty$.

(i). $\sqrt{n + \sqrt{n}} - \sqrt{n}$.

(ii). $\left(\frac{2n-1}{3n+1}\right)^n$.

(iii). $\frac{1}{(n+1)^2} + \frac{1}{(n+2)^2} + \cdots + \frac{1}{(n+n)^2}$.

(iv). $\sqrt{\frac{2n^2+1}{3n^2+n+1}}$.

(v). $n^2\left[\left(2 + \frac{1}{n^2}\right)^3 - 8\right]$.

(vi). $\frac{2^n+3^n}{2^{n+1}+3^{n+1}}$.

(vii). $\left(1 - \frac{1}{2^2}\right)\left(1 - \frac{1}{3^2}\right)\cdots\left(1 - \frac{1}{n^2}\right)$.

7.6. (i). Let $\{u_n\}$ be a bounded sequence of positive numbers. Suppose $u_n \geq 1$ for all $n \in \mathcal{N}$. Prove that $\lim_{n\to\infty} \sqrt[n]{u_n} = 1$.

(ii). Let a and b be positive numbers. Show that $\lim_{n\to\infty} \sqrt[n]{a^n + b^n} = \max\{a, b\}$.

7.7. (*Ratio Test*). Let $\{u_n\}$ be a sequence of nonzero terms such that

$$\lim_{n\to\infty} \left|\frac{u_{n+1}}{u_n}\right| = \alpha.$$

Show that

(i). If $\alpha < 1$, then $\lim u_n = 0$.

(ii). If $\alpha > 1$, then $\lim |u_n| = +\infty$.

(iii). If $\alpha = 1$, then $\{u_n\}$ may converge, diverge to $+\infty$ or $-\infty$, or oscillate.

In particular, use the ratio test to show that for any $|a| < 1$ and $k > 0$, $\lim_{n\to\infty} n^k a^n = 0$.

7.8. Let $\{u_n\}$ be a sequence. Define the sequence $\{\bar{u}_n\}$ of *Cesàro means* of $\{u_n\}$ by $\bar{u}_n = (u_1 + u_2 + \cdots + u_n)/n$ for every $n \in \mathcal{N}$. Prove that if $\lim_{n\to\infty} u_n = u$, then $\lim_{n\to\infty} \bar{u}_n = u$; however, the converse is not true.

7.9. Show that

(i). Every Cauchy sequence is bounded.

(ii). A subsequence of a Cauchy sequence is a Cauchy sequence.

7.10. Show that the sequence $\{u_n\}$, where

(i). $u_n = \sum_{i=0}^n \frac{1}{i!}$ is a Cauchy sequence.

(ii). $u_n = \sum_{i=0}^n \frac{1}{i}$ is not a Cauchy sequence.

7.11. A sequence $\{u_n\}$ is said to be *contractive* iff there exists a constant $k \in (0,1)$ such that $|u_{n+2} - u_{n+1}| \leq k|u_{n+1} - u_n|$ for all $n \in \mathcal{N}$. Show that every contractive sequence is a Cauchy sequence, and hence converges.

7.12. Let $\{u_n\}$ be a sequence with $u_n > 0$, $n \in \mathcal{N}$. Show that $\{u_n\}$ diverges to $+\infty$ iff the sequence $\{1/u_n\}$ converges to zero.

Answers or Hints

7.1. (i). $\left|\frac{n^2-n+5}{5n^2+2n-3} - \frac{1}{5}\right| = \left|\frac{-7n+28}{5(5n^2+2n-3)}\right| < \frac{7n}{5(5n^2)} < \frac{1}{n} < \epsilon$ if $n > N = \max\left\{4, \left[\frac{1}{\epsilon}\right]\right\}$.

(ii). Let $\epsilon > 0$ be given, choose N such that $N > 1/(3\epsilon^2)$. Then, for $n > N$ we have $|\sqrt{3n+1} - \sqrt{3n} - 0| = \left|\frac{(\sqrt{3n+1}-\sqrt{3n})(\sqrt{3n+1}+\sqrt{3n})}{(\sqrt{3n+1}+\sqrt{3n})}\right| = \left|\frac{1}{\sqrt{3n+1}+\sqrt{3n}}\right| < \frac{1}{\sqrt{3n}} < \frac{1}{\sqrt{3N}} < \epsilon$.

(iii). Let $\epsilon > 0$ be given, choose N such that $N > 4/\epsilon$. Then, for $n > N$, we have $\left|\frac{2^n}{n!}\right| = \left|\frac{2}{1}\frac{2}{2}\cdots\frac{2}{n}\right| < \frac{2}{1}\cdot\frac{2}{n} = \frac{4}{n} < \frac{4}{N} < \epsilon$.

(iv). Since $\frac{1}{k} - \frac{1}{k+1} = \frac{1}{k(k+1)}$ it follows that $\frac{1}{1\cdot2} + \frac{1}{2\cdot3} + \cdots + \frac{1}{(n-1)n} = \frac{1}{1} - \frac{1}{n} = \frac{n-1}{n}$. Now $\left|\frac{n-1}{n} - 1\right| = \frac{1}{n} < \epsilon$ if $n > \left[\frac{1}{\epsilon}\right]$.

(v). $\left|\frac{n+(-1)^n}{n^2+1} - 0\right| \leq \frac{2n}{n^2} = \frac{2}{n} < \epsilon$ if $n > \left[\frac{2}{\epsilon}\right]$.

(vi). $\left|\frac{\sqrt{n^2+n}}{n} - 1\right| = \left|\sqrt{1+\frac{1}{n}} - 1\right| = \left|\frac{1/n}{\sqrt{1+1/n}+1}\right| < \frac{1}{n} < \epsilon$ if $n > \left[\frac{1}{\epsilon}\right]$.

7.2. (i). False, let $u_n = (-1)^n$, $v_n = -(-1)^n$.

(ii). False, let $u_n = v_n = (-1)^n$.

(iii). True, for $\epsilon > 0$ there exists N such that $||u_n| - 0| < \epsilon$ if $n > N$. Since $|u_n - 0| = ||u_n| - 0|$, we have the result.

(iv). False, let $u_n = (-1)^n$.

(v). True, $\lim_{n\to\infty} u_n = A$ implies $\lim_{n\to\infty} u_n^2 = A^2$.

(vi). False, let $u_n = (-1)^n$.

(vii). False, let $u_n = -1$, $\lim_{n\to\infty} u_n^2 = 1$ and $\lim_{n\to\infty} u_n = -1$.

7.3. Since $\lim_{n\to\infty} u_n = u$ for every $\epsilon > 0$ there exists an $N \in \mathcal{N}$ such that $|u_n - u| < \epsilon$ for all $n \geq N$. Hence, for all $n \geq N$, $||u_n| - |u|| \leq |u_n - u| < \epsilon$ and hence $\lim_{n\to\infty} |u_n| = |u|$. (i). True, see Problem 7.2(iii). (ii). False, see Problem 7.2(iv).

7.4. Suppose $u > v$ and set $\epsilon = (u - v)/2$. Choose $N_1 > N$ such that $|u_n - u| < \epsilon$ and $|v_n - v| < \epsilon$ for $n \geq N_1$. Then, for such an n, $u_n > u - \epsilon = u - (u-v)/2 = v + (u-v)/2 = v + \epsilon > v_n$, which is a contradiction. Clearly, $1/n^2 < 1/n$ but $\lim_{n\to\infty} 1/n^2 = \lim_{n\to\infty} 1/n = 0$.

7.5. (i). $\sqrt{n+\sqrt{n}} - \sqrt{n} = \frac{\sqrt{n}}{\sqrt{n+\sqrt{n}}+\sqrt{n}} = \frac{1}{\sqrt{1+(1/\sqrt{n})}+1} \to \frac{1}{2}$.

(ii). $0 < \left(\frac{2n-1}{3n+1}\right)^n < \left(\frac{2}{3}\right)^n \to 0.$

(iii). $0 < \frac{1}{(n+1)^2} + \cdots + \frac{1}{(n+n)^2} < \frac{n}{(n+1)^2} \to 0.$

(iv). $\frac{2n^2+1}{3n^2+n+1} = \frac{2+1/n^2}{3+1/n+1/n^2} \to \frac{2}{3}.$ Therefore, $\sqrt{\frac{2n^2+1}{3n^2+n+1}} \to \sqrt{\frac{2}{3}}.$

(v). $n^2\left[\left(2+\frac{1}{n^2}\right)^3 - 8\right] = 12 + \frac{6}{n^2} + \frac{1}{n^4} \to 12.$

(vi). $\frac{2^n+3^n}{2^{n+1}+3^{n+1}} = \frac{(2/3)^n+1}{2(2/3)^n+3} \to \frac{1}{3}.$

(vii). Since $1 - \frac{1}{k^2} = \frac{k^2-1}{k^2} = \frac{(k-1)(k+1)}{k^2}$ it follows that $\left(1 - \frac{1}{2^2}\right)\cdots\left(1 - \frac{1}{n^2}\right)$
$= \frac{1\cdot3}{2^2}\frac{2\cdot4}{3^2}\frac{3\cdot5}{4^2}\cdots\frac{(n-2)n}{(n-1)^2}\frac{(n-1)(n+1)}{n^2} = \frac{n+1}{2n} \to \frac{1}{2}.$

7.6. (i). Let $u_n \leq C$ for all $n \in \mathcal{N}$. Then, we have $1 \leq \sqrt[n]{u_n} \leq \sqrt[n]{C}$. However, since $\sqrt[n]{C} \to 1$ as $n \to \infty$ it follows that $1 \leq \lim_{n\to\infty} \sqrt[n]{u_n} \leq \lim_{n\to\infty} \sqrt[n]{C} = 1$, i.e., $\lim_{n\to\infty} \sqrt[n]{u_n} = 1$.

(ii). If $a = b$, then $\lim_{n\to\infty} \sqrt[n]{a^n + b^n} = \lim_{n\to\infty} \sqrt[n]{2a^n} = a \lim_{n\to\infty} \sqrt[n]{2} = a$. If $a > b$, then $\lim_{n\to\infty} \sqrt[n]{a^n + b^n} = a \lim_{n\to\infty} \sqrt[n]{1 + (b/n)^n} = a$. If $b > a$, then $\lim_{n\to\infty} \sqrt[n]{a^n + b^n} = b$. Combining, we get $\lim_{n\to\infty} \sqrt[n]{a^n + b^n} = \max\{a,b\}$.

7.7. (i). If $0 \leq \alpha < 1$, then there exists an $N \in \mathcal{N}$ such that $|u_{n+1}/u_n| < \beta$, $n \geq N$ for some $\beta \in (\alpha, 1)$. But, then $|u_{n+1}| < \beta|u_n| < \beta^2|u_{n-1}| < \cdots < \beta^{n+1-N}|u_N| \to 0$. Hence, $u_n \to 0$.

(ii). If $\alpha > 1$, then $|u_{n+1}| > \beta|u_n|$, $n \geq N$ for some $\beta \in (1, \alpha)$.

(iii). Consider the sequences $\{1/n\}$, $\{n\}$, $\{-n\}$, $\{(-1)^n\}$.
Note that $\left|\frac{a_{n+1}}{a_n}\right| = \left|\frac{(n+1)^k a^{n+1}}{n^k a^n}\right| = \left|\left(1 + \frac{1}{n}\right)^k a\right| \to |a| < 1.$

7.8. Since $\{u_n\}$ converges, there exists a constant $C > 0$ such that $|u_n| \leq C$ for all $n \geq 1$, and there exists an $N \in \mathcal{N}$ such that $|u_n - u| < \epsilon/2$ for all $n \geq N$. Choose $N_1 = \max\{2N(C + |u|)/\epsilon, N\}$. Then, for all $n \geq N_1$ we have
$$\left|\frac{u_1+\cdots+u_n}{n} - u\right| = \frac{1}{n}|(u_1 - u) + \cdots + (u_n - u)|$$
$$\leq \frac{1}{n}[|u_1 - u| + \cdots + |u_N - u| + |u_{N+1} - u| + \cdots + |u_n - u|]$$
$$\leq \frac{1}{n}[(|u_1| + |u|) + \cdots + (|u_n| + |u|) + (n - N)\epsilon/2]$$
$$\leq \frac{1}{n}[N(C + |u|)] + \frac{1}{n}(n - N)\frac{\epsilon}{2} < \frac{\epsilon}{2} + \frac{\epsilon}{2} = \epsilon.$$
The converse is not true. For example, consider the divergent sequence $\{(-1)^n\}$. For this, $\bar{u}_1 = -1$, $\bar{u}_2 = 0$, $\bar{u}_3 = -1/3$, $\bar{u}_4 = 0$, $\bar{u}_5 = -1/5, \cdots$ and hence $\{\bar{u}_n\}$ converges to zero.

7.9. (i). See the proof of Theorem 7.2.

(ii). Let $\{u_n\}$ be a Cauchy sequence. Then, for every $\epsilon > 0$ there exists an $N \in \mathcal{N}$ such that $|u_n - u_m| < \epsilon$ for all $n, m \geq N$. Let $\{u_{n_k}\}$ be a subsequence of $\{u_n\}$, then for all $k, \ell \geq N$ implies $n_k \geq k \geq N$ and $n_\ell \geq \ell \geq N$, and hence $|u_{n_k} - u_{n_\ell}| < \epsilon$.

7.10. (i). Given $\epsilon > 0$ let $N \in \mathcal{N}$ be such that $1/N < \epsilon$. Then, for $n > m \geq N$ we have

$$|u_n - u_m| = \frac{1}{(m+1)!} + \cdots + \frac{1}{n!}$$

$$\leq \frac{1}{2^m} + \cdots + \frac{1}{2^{n-1}} \quad (\text{using } n! \geq 2^{n-1} \text{ for all } n \geq 1)$$

$$< \frac{1}{2^m}\left(1 + \frac{1}{2} + \frac{1}{2^2} + \cdots\right) = \frac{1}{2^{m-1}} \leq \frac{1}{m} \leq \frac{1}{N} < \epsilon.$$

(ii). $|u_{2n} - u_n| = \frac{1}{n+1} + \frac{1}{n+2} + \cdots + \frac{1}{2n} > \frac{1}{n+n} + \frac{1}{n+n} + \cdots + \frac{1}{n+n} = \frac{n}{2n} = \frac{1}{2}.$

7.11. Clearly, $|u_{n+2} - u_{n+1}| \leq k|u_{n+1} - u_n| \leq k^2|u_n - u_{n-1}| \leq \cdots \leq k^n|u_2 - u_1|$. Now for $m > n$, we have $|u_m - u_n| \leq |u_m - u_{m-1}| + \cdots + |u_{n+1} - u_n| \leq (k^{m-2} + \cdots + k^{n-1})|u_2 - u_1| < [k^{n-1}/(1-k)]|u_2 - u_1| \to 0.$

7.12. Suppose that $u_n \to +\infty$. Then, there exists an $N \in \mathcal{N}$ such that for every $1/M = \epsilon > 0$, $u_n > 1/\epsilon$, $\forall\, n \geq N$. But, then $|1/u_n - 0| = 1/u_n < \epsilon$, $\forall\, n \geq N$. The converse is proved similarly.

Chapter 8

Real Sequences (Contd.)

In this chapter we will show that it is always possible to extract a convergent sequence from any bounded sequence. This result supplements Theorem 7.3 and is more transparent than Theorem 4.1. For this, as a first step we shall examine the convergence of monotone sequences. Then, we will introduce $\lim\sup$ and $\lim\inf$ of a given sequence. These notations provide necessary and sufficient conditions for the convergence of sequences.

We begin this chapter with the definition of monotone sequences. A sequence $\{u_n\}$ of real numbers is called *increasing* (*strictly increasing*) if $u_n \le u_{n+1}$ ($u_n < u_{n+1}$), $\forall\, n \in \mathcal{N}$. It is called *decreasing* (*strictly decreasing*) if $u_n \ge u_{n+1}$ ($u_n > u_{n+1}$), $\forall\, n \in \mathcal{N}$, and it is called *monotone* if it is either increasing or decreasing. If $\{u_n\}$ is increasing (decreasing) and converges to u, we shall write $u_n \uparrow u$ ($u_n \downarrow u$). It is clear that $\{u_n\}$ is increasing iff $\{-u_n\}$ is decreasing. A sequence may be monotonic after certain number of terms.

The following results are fundamental for monotone sequences.

Theorem 8.1. (1). If $\{u_n\}$ is increasing and bounded above and $u = \sup\{u_n : n \in \mathcal{N}\}$, then $u_n \uparrow u$.

(2). If $\{u_n\}$ is decreasing and bounded below and $\ell = \inf\{u_n : n \in \mathcal{N}\}$, then $u_n \downarrow \ell$.

Proof. (1). By the completeness axiom the supremum u exists. Thus, for any $\epsilon > 0\, \exists\, N \in \mathcal{N}$ such that $u - \epsilon < u_N \le u$. Since $u_N \le u_n$, $\forall\, n \ge N$ and $u_n \le u$, $\forall\, n \in \mathcal{N}$ it follows that $u - \epsilon < u_n \le u$, $\forall\, n \in \mathcal{N}, n \ge N$. But this means that $u_n \uparrow u$.

(2). It suffices to note that the sequence $\{-u_n\}$ is increasing with supremum $-\ell$.

It is clear that an increasing sequence which is not bounded above must diverge to ∞, and a decreasing sequence which is not bounded below must diverge to $-\infty$.

Example 8.1. As an application of Theorem 8.1, we shall show that each decimal of the form $0.d_1 d_2 d_3 \cdots$ where $d_i \in \{0, 1, \cdots, 9\}$, $i \in \mathcal{N}$

represents a unique number in the interval $[0, 1]$. For this, consider the sequence $\{u_n\}$ where $u_n = 0.d_1 d_2 \cdots d_n$. This sequence is increasing because $u_{n+1} - u_n = 0.00 \cdots 0 d_{n+1} \geq 0$, and bounded above by 1. Thus, it has a unique limit, which is the real number in $[0, 1]$.

Example 8.2. The *Fibonacci sequence* $\{F_n\}$ (see Problem 1.4) is strictly increasing, and diverges to ∞. Here we shall show that the sequence of ratios $u_n = F_{n+1}/F_n$ does converge. The first few terms of the sequence $\{u_n\}$ are $1, 2, 3/2, 5/3, 8/5, 13/8$. The following steps for this sequence are immediate: 1. $u_{n+1} = 1 + 1/u_n$. 2. $1 \leq u_n \leq 2$. 3. $u_n = 1 + u_{n-2}/(1 + u_{n-2})$. 4. $u_{n+2} - u_n = (u_n - u_{n-2})/(1 + u_n)(1 + u_{n-2})$. 5. $(u_{n+2} - u_n) = (u_n - u_{n-2})$. 6. $(u_{2n+1} - u_{2n-1}) = (u_3 - u_1)$. 7. $(u_{2n} - u_{2n-2}) = (u_4 - u_2)$. 8. $\{u_{2n+1}\}$ is increasing and bounded above by 2, and hence converges to u. 9. $\{u_{2n}\}$ is decreasing and bounded below by 1, and hence converges to v. 10. $u = 1 + u/(1 + u)$ and $v = 1 + v/(1 + v)$. 11. $u = v = (1 + \sqrt{5})/2 = \Phi$. In the literature Φ is called the *golden ratio*.

Example 8.3. To find an approximate value of $\sqrt{A}$, $A \in \mathcal{R}^+$ we solve the equation $x^2 - A = 0$ by Newton's method, which requires the construction of the sequence recursively by

$$u_{n+1} = u_n - \frac{u_n^2 - A}{2u_n} = \frac{1}{2}\left(u_n + \frac{A}{u_n}\right),$$

where $u_0 > 0$ is an initial approximation of $\sqrt{A}$. The convergence of this sequence to $\sqrt{A}$ immediately follows from the relation

$$\frac{u_n - \sqrt{A}}{u_n + \sqrt{A}} = \left(\frac{u_{n-1} - \sqrt{A}}{u_{n-1} + \sqrt{A}}\right)^2 = \cdots = \left(\frac{u_0 - \sqrt{A}}{u_0 + \sqrt{A}}\right)^{2^n}$$

and the fact that $|u_0 - \sqrt{A}|/|u_0 + \sqrt{A}| < 1$. Now from the above relation, we note that $u_n - \sqrt{A} > 0$ for all $n \in \mathcal{N}$. Hence, the sequence $\{u_n\}$ is bounded below by $\sqrt{A}$. Further, since

$$u_{n+1} - u_n = \frac{1}{2u_n}(A - u_n^2) < 0$$

the sequence $\{u_n\}$ is decreasing. Hence, the convergence of $\{u_n\}$ also follows from Theorem 8.1.

Theorem 8.2. (1). A sequence monotonic after $m-1$ terms converges iff it is bounded, and it converges to the supremum or infimum of the range set $S = \{u_n : n \geq m\}$ according as the sequence is increasing or decreasing.

(2). If $a_n \leq a_{n+1} \leq b_{n+1} \leq b_n$, $\forall n \geq m$ and $\lim(b_n - a_n) = 0$, then $\lim b_n = \lim a_n$.

Proof. (1). It follows from Theorems 7.2 and 8.1.

(2). Clearly the sequences $\{a_n\}$ and $\{b_n\}$ are bounded, and increasing and decreasing after $m-1$ terms. Thus from (1), $a_n \to \sup\{a_n : n \geq m\}$ and $b_n \to \inf\{b_n : n \geq m\}$. Now since $\{a_n\}$ and $\{b_n\}$ are convergent, we have $\lim b_n - \lim a_n = \lim(b_n - a_n) = 0$ and hence $\lim b_n = \lim a_n = \sup\{a_n : n \geq m\} = \inf\{b_n : n \geq m\}$.

For $m = 1$, Theorem 8.2(2) can be expressed as follows: If a sequence of nonempty, closed, and bounded intervals $I_n = [a_n, b_n]$ is such that $I_{n+1} \subseteq I_n$, $\forall n \in \mathcal{N}$, then $K = \cap_{n \in \mathcal{N}} I_n \neq \emptyset$, and if $\lim |b_n - a_n| = 0$ then K consists of exactly one point. This property is called the *nested interval property*. This property enables us to prove the *sequential analog of the Bolzano-Weierstrass theorem.*

Theorem 8.3. Every bounded sequence has a convergent subsequence.

Proof. Let $\{u_n\}$ be a bounded sequence. Then there is a positive real number M such that $|u_n| \leq M$, $\forall n \in \mathcal{N}$. Hence, $u_n \in [-M, M]$, $\forall n \in \mathcal{N}$. Consider the intervals $[-M, 0]$ and $[0, M]$. Clearly, at least one of these intervals must contain u_n for infinitely many values of n (note that the u_n are not necessarily distinct). We call such an interval I_0. Now divide this interval I_0 into two subintervals of equal length. At least one of these subintervals must contain u_n for infinitely many values of n. We continue this process to obtain a sequence of intervals $I_0, I_1, I_2, \cdots$ with $I_0 \supset I_1 \supset I_2 \supset \cdots$. The length of I_n is $M/2^n$, which goes to 0 as $n \to \infty$. By the nested interval property there is only one point, say, u common to all these intervals. Now choose $u_{n_1} \in I_1$, $u_{n_2} \in I_2$ with $n_2 > n_1$, $u_{n_3} \in I_3$ with $n_3 > n_2$, and so on. This selection is always possible because each I_n contains u_n for infinitely many values of n. Then $\{u_{n_k}\}$ is a subsequence of $\{u_n\}$, and both u_{n_k} and u are in I_k. Thus, $|u_{n_k} - u| < M/2^k$ and hence $u_{n_k} \to u$.

It is clear that unbounded sequences may not have convergent subsequences, e.g., $\{\sqrt{n}\}$. Further, a bounded sequence may have more than one subsequence which converge to different limits. For example, the sequence $\{(-1)^n\}$ has subsequences $\{1, 1, \cdots\}$ and $\{-1, -1, \cdots\}$ which converge to 1 and -1, respectively. The sequence $\{1, 1, 2, 1, 2, 3, 1, 2, 3, 4, \cdots\}$ has subsequences which converge to each natural number. Now let E be the set of all limits of subsequences of a given sequence $\{u_n\}$. The set E may be empty, finite, or an infinite subset of $\mathcal{R}$. If $\{u_n\}$ is bounded, E is necessarily nonempty and bounded, and therefore $\sup E$ and $\inf E$ exist, and belong to E (see Theorem 4.2). In this case we write $\limsup_{n \to \infty} u_n = \sup E$ and $\liminf_{n \to \infty} u_n = \inf E$. For example, the sequence $\{u_n\}$ defined by $u_n = (-1)^n(3 + 1/n)$ is bounded with lower bound -4 and upper bound $7/2$. For this sequence, $E = \{-3, 3\}$, and hence $\limsup_{n \to \infty} u_n = 3$ and $\liminf_{n \to \infty} u_n = -3$. If the sequence $\{u_n\}$ is not bounded above, we write $\limsup_{n \to \infty} u_n = \infty$, and if $\{u_n\}$ is not bounded below, we write $\liminf_{n \to \infty} u_n = -\infty$. Thus, roughly the limit superior (limit inferior) is a measure of how big (small) u_n can be when n is large. The following properties of $L = \limsup_{n\to\infty} u_n$ and $\ell = \liminf_{n\to\infty} u_n$ can be proved easily.

(P1). $\ell \leq L$.

(P2). Given any $\epsilon > 0$ there exists a $N \in \mathcal{N}$ such that for all $n \geq N$, $\ell - \epsilon < u_n < L + \epsilon$.

(P3). $\lim_{n\to\infty} u_n = u$ iff $\ell = u = L$.

(P4). $\lim_{n\to\infty} u_n = \infty$ iff $\ell = \infty = L$.

(P5). $\lim_{n\to\infty} u_n = -\infty$ iff $\ell = -\infty = L$.

To illustrate we shall prove (P2). Since L and ℓ are real numbers, $\{u_n\}$ is a bounded sequence. Let $\epsilon > 0$ be given and suppose $u_n \geq L + \epsilon$ for infinitely many natural numbers n. Then there exists natural numbers $n_1 < n_2 < \cdots$ such that $u_{n_k} \geq L + \epsilon$ for $k = 1, 2, \cdots$. Now since $\{u_{n_k}\}$ is a subsequence of $\{u_n\}$ it is bounded. Hence, by Theorem 8.3, $\{u_{n_k}\}$ has a subsequence which converges to, say, λ. But, then $u_{n_k} \geq L + \epsilon$ implies that $\lambda \geq L + \epsilon$. This contradicts the definition of L. Thus, $u_n \geq L + \epsilon$ is possible only for a finite number of natural numbers n. Therefore, there exists a $N_1 \in \mathcal{N}$ such that $u_n < L + \epsilon$ for all $n \geq N_1$. Similarly, it follows that there exists a $N_2 \in \mathcal{N}$ such that $\ell - \epsilon < u_n$ for all $n \geq N_2$. Now let $N = \max\{N_1, N_2\}$. Then, $\ell - \epsilon < u_n < L + \epsilon$ for all $n \geq N$.

Example 8.4. For a bounded sequence $\{u_n\}$ of positive terms we shall show that

$$\liminf_{n\to\infty} \frac{u_{n+1}}{u_n} \leq \liminf_{n\to\infty} \sqrt[n]{u_n} \leq \limsup_{n\to\infty} \sqrt[n]{u_n} \leq \limsup_{n\to\infty} \frac{u_{n+1}}{u_n}.$$

Let $\ell_1 = \liminf_{n\to\infty} u_{n+1}/u_n$, and $\ell_2 = \liminf_{n\to\infty} \sqrt[n]{u_n}$. We shall only prove that $\ell_1 \leq \ell_2$. Since $\sqrt[n]{u_n} > 0$, $\forall n \in \mathcal{N}$, Problem 8.13(i) implies that $\ell_2 \geq 0$, and hence the inequality $\ell_1 \leq \ell_2$ is obvious if $\ell_1 = 0$. If $\ell_1 > 0$, then for every $0 < \lambda < \ell_1 \exists N \in \mathcal{N}$ such that $\lambda \leq u_{n+1}/u_n$, $n \geq N$ i.e., $u_{n+1} \geq \lambda u_n$, which implies $u_n \geq \lambda^{n-N} u_N$, $n \geq N + 1$. Hence, $u_n^{1/n} \geq \lambda^{1-N/n} u_N^{1/n}$, $n \geq N + 1$. Thus, from Example 7.7 it follows that $\ell_2 \geq \lambda$, which in turn implies that $\ell_2 \geq \ell_1$.

Finally, in this chapter we shall prove the following result.

Theorem 8.4 (Cesaro-Stolz's Theorem). Let $\{a_n\}$ and $\{b_n\}$ be two sequences. If $\{b_n\}$ is a divergent strictly monotone sequence (i.e., diverges to $-\infty$ or ∞) and $\lim_{n\to\infty} \frac{a_{n+1}-a_n}{b_{n+1}-b_n}$ exists and is equal to $L \in [-\infty, \infty]$, then $\lim_{n\to\infty} \frac{a_n}{b_n} = L$.

Proof. If L is finite and $\{b_n\}$ is strictly increasing, then for all $\epsilon > 0$ there exists an n_ϵ such that for all $n \geq n_\epsilon$, we have $L - \epsilon < \frac{a_{n+1}-a_n}{b_{n+1}-b_n} < L + \epsilon$. Since $b_{n+1} - b_n > 0$, it follows that

$$(L - \epsilon)(b_{n+1} - b_n) < a_{n+1} - a_n < (b_{n+1} - b_n)(L + \epsilon). \tag{8.1}$$

Let $k \geq n_\epsilon$ be a natural number. Summing the inequality (8.1) for $n_\epsilon, n_\epsilon + 1, ..., k$, we get

$$(L - \epsilon) \sum_{n=n_\epsilon}^{k} (b_{n+1} - b_n) < \sum_{n=n_\epsilon}^{k} (a_{n+1} - a_n) < (L + \epsilon) \sum_{n=n_\epsilon}^{k} (b_{n+1} - b_n),$$

which gives

$$(L - \epsilon)(b_{k+1} - b_{n_\epsilon}) < a_{k+1} - a_{n_\epsilon} < (L + \epsilon)(b_{k+1} - b_{n_\epsilon}). \qquad (8.2)$$

Dividing (8.2) by b_{k+1}, we obtain

$$(L - \epsilon)\left(1 - \frac{b_{n_\epsilon}}{b_{k+1}}\right) + \frac{a_{n_\epsilon}}{b_{k+1}} < \frac{a_{k+1}}{b_{k+1}} < (L + \epsilon)\left(1 - \frac{b_{n_\epsilon}}{b_{k+1}}\right) + \frac{a_{n_\epsilon}}{b_{k+1}}.$$

Since $\lim_{k \to \infty} b_k = \infty$, the above inequality immediately gives

$$(L - \epsilon) < \lim_{k \to \infty} \frac{a_{k+1}}{b_{k+1}} < (L + \epsilon)$$

and hence, $\lim_{n \to \infty} \frac{a_n}{b_n} = L$.

If $L = +\infty$, then for all $\epsilon > 0$ there exists an n_ϵ such that for all $n \geq n_\epsilon$, we have $\frac{a_{n+1} - a_n}{b_{n+1} - b_n} > \epsilon$. Thus, it follows that $a_{n+1} - a_n > \epsilon(b_{n+1} - b_n)$. Let $k \geq n_\epsilon$ be a natural number. Adding the last inequality for $n_\epsilon, n_\epsilon + 1, ..., k$, we get $a_{k+1} - a_{n_\epsilon} > \epsilon(b_{k+1} - b_{n_\epsilon})$. Dividing this inequality by b_{k+1}, we obtain

$$\frac{a_{k+1}}{b_{k+1}} - \frac{a_{n_\epsilon}}{b_{k+1}} > \epsilon\left(1 - \frac{b_{n_\epsilon}}{b_{k+1}}\right).$$

Now since $\frac{b_{n_\epsilon}}{b_{k+1}}$ and $\frac{a_{n_\epsilon}}{b_{k+1}}$ tend to zero as $k \to \infty$, we find $\frac{a_{k+1}}{b_{k+1}} > \epsilon$, which implies that $\lim_{n \to \infty} \frac{a_n}{b_n} = \infty$.

We complete this chapter with the remark that the completeness axiom, Dedekind's property, Bolzano-Weierstrass property, Cauchy's convergence criterion and the nested interval property are equivalent to each other.

Problems

8.1. Use Theorem 8.1 to show that the following limits exist.

(i). $\lim_{n \to \infty} \frac{1 \cdot 3 \cdot 5 \cdots (2n-1)}{2 \cdot 4 \cdot 6 \cdots (2n)}$.

(ii). $\lim_{n \to \infty} \left(\frac{1}{1 \cdot 2} + \frac{1}{3 \cdot 4} + \cdots + \frac{1}{(2n-1)(2n)}\right)$.

(iii). $\lim_{n \to \infty} \sqrt[n]{n}$.

8.2. Determine which of the following sequences are monotonic and bounded.

(i). $\left\{ \frac{3n-1}{4n+5} \right\}$.

(ii). $\left\{ \frac{n^2}{2^n} \right\}$.

(iii). $\left\{ \frac{n^n}{n!} \right\}$.

(iv). $\left\{ \frac{1\cdot3\cdot5\cdots(2n-1)}{2^n\, n!} \right\}$.

8.3. Prove that if n is any positive integer, then

(i). $\left(1+\frac{1}{n}\right)^n < \left(1+\frac{1}{n+1}\right)^{n+1}$.

(ii). $2 < \left(1+\frac{1}{n}\right)^n < 3$.

8.4. In view of Problem 8.3 and Theorem 8.1, $\lim_{n\to\infty}\left(1+\frac{1}{n}\right)^n$ exists and we denote it by e. Establish the following limits.

(i). $\left(1+\frac{1}{n^2}\right)^{n^2} \to e$.

(ii). $\left(1+\frac{1}{2n}\right)^n \to e^{1/2}$.

(iii). $\left(1+\frac{1}{n!}\right)^{n!} \to e$.

(iv). $\left(1+\frac{1}{n+3}\right)^{2n} \to e^2$.

(v). $\left(1-\frac{1}{n}\right)^n \to \frac{1}{e}$.

(vi). $\left(1+\frac{1}{n^2}\right)^n \to 1$.

(vii). $\left(1+\frac{1}{n}+\frac{1}{n^2}\right)^n \to e$.

8.5. Show that

(i).   If $0 < a < 2$, then $a < \sqrt{2a} < 2$.

(ii).   The sequence $\{u_n\}$ defined by $u_1 = \sqrt{2}$, $u_{n+1} = \sqrt{2u_n}$, $n \in \mathcal{N}$ converges, and $u_n \to 2$.

8.6. Prove that the sequence given by $\sqrt{2}, \sqrt{2+\sqrt{2}}, \sqrt{2+\sqrt{2+\sqrt{2}}}, \cdots$ tends to a limit and find its value.

8.7. Suppose that a sequence $\{u_n\}$ is defined by

$$u_1 = 1, \quad u_{n+1} = 2(2u_n + 1)/(u_n + 3), \quad n \geq 1.$$

Prove that $1 \leq u_n < 2$ and $\{u_n\}$ is strictly increasing. Deduce that $\{u_n\}$ tends to 2.

8.8. Suppose that a sequence $\{u_n\}$ is defined by

$$u_2 > u_1 > 0, \quad u_{n+2} = (u_{n+1}^2 u_n)^{1/3}, \quad n \geq 1.$$

Show that both the sequences $\{u_{2n}\}$, $\{u_{2n+1}\}$ are bounded and monotonic, and $\lim u_{2n} = \lim u_{2n+1} = \lim u_n = (u_2^3 u_1)^{1/4}$.

8.9. Let $u_n = \frac{1}{n+1} + \frac{1}{n+2} + \cdots + \frac{1}{n+n}$, $n \in \mathcal{N}$. Since $1/(n+k) \to 0$, $1 \le k \le n$ can we conclude that $\lim_{n \to \infty} u_n = 0$? Give reasons for your answer.

8.10. Give examples to show that nested interval property is not true if

(i). Intervals are not closed.

(ii). Intervals are not bounded.

8.11. Use the nested interval property to show that the set of all real numbers contained in the interval $[0, 1]$ is uncountable.

8.12. Use the nested interval property to prove the Heine-Borel theorem: If $\mathcal{O} = \{O_i\}$ is an open cover of the closed bounded set S then there is a finite subcover of $\mathcal{O}$ which is also an open cover of S.

8.13. Let $\{u_n\}$ and $\{v_n\}$ be bounded sequences of real numbers. Show that

(i). If $u_n \le v_n$, $n \in \mathcal{N}$, then $\limsup_{n \to \infty} u_n \le \limsup_{n \to \infty} v_n$ and $\liminf_{n \to \infty} u_n \le \liminf_{n \to \infty} v_n$.

(ii). $\limsup_{n \to \infty}(u_n + v_n) \le \limsup_{n \to \infty} u_n + \limsup_{n \to \infty} v_n$.

(iii). $\liminf_{n \to \infty}(u_n + v_n) \ge \liminf_{n \to \infty} u_n + \liminf_{n \to \infty} v_n$.

(iv). Give an example to show that in (ii), an inequality cannot be replaced by equality.

8.14. Show that if the sequence $\{u_n\}$ has no convergent subsequence, then $\{|u_n|\}$ diverges to infinity.

8.15. Let the sequences $\{u_n\}$ and $\{v_n\}$ be such that $v_n > 0$, $v_{n+1} < v_n$ for all large n, and $\lim_{n \to \infty} u_n = \lim_{n \to \infty} v_n = 0$. Show that

$$\liminf \frac{u_{n+1} - u_n}{v_{n+1} - v_n} \le \liminf \frac{u_n}{v_n} \le \limsup \frac{u_n}{v_n} \le \limsup \frac{u_{n+1} - u_n}{v_{n+1} - v_n}.$$

Answers or Hints

8.1. (i). $\frac{1 \cdot 3 \cdot 5 \cdots (2n-1)}{2 \cdot 4 \cdot 6 \cdots (2n)} > \frac{1 \cdot 3 \cdot 5 \cdots (2n-1)}{2 \cdot 4 \cdot 6 \cdots (2n)} \left(\frac{2(n+1)-1}{2(n+1)} \right) \ge 0$.

(ii). $\left\{ \frac{1}{1 \cdot 2} + \frac{1}{3 \cdot 4} + \cdots + \frac{1}{(2n-1)(2n)} \right\}$ is clearly monotone increasing. Further, $\frac{1}{1 \cdot 2} + \frac{1}{3 \cdot 4} + \cdots + \frac{1}{(2n-1)(2n)} < \frac{1}{1 \cdot 2} + \frac{1}{2 \cdot 3} + \cdots + \frac{1}{(2n-2)(2n-1)} + \frac{1}{(2n-1)(2n)}$ $= \left(1 - \frac{1}{2}\right) + \left(\frac{1}{2} - \frac{1}{3}\right) + \cdots + \left(\frac{1}{2n-1} - \frac{1}{2n}\right) = 1 - \frac{1}{2n} \le 1$.

(iii). $\{\sqrt[n]{n}\}$ is bounded below by 1. From (1.1) we have $n^{1/n} > 1 + \frac{1}{n}$, or $n^{(n+1)/n} > (n+1)$, or $n^{1/n} > (n+1)^{1/(n+1)}$. Hence, $\{\sqrt[n]{n}\}$ is monotone decreasing for $n \ge 3$.

8.2. (i). Since $u_n = \frac{3n-1}{4n+5}$ we have $\frac{u_{n+1}}{u_n} = \frac{3n+2}{4n+9} \cdot \frac{4n+5}{3n-1} = \frac{12n^2+23n+10}{12n^2+23n-9} > 1$, $n \in \mathcal{N}$. Thus, $\{u_n\}$ is increasing. Also, $0 < \frac{3n-1}{4n+5} < 1$, $n \in \mathcal{N}$. Clearly, $u_n \to \frac{3}{4}$.

(ii). Since $u_n = \frac{n^2}{2^n}$ we have $\frac{u_{n+1}}{u_n} = \frac{(n+1)^2}{n^2} \frac{1}{2} < 1$ if $2n^2 > (n+1)^2$, i.e., $n(n-2) > 1$, which is true if $n \geq 3$. Thus, $\{u_n\}$ is decreasing. Also, $0 < \frac{n^2}{2^n} < \frac{9}{8}$. Clearly, $u_n \to 0$.

(iii). Since $u_n = \frac{n^n}{n!}$ we have $\frac{u_{n+1}}{u_n} = \frac{(n+1)^{n+1}}{(n+1)!} \frac{n!}{n^n} = \left(1 + \frac{1}{n}\right)^n > 1$, $n \in \mathcal{N}$. Thus, $\{u_n\}$ is increasing. Also, $u_n = \frac{n \cdot n \cdots n}{1 \cdot 2 \cdots n} \geq n$ implies $\{u_n\}$ is bounded below by 1, but unbounded above.

(iv). Since $u_n = \frac{1 \cdot 3 \cdot 5 \cdots (2n-1)}{2^n \, n!}$ we have $\frac{u_{n+1}}{u_n} = \frac{2n+1}{2n+2} < 1$, $n \in \mathcal{N}$. Thus, $\{u_n\}$ is decreasing and bounded below by 0 and above by 1/2.

8.3. (i). $\left(1 + \frac{1}{n}\right)^n = 1 + n\left(\frac{1}{n}\right) + \frac{n(n-1)}{2!}\left(\frac{1}{n}\right)^2 + \cdots + \left(\frac{1}{n}\right)^n$

$$= 1 + 1 + \frac{1}{2!}(1)\left(1 - \frac{1}{n}\right) + \frac{1}{3!}(1)\left(1 - \frac{1}{n}\right)\left(1 - \frac{2}{n}\right) + \cdots$$
$$+ \frac{1}{n!}(1)\left(1 - \frac{1}{n}\right)\cdots\left(1 - \frac{n-1}{n}\right)$$
$$< 1 + 1 + \frac{1}{2!}(1)\left(1 - \frac{1}{n+1}\right) + \frac{1}{3!}(1)\left(1 - \frac{1}{n+1}\right)\left(1 - \frac{2}{n+1}\right) + \cdots$$
$$+ \frac{1}{n!}(1)\left(1 - \frac{1}{n+1}\right)\cdots\left(1 - \frac{n-1}{n+1}\right) < \left(1 + \frac{1}{n+1}\right)^{n+1}.$$

(ii). $\left(1 + \frac{1}{n}\right)^n = 1 + 1 + \frac{1}{2!}n(n-1)\frac{1}{n^2} + \cdots + \frac{1}{n^n}$ $\left(\left(1 + \frac{1}{n}\right)^n > 2 \text{ is obvious}\right)$

$$\leq 1 + 1 + \frac{1}{1 \cdot 2} + \cdots + \frac{1}{1 \cdot 2 \cdot 3 \cdots n}$$
$$\leq 1 + 1 + \frac{1}{2} + \frac{1}{2^2} + \cdots + \frac{1}{2^{n-1}} = 1 + \frac{1 - (1/2)^n}{1 - (1/2)} < 3.$$

8.4. (i). $\left(1 + \frac{1}{n^2}\right)^{n^2}$ is a subsequence of $\left(1 + \frac{1}{n}\right)^n$.

(ii). $\left(1 + \frac{1}{2n}\right)^{2n}$ is a subsequence of $\left(1 + \frac{1}{n}\right)^n$, and hence $\left(1 + \frac{1}{2n}\right)^n = \sqrt{\left(1 + \frac{1}{2n}\right)^{2n}} \to e^{1/2}$.

(iii). $\left(1 + \frac{1}{n!}\right)^{n!}$ is a subsequence of $\left(1 + \frac{1}{n}\right)^n$, and hence $\left(1 + \frac{1}{n!}\right)^{n!} \to e$.

(iv). $\left(1 + \frac{1}{n+3}\right)^{2n} = \left(1 + \frac{1}{n+3}\right)^{2(n+3)}\left(1 + \frac{1}{n+3}\right)^{-6} \to e^2 \times 1 = e^2$.

(v). $\left(1 - \frac{1}{n}\right)^n = \left(\frac{n}{n-1}\right)^{-n} = \left(1 + \frac{1}{n-1}\right)^{-(n-1)}\left(1 + \frac{1}{n-1}\right)^{-1} \to \frac{1}{e \times 1} = \frac{1}{e}$.

(vi). Since $1 \leq \left(1 + \frac{1}{n^2}\right)^n \leq \sqrt[n]{\left(1 + \frac{1}{n^2}\right)^{n^2}} \leq \sqrt[n]{e}$ and $\sqrt[n]{e} \to 1$ as $n \to \infty$ it follows that $\left(1 + \frac{1}{n^2}\right)^n \to 1$.

(vii). Since $\left(1 + \frac{1}{n} + \frac{1}{n^2}\right)^n = \left(1 + \frac{1}{n}\right)^n \left(1 + \frac{1}{n(n+1)}\right)^n$ it follows that $\left(1 + \frac{1}{n}\right)^n \leq \left(1 + \frac{1}{n} + \frac{1}{n^2}\right) \leq \left(1 + \frac{1}{n}\right)^n \left(1 + \frac{1}{n^2}\right)^n$. Hence from part (vi), $\left(1 + \frac{1}{n} + \frac{1}{n^2}\right)^n \to e$.

8.5. (i). $0 < a < 2$ implies $a = \sqrt{a}\sqrt{a} < \sqrt{2}\sqrt{a} = \sqrt{2a} < \sqrt{2}\sqrt{2} = 2$.

(ii). $0 < u_1 = \sqrt{2} < 2$ implies $u_1 = \sqrt{2} < \sqrt{2u_1} = a_2 < 2$ by (i). Thus, by induction $u_n < \sqrt{2u_n} = u_{n+1} < 2$. Hence, $\{u_n\}$ is increasing and bounded above. Let $u_n = u$, then $u = \sqrt{2u}$ implies $u = 2$.

8.6. Let $u_1 = \sqrt{2}$ and $u_{n+1} = \sqrt{2 + u_n}$, $n \geq 1$. By induction we shall show that $u_n < 2$, $n \geq 1$. Indeed, $u_1 < 2$, if $u_k < 2$ then $u_{k+1} = \sqrt{2 + u_k} < \sqrt{2 + 2} = 2$. It is obvious that $\{u_n\}$ is strictly increasing. Hence, $\lim_{n \to \infty} u_n$ exists. Let $u_n \to u$, then $u = \sqrt{2 + u}$. Taking the positive root, we find that $u = 2$.

8.7 Obviously, $u_1 < 2$. Since $2 - u_{n+1} = 2(2 - u_n)/(u_n + 3)$ and $u_n > 0$, if $u_n < 2$ then $u_{n+1} < 2$. Now $u_{n+1} - u_n = (2 - u_n)(1 + u_n)/(u_n + 3)$, since $0 < u_n < 2$ we have $u_{n+1} > u_n$. Thus, $\{u_n\}$ is strictly increasing and $1 \leq u_n < 2$. Let $u_n \to u$, then $u = 2(2u + 1)/(u + 3)$. By taking the positive root, we get $u = 2$.

8.8. Use induction to show that sequences $\{u_{2n}\}$, $\{u_{2n+1}\}$ are monotonic. If $u_{2n} \to \ell_1$ and $u_{2n+1} \to \ell_2$, then $u_{2n+1} = (u_{2n}^2 u_{2n-1})^{1/3}$ implies $\ell_2 = (\ell_1^2 \ell_2)^{1/3}$ and hence $\ell_1 = \ell_2$. Observe that $u_{n+2}^3 u_{n+1} = u_{n+1}^3 u_n = \cdots = u_2^3 u_1$.

8.9. No. $u_n \geq \frac{1}{n+n} + \frac{1}{n+n} + \cdots + \frac{1}{n+n} = \frac{n}{2n} = \frac{1}{2}$. Hence, $\lim u_n \geq 1/2$ (if exists). In fact, $\lim u_n = \ln 2$.

8.10. (i). Consider $I_n = (0, 1/n)$. If $x \in (0, 1/n)$, $\forall n \in \mathcal{N}$, then $0 < x < 1/n$, $\forall n \in \mathcal{N}$. But this contradicts the Archimedean property.
(ii). Consider $I_n = [n, \infty)$.

8.11. Suppose to the contrary that there is a bijective mapping f from $\mathcal{N}$ onto the interval $[0, 1]$. We write these numbers as $f(1), f(2), \cdots$. Now divide the interval $[0, 1]$ into three equal parts. We choose from the subintervals obtaining the one which does not contain $f(1)$, and denote it by $[a_1, b_1]$. If there were two subintervals which do not contain $f(1)$ we choose either of them. Next we divide the interval $[a_1, b_1]$ into three parts and choose the interval, say, $[a_2, b_2]$ which does not contain $f(2)$. Continuing this process we obtain a sequence of intervals $[a_n, b_n]$, $n \in \mathcal{N}$ such that each of them (starting from the second) is one third of the preceding interval, and $f(n) \notin [a_n, b_n]$, $\forall n \in \mathcal{N}$. Now by the nested interval property, there exists a number y belonging to all intervals $[a_n, b_n]$. Since $0 \leq y \leq 1$ from our assumption there must be some $m \in \mathcal{N}$ such that $y = f(m)$. However, on the other hand, $f(m) \notin [a_m, b_m]$. This contradiction completes the proof.

8.12. Since S is bounded, there exists $M > 0$ such that $S \subseteq [-M, M]$. Consider two intervals $[-M, 0]$ and $[0, M]$; at least one of these intervals, call it I_0, must contain that portion of S which cannot be covered by a finite number of sets from $\mathcal{O}$. Follow Theorem 8.3 to construct intervals I_n with the same property as I_0. By the nested interval property there is only one point, say, u common to all these intervals. We shall show that u is a limit point of S. Let $\epsilon > 0$ be given. Choose n so large that $M/2^n < \epsilon$. Then, $I_n \subset N(u, \epsilon)$. But, I_n contains infinitely many points of S (if $S \cap I_n$ were a finite set then it could be covered by finitely many sets from $\mathcal{O}$), and hence there is an $x \in S$ with $x \neq u$ and $|x - u| < \epsilon$. Thus, u is a limit point of S, and $u \in S$ since S closed.

Now since $\mathcal{O}$ is an open cover of S, there is a set $O_p \in \mathcal{O}$ such that $u \in O_p$. But since O_p is an open set, there is an $\eta > 0$ such that $N(u, \eta) \subseteq O_p$. Now choose $m > 0$ so large that $I_m \subset N(u, \eta)$. Then, $I_m \subset O_p$, i.e., I_m is covered by a finite number of sets (namely one) from $\mathcal{O}$, and so clearly that portion of S in I_m is covered by a finite number of sets of $\mathcal{O}$. But this contradicts our construction of the interval I_m.

8.13. (i). Since $\sup_{n \geq m} u_n \leq \sup_{n \geq m} v_n$ and $\inf_{n \geq m} u_n \leq \inf_{n \geq m} v_n$, results follow as $m \to \infty$.

(ii) and (iii). For $m \geq 1$, we have $\inf_{n \geq m} u_n + \inf_{n \geq m} v_n \leq u_i + v_i \leq \sup_{n \geq m} u_n + \sup_{n \geq m} v_n$, for all $i \geq m$. From $\inf_{n \geq m} u_n + \inf_{n \geq m} v_n \leq u_i + v_i$, we have $\inf_{n \geq m} u_n + \inf_{n \geq m} v_n \leq \inf_{n \geq m}(u_n + v_n)$. Also from $u_i + v_i \leq \sup_{n \geq m} u_n + \sup_{n \geq m} v_n$, we have $\sup_{n \geq m}(u_n + v_n) \leq \sup_{n \geq m} u_n + \sup_{n \geq m} v_n$. Results now follow as $i \to \infty$.

(iv). Consider $u_n = (-1)^n$ and $v_n = (-1)^{n+1}$.

8.14. If $\{|u_n|\}$ does not diverge to ∞, then there exists an $M > 0$ such that $|u_m| \leq M$ for all $m > n \geq n_0$. Thus, we can construct a subsequence $\{u_{n_k}\}$ such that $|u_{n_k}| \leq M$. But, then this subsequence is bounded and from Theorem 8.3 has a convergent subsequence, which is a contradiction.

8.15. Let $N \in \mathcal{N}$ be sufficiently large so that for all $n \geq N$, $v_n > 0$ and $v_{n+1} < v_n$. We assume that $(u_{n+1} - u_n)/(v_{n+1} - v_n) \geq c$ for all $n \geq N$, where $c \in \mathcal{R}$. Then, $(u_{n+1} - u_n) \leq c(v_{n+1} - v_n)$ and by summation $u_{n+p} - u_n \leq c(v_{n+p} - v_n)$ for all $n \geq N$ and $p \in \mathcal{N}$. Letting $p \to \infty$, we find $-u_n \leq -cv_n$, which is the same as $u_n/v_n \geq c$ for all $n \geq N$. The same holds with the inequalities reversed.

Chapter 9

Infinite Series

Infinite series occur very frequently in all types of problems, and hence the study of their convergence or divergence is of fundamental importance. Unless a series employed in an investigation is convergent, it may lead to absurd conclusions. Therefore, in this and the next chapters we shall provide several easily verifiable tests which guarantee the convergence or divergence of a given series.

Let $\{u_n\}$ be a real sequence. An expression of the form $\sum_{n=1}^{\infty} u_n = u_1 + u_2 + \cdots$ is called an *infinite series* with terms u_n. Associated with this series we define a new sequence $\{U_n\}$ of *partial sums*, where $U_n = \sum_{k=1}^{n} u_k = u_1 + u_2 + \cdots + u_n$. The series is said to *converge* if its sequence of partial sums converges to some $U \in \mathcal{R}$. In this case we write $\sum_{n=1}^{\infty} u_n = U$, and call U the *sum* or *value* of the series. The series is said to *diverge* if its sequence of partial sums diverges. When $U_n \to +\infty \ (-\infty)$, we write $\sum_{n=1}^{\infty} u_n = +\infty \ (-\infty)$. The series is said to *oscillate* if its sequence of partial sums oscillates. The following two properties of series are fundamental.

(P1). Convergence of a series remains unchanged by the replacement, inclusion, or omission of a finite number of terms.

(P2). A series remains convergent, divergent, or oscillatory when each term is multiplied by a fixed number other than zero.

Example 9.1. The *geometric series* $\sum_{n=1}^{\infty} r^{n-1} = 1 + r + r^2 + \cdots$ converges to $1/(1 - r)$ iff $|r| < 1$, diverges to $+\infty$ if $r \geq 1$, oscillates finitely if $r = -1$ and infinitely if $r < -1$. Clearly, we have

$$U_n = \sum_{k=1}^{n} r^{k-1} = \begin{cases} \frac{1-r^n}{1-r} & \text{if } r \neq 1 \\ n & \text{if } r = 1. \end{cases}$$

Thus, if $|r| < 1$, $r^n \to 0$, and so $U_n \to 1/(1 - r)$. If $r = 1$, $U_n = n \to +\infty$. If $r > 1$, $r^n \to +\infty$, and so $U_n \to +\infty$. If $r = -1$, r^n oscillates finitely, and so $\{U_n\}$ oscillates finitely. If $r < -1$, r^n oscillates infinitely, and so $\{U_n\}$ oscillates infinitely.

Example 9.2. The *harmonic series* $\sum_{n=1}^{\infty} 1/n = 1 + 1/2 + 1/3 + \cdots$ is divergent. Clearly, for $U_n = \sum_{k=1}^{n} 1/n$ we have

$$
\begin{aligned}
U_1 &= 1 \\
U_2 &= 1 + \tfrac{1}{2} = \tfrac{3}{2} \\
U_4 &= U_2 + \tfrac{1}{3} + \tfrac{1}{4} > \tfrac{3}{2} + \tfrac{1}{4} + \tfrac{1}{4} = 2 \\
U_8 &= U_4 + \tfrac{1}{5} + \tfrac{1}{6} + \tfrac{1}{7} + \tfrac{1}{8} > 2 + \tfrac{1}{8} + \tfrac{1}{8} + \tfrac{1}{8} + \tfrac{1}{8} = \tfrac{5}{2}
\end{aligned}
$$

and in general, by induction, it follows that $U_{2^n} > (n+2)/2$. Thus, the subsequence $\{U_{2^n}\}$ of $\{U_n\}$ diverges to $+\infty$, which in turn implies that $U_n \to \infty$.

Our first result provides a necessary condition for the convergence of a series.

Theorem 9.1. If the series $\sum_{n=1}^{\infty} u_n$ converges, then it is necessary that $u_n \to 0$.

Proof. Suppose $\sum_{n=1}^{\infty} u_n = U$, i.e., $U_n \to U$. Then, since $u_n = U_n - U_{n-1}$ it follows that $\lim u_n = \lim(U_n - U_{n-1}) = (U - U) = 0$.

The condition $u_n \to 0$ in the above theorem is not sufficient to ensure the convergence of $\sum_{n=1}^{\infty} u_n$. In fact, in Example 9.2 we have seen that the harmonic series $\sum_{n=1}^{\infty} 1/n$ diverges, even though $u_n = 1/n \to 0$. However, from Theorem 9.1 it is clear that if $u_n \to u \neq 0$, then $\sum_{n=1}^{\infty} u_n$ must diverge. Further, if $u > 0$ the series must diverge to $+\infty$, whereas if $u < 0$ it should diverge to $-\infty$. Thus, the series $\sum_{n=1}^{\infty} \cos(x/n)$, $x \in \mathcal{R}$ diverges to $+\infty$, whereas the series $\sum_{n=1}^{\infty}(3 - n^2)/(2n^2 + 7)$ diverges to $-\infty$. From Theorem 9.1 it also follows that the series $\sum_{n=1}^{\infty}(-1)^n$ diverges.

Now we state two results which are immediate consequences of Theorems 7.5(1) and 7.7, respectively.

Theorem 9.2. If $\sum_{n=1}^{\infty} u_n$ converges to U and $\sum_{n=1}^{\infty} v_n$ converges to V, then $\sum_{n=1}^{\infty}(u_n \pm v_n)$ converges to $U \pm V$.

The series $(1 - 1) + (1 - 1) + \cdots$ obviously converges to zero, however $\sum_{n=1}^{\infty} 1 - \sum_{n=1}^{\infty} 1$ does not make sense. Thus, the converse of Theorem 9.2 does not hold.

Theorem 9.3 (Cauchy's Convergence Criterion). The series $\sum_{n=1}^{\infty} u_n$ converges iff given $\epsilon > 0$ there is an $N \in \mathcal{N}$ such that $m \geq n \geq N$ implies $|\sum_{k=n}^{m} u_k| < \epsilon$.

In particular, if we choose $m = n$, the Cauchy criterion reduces to $|u_n| < \epsilon$, $n \geq N$ which is the same as Theorem 9.1.

The following result provides a necessary and sufficient condition for the convergence of the series $\sum_{n=1}^{\infty} u_n$ when the sequence $\{u_n\}$ is nonnegative.

Theorem 9.4. The series $\sum_{n=1}^{\infty} u_n$, where $u_n \geq 0$, $n \geq N \in \mathcal{N}$ converges iff its sequence of partial sums $\{U_n\}$ is bounded, in which case, $U = \sup\{U_n : n \geq N\} = \sum_{n=1}^{\infty} u_n$.

Proof. If $\sum_{n=1}^{\infty} u_n$ converges, then $\{U_n\}$ converges. Since in view of Theorem 7.2 every convergent sequence is bounded, $\sum_{n=1}^{\infty} u_n$ has bounded partial sums. On the other hand, suppose $|U_n| \leq M$, $n \in \mathcal{N}$. Since $u_n \geq 0$ for $n \geq N$, U_n is an increasing sequence for $n \geq N$. Now in view of Theorem 8.1(1) every increasing bounded sequence converges to its supremum, it follows that $\sum_{n=1}^{\infty} u_n$ converges to U.

In many applications, the sequence $\{u_n\}$ is not only nonnegative, but decreases also. In such a case a "thin" subsequence of $\{u_n\}$ determines the convergence or divergence of the series.

Theorem 9.5 (Cauchy's Condensation Criterion). The series $\sum_{n=1}^{\infty} u_n$, where $u_n \geq 0$ and $u_n \geq u_{n+1}$, $n \in \mathcal{N}$ converges iff the series $\sum_{n=0}^{\infty} 2^n u_{2^n}$ converges.

Proof. Let $U_n = \sum_{k=1}^{n} u_k$ and $T_m = \sum_{k=0}^{m} 2^k u_{2^k}$. For $n \leq 2^m$, we have

$$
\begin{aligned}
U_n &= u_1 + u_2 + \cdots + u_n \\
&\leq u_1 + (u_2 + u_3) + (u_4 + u_5 + u_6 + u_7) + \cdots + (u_{2^m} + \cdots + u_{2^{m+1}-1}) \\
&\leq u_1 + 2u_2 + 4u_4 + \cdots + 2^m u_{2^m} = T_m.
\end{aligned}
$$

If $n \geq 2^m$, we find

$$
\begin{aligned}
U_n &= u_1 + u_2 + \cdots + u_n \\
&\geq u_1 + u_2 + (u_3 + u_4) + \cdots + (u_{2^{m-1}+1} + \cdots + u_{2^m}) \\
&\geq \tfrac{1}{2}u_1 + u_2 + 2u_4 + \cdots + 2^{m-1}u_{2^m} = \tfrac{1}{2}T_m.
\end{aligned}
$$

Thus, the sequences of partial sums of $\{U_n\}$ and $\{T_m\}$ are either both bounded, or both unbounded. The result now follows from Theorem 9.4.

Example 9.3. The divergence of the *p-series* $\sum_{n=1}^{\infty} 1/n^p$ for $p \leq 0$ follows from Theorem 9.1, whereas for $p = 1$ its divergence is shown in Example 9.2. If $p > 0$, Theorem 9.5 is applicable, and we need to consider the series $\sum_{n=0}^{\infty} 2^n (1/2^n)^p = \sum_{n=0}^{\infty} (2^{(1-p)})^n$, which in view of Example 9.1 converges iff $2^{1-p} < 1$, i.e., $1 - p < 0$ or $p > 1$, and diverges if $2^{1-p} \geq 1$, i.e., $1 - p \geq 0$, or $p \leq 1$. Thus, the *p*-series converges if $p > 1$ and diverges if $p \leq 1$.

In what follows, if $u_n \geq 0$ for all large n, we shall write $\sum_{n=1}^{\infty} u_n < \infty$ provided the series converges, and $\sum_{n=1}^{\infty} u_n = \infty$ if the series diverges.

Next, we shall provide some widely applicable tests which ensure the convergence of a given series with nonnegative terms.

Theorem 9.6 (Comparison Test). Suppose $0 \leq u_n \leq v_n$ for large $n \in \mathcal{N}$.

(1). If $\sum_{n=1}^{\infty} v_n < \infty$, then $\sum_{n=1}^{\infty} u_n < \infty$.

(2). If $\sum_{n=1}^{\infty} u_n = \infty$, then $\sum_{n=1}^{\infty} v_n = \infty$.

Proof. Let $N \in \mathcal{N}$ be so large that $0 \leq u_n \leq v_n$, $n > N$. Then, for the partial sums $U_n = \sum_{k=1}^{n} u_k$ and $V_n = \sum_{k=1}^{n} v_k$, we have $0 \leq U_n - U_N \leq V_n - V_N$, $n \geq N$. Since N is fixed, U_n is bounded if V_n is bounded, and V_n is unbounded if U_n is unbounded. The result now follows from Theorem 9.4.

Example 9.4. Since $n! \geq 2^{n-1}$, $n \in \mathcal{N}$, the convergence of the series $\sum_{n=1}^{\infty} 1/n!$ immediately follows from Theorem 9.6 and Example 9.1. Similarly, the divergence of the series $\sum_{n=1}^{\infty} 1/n^{\epsilon}$, $0 \leq \epsilon < 1$ follows by comparing it with the harmonic series.

Theorem 9.7 (Limit Comparison Test). Suppose $u_n, v_n > 0$ for large $n \in \mathcal{N}$. If $0 < \lim_{n \to \infty} u_n/v_n < \infty$, then $\sum_{n=1}^{\infty} u_n$ converges iff $\sum_{n=1}^{\infty} v_n$ converges.

Proof. Let $\ell = \lim_{n \to \infty} u_n/v_n$. Then there is a large $N \in \mathcal{N}$ such that $(\ell/2)v_n < u_n < (3\ell/2)v_n$ for $n \geq N$. The result now follows from Theorem 9.6 and the property (P2).

Example 9.5. As an application to Theorem 9.7 we shall show that $\sum_{n=1}^{\infty} [(n^3+1)^{1/3} - n]$ converges. For this, it suffices to consider the convergent series $\sum_{n=1}^{\infty} 1/n^2$, and note that $u_n = [(n^3 + 1)^{1/3} - n]$ and $v_n = 1/n^2$ both are positive for all $n \in \mathcal{N}$, and

$$\frac{u_n}{v_n} = n^2[(n^3 + 1)^{1/3} - n] = \frac{n^2[(n^3+1)-n^3]}{(n^3+1)^{2/3}+n(n^3+1)^{1/3}+n^2}$$
$$= \frac{1}{(1+1/n^3)^{2/3}+(1+1/n^3)^{1/3}+1} \to \frac{1}{3}.$$

Theorem 9.8 (Ratio Test). Suppose $u_n > 0$, $n \in \mathcal{N}$. Let $\limsup u_{n+1}/u_n = L$, and $\liminf u_{n+1}/u_n = \ell$. Then,

(1). $L < 1$ implies that the series $\sum_{n=1}^{\infty} u_n$ converges.

(2). $\ell > 1$ implies that the series $\sum_{n=1}^{\infty} u_n$ diverges.

Proof. (1). If $L < 1$, then for any given ϵ, $0 < \epsilon < 1 - L$ there exists $N \in \mathcal{N}$ such that $u_{n+1}/u_n < L+\epsilon < 1$, $n \geq N$. Thus, $u_{n+1} < (L+\epsilon)u_n$, $n \geq N$ which implies $u_n < (L + \epsilon)^{n-N}u_N$, $n \geq N + 1$. Therefore, $\sum_{n=N+1}^{\infty} u_n < u_N \sum_{n=N+1}^{\infty} (L + \epsilon)^{n-N}$. However, since $(L + \epsilon) < 1$ the convergence of the series $\sum_{n=1}^{\infty} u_n$ follows from Example 9.1, Theorem 9.6(1), and (P1).

(2). If $\ell > 1$ there exists $N \in \mathcal{N}$ such that $u_{n+1}/u_n > 1$, $n \geq N$. Thus, $u_n > u_N > 0$, $n \geq N + 1$. But, this implies $\lim u_n \neq 0$. Therefore, $\sum_{n=1}^{\infty} u_n$ must diverge.

If in Theorem 9.8, $\ell = L$, i.e., $\lim u_{n+1}/u_n = L$, then the series $\sum_{n=1}^{\infty} u_n$ converges if $L < 1$, and diverges if $L > 1$. The test fails if $L = 1$, e.g., $\sum_{n=1}^{\infty} 1/n^2$ converges, whereas $\sum_{n=1}^{\infty} 1/n$ diverges.

Example 9.6. For the series $\sum_{n=1}^{\infty} x^n n!/n^n$, $x > 0$ we have $\lim u_{n+1}/u_n = x/e$, and hence Theorem 9.8 ensures it converges or diverges according as $0 < x < e$, or $x > e$. For $x = e$, Theorem 9.8 is not applicable; however, since $u_{n+1}/u_n = e/[(1 + 1/n)^n] > 1$ for all $n \in \mathcal{N}$ it diverges.

Theorem 9.9 (Root Test). Suppose $u_n \geq 0$, $n \in \mathcal{N}$ and $\limsup \sqrt[n]{u_n} = L$. Then,

(1). $L < 1$ implies that the series $\sum_{n=1}^{\infty} u_n$ converges.

(2). $L > 1$ implies that the series $\sum_{n=1}^{\infty} u_n$ diverges.

Proof. (1). For $L < (1 + L)/2 < 1$ there exists $N \in \mathcal{N}$ such that $0 < \sqrt[n]{u_n} < (1 + L)/2$, $n \geq N$. Thus, $u_n < [(1 + L)/2]^n$, $n \geq N$. Now since $\sum_{n=N}^{\infty} [(1 + L)/2]^n$ is convergent, the convergence of $\sum_{n=1}^{\infty} u_n$ follows from Theorem 9.6(1), and (P1).

(2). It suffices to note that if $L > 1$ there exists $N \in \mathcal{N}$ such that $u_n > 1$, $n \geq N$.

If in Theorem 9.9, $\lim \sqrt[n]{u_n} = L$, then the series $\sum_{n=1}^{\infty} u_n$ converges if $L < 1$, and diverges if $L > 1$. The test fails if $L = 1$.

Example 9.7. For the series $\sum_{n=1}^{\infty} 2^{-n+(-1)^n}$, we have $\limsup u_{n+1}/u_n = 2$ and $\liminf u_{n+1}/u_n = 1/8$ and hence the ratio test is not applicable; however, since $\lim \sqrt[n]{u_n} = \lim 2^{-1+(-1)^n/n} = 1/2$ the root test ensures the convergence of the series. Similarly, for the series $\sum_{n=1}^{\infty} 2^{n-(-1)^n}$ we find $\limsup u_{n+1}/u_n = 8$ and $\liminf u_{n+1}/u_n = 1/2$ and hence the ratio test is not applicable; however, since $\lim \sqrt[n]{u_n} = \lim 2^{1-(-1)^n/n} = 2$ the root test ensures the divergence of the series.

From Examples 8.4 and 9.7 it is clear that the root test is stronger than the ratio test.

Example 9.8. For the series $\sum_{n=1}^{\infty} [(5 + (-1)^n)/2]^{-n}$, we have $\limsup \sqrt[n]{u_n} = 1/2$ and $\liminf \sqrt[n]{u_n} = 1/3$, and hence Theorem 9.9(1) ensures its convergence. Similarly, for the series $\sum_{n=1}^{\infty} [(5 + (-1)^n)/2]^n$, we find $\limsup \sqrt[n]{u_n} = 3$ and $\liminf \sqrt[n]{u_n} = 2$, and therefore Theorem 9.9(2) guarantees its divergence.

Problems

9.1. Show that the series $\sum_{n=1}^{\infty} u_n$ converges iff $R_n = \sum_{k=n+1}^{\infty} u_k \to 0$.

9.2. Let $\sum_{n=1}^{\infty} u_n$ be a divergent series of positive numbers. Show that there exists a sequence $\{\epsilon_n\}$ of positive numbers which converges to zero, but $\sum_{n=1}^{\infty} \epsilon_n u_n$ diverges.

9.3. Let $\{u_n\}$ be a nonincreasing sequence of positive numbers and $\sum_{n=1}^{\infty} u_n$ converges. Show that $\lim_{n \to \infty} n u_n = 0$. Further, give an example to show that if the sequence $\{u_n\}$ is not nonincreasing then the result is false.

9.4. Suppose u_n, $v_n > 0$, $n \in \mathcal{N}$, and $\{u_n/v_n\}$, $\{v_n/u_n\}$ are both bounded sequences. Show that the series $\sum_{n=1}^{\infty} u_n$ and $\sum_{n=1}^{\infty} v_n$ either both converge or both diverge.

9.5. Suppose that $\{u_n\}$ and $\{v_n\}$ are sequences of positive real numbers, and there exists an $N \in \mathcal{N}$ such that $u_{n+1}/u_n \le v_{n+1}/v_n$ for all $n \ge N$. Show that

(i).　If $\sum_{n=1}^{\infty} v_n$ converges, then $\sum_{n=1}^{\infty} u_n$ converges.

(ii).　If $\sum_{n=1}^{\infty} u_n$ diverges, then $\sum_{n=1}^{\infty} v_n$ diverges.

9.6. Suppose that $\{u_n\}$ is a sequence of positive real numbers, and the series $\sum_{n=1}^{\infty} u_n$ diverges. Show that the series

(i).　$\sum_{n=1}^{\infty} u_n/(1 + n^2 u_n)$ converges.

(ii).　$\sum_{n=1}^{\infty} u_n/(1 + n u_n)$ diverges.

(iii).　$\sum_{n=1}^{\infty} u_n/(1 + u_n^2)$ diverges.

9.7. Test the given series $\sum_{n=1}^{\infty} u_n$ for convergence or divergence, where

(i)　$u_n = \frac{1}{n(\ln n)}$.

(ii).　$u_n = \frac{1}{n(\ln n)^2}$.

(iii).　$u_n = \frac{n^{n^2}}{(n+1)^{n^2}}$.

(iv).　$u_n = \frac{2^{n+1}-2}{2^{n+1}+1} x^n$, $x > 0$.

(v).　$u_{2n-1} = 1/3^{n+1}$ and $u_{2n} = 1/3^n$.

9.8　(Kummer's Test). Let $\sum_{n=1}^{\infty} u_n$ be a series with positive terms, and let $\{P_n\}$ be a sequence of positive constants. Assume that the sequence $\{v_n\}$ defined by $v_n = P_n(u_n/u_{n+1}) - P_{n+1}$ has a limit L, possibly infinite. Show that

(i).　If $L > 0$, then $\sum_{n=1}^{\infty} u_n$ converges.

(ii).　If $L < 0$, and $\sum_{n=1}^{\infty} 1/P_n$ diverges, then $\sum_{n=1}^{\infty} u_n$ diverges.

(iii).　If $L = 0$ the test fails.

Further, note that when $P_n = 1$, Kummer's test reduces to the ratio test.

9.9 (Raabe-Duhamel's Test). Let $\sum_{n=1}^{\infty} u_n$ be a series with positive terms, and let $\lim_{n\to\infty} n(u_n/u_{n+1} - 1) = L$. Show that

(i). If $L > 1$, then $\sum_{n=1}^{\infty} u_n$ converges.

(ii). If $L < 1$, then $\sum_{n=1}^{\infty} u_n$ diverges.

(iii). If $L = 1$ the test fails.

Further, show that Raabe-Duhamel's test is stronger than the ratio test.

9.10 (Logarithmic Ratio Test). Let $\sum_{n=1}^{\infty} u_n$ be a series with positive terms, and let $\lim_{n\to\infty}[n \ln u_n/u_{n+1}] = L$. Show that

(i). If $L > 1$, then $\sum_{n=1}^{\infty} u_n$ converges.

(ii). If $L < 1$, then $\sum_{n=1}^{\infty} u_n$ diverges.

(iii). If $L = 1$ the test fails.

Further, show that logarithmic ratio test is stronger than the ratio test.

Answers or Hints

9.1. Let $\sum_{n=1}^{\infty} u_n = U$. Then, $U_n \to U$, so that $U = U_n + R_n$ implies $R_n \to 0$. Now if $R_n \to 0$, then for $\epsilon > 0 \, \exists \, N \in \mathcal{N}$ such that $|R_n| < \epsilon/2, \, \forall \, n \geq N$. Thus, $|u_{n+1} + u_{n+2} + \cdots + u_{n+p}| = |R_n - R_{n+p}| \leq |R_n| + |R_{n+p}| \leq \epsilon, \, \forall \, n \geq N$ and $p \geq 0$. The convergence of $\sum_{n=1}^{\infty} u_n$ now follows from Theorem 9.3.

9.2. For any fixed $N \in \mathcal{N}$ choose $m \in \mathcal{N}$ so that $U_{n+1} > 2U_N$ for all $n \geq m$. This is always possible because $\{U_n\}$ diverges to infinity. Now since $\{U_n\}$ is nondecreasing, we have $\sum_{k=N}^{n}(U_{k+1} - U_k)/U_{k+1} \geq \sum_{k=N}^{n}(U_{k+1} - U_k)/U_{n+1} = (U_{n+1} - U_N)/U_{n+1} > (U_{n+1} - U_{n+1}/2)/U_{n+1} = 1/2$. Thus, the series $\sum_{n=1}^{\infty}(U_{n+1} - U_n)/U_{n+1}$ does not satisfy Cauchy's criterion, and hence $\sum_{n=1}^{\infty}(U_{n+1} - U_n)/U_{n+1} = \sum_{n=1}^{\infty} u_{n+1}/U_{n+1} = \infty$. Now let $\epsilon_n = 1/U_n$. Obviously $\epsilon_n \to 0$.

9.3. If $\sum_{n=1}^{\infty} u_n = U$, then $\lim U_n = U = \lim U_{2n}$. Thus, $\lim(U_{2n} - U_n) = 0$. Now $U_{2n} - U_n = u_{n+1} + u_{n+2} + \cdots + u_{2n} \geq u_{2n} + u_{2n} + \cdots + u_{2n} = n u_{2n} \geq 0$. Thus $\lim(n u_{2n}) = 0$ and so $\lim(2n u_{2n}) = 0$. Next since $(2n+1)u_{2n+1} \leq (2n+1)u_{2n} = \frac{2n+1}{2n}2n u_{2n}$ and hence $(2n + 1)u_{2n+1} = 0$. Therefore, $\lim(n u_n) = 0$. To show the remaining part consider the sequence $u_n = 1/n$ if n is a perfect square, and $1/n^2$ if n is not a perfect square. Clearly, $n a_n = 1$ whenever n is not a perfect square. However, $\frac{1}{1} + \frac{1}{2^2} + \frac{1}{3^2} + \frac{1}{4} + \frac{1}{5^2} + \cdots = \left(\frac{1}{2^2} + \frac{1}{3^2} + \frac{1}{5^2} + \cdots\right) + \left(1 + \frac{1}{4} + \frac{1}{9} + \cdots\right) \leq 2\sum_{n=1}^{\infty} 1/n^2$, and hence $\sum_{n=1}^{\infty} u_n$ converges.

9.4. If $u_n/v_n \leq M_1$ and $v_n/u_n \leq M_2$, then $u_n \leq M_1 v_n \leq M_1 M_2 u_n$. Now apply Theorem 9.6.

9.5. Since $u_{n+1}/v_{n+1} \leq u_n/v_n$, $n \geq N$ the sequence $\{u_n/v_n\}$ is decreasing for all $n \geq N$. Thus, $u_n/v_n \leq u_N/v_N = M$, which implies $u_n \leq Mv_n$, $n \geq N+1$.

9.6. (i). For all n, $u_n/(1+n^2u_n) \leq 1/n^2$.
(ii). Since $u_n \to \ell > 0$, for all large n, $u_n > \ell/2$ and $u_n/(1+nu_n) > 1/(n+2/\ell)$.
(iii). $u_n/(1+u_n^2) \to \ell/(1+\ell^2) > 0$.

9.7. (i). Since $\sum_{n=2}^{\infty} 2^n u_{2^n} = 1/(\ln 2)\sum_{n=2}^{\infty} 1/n$ diverges, from Theorem 9.5 the series diverges.
(ii). Since $\sum_{n=2}^{\infty} 2^n u_{2^n} = 1/(\ln 2)^2 \sum_{n=2}^{\infty} 1/n^2$ converges, from Theorem 9.5 the series converges.
(iii). $\sqrt[n]{u_n} = n^n/(n+1)^n = (1+1/n)^{-n} \to 1/e < 1$ thus from root test the series converges.
(iv). $\frac{u_{n+1}}{u_n} = \frac{2^{n+1}+1}{2^{n+2}+1}\frac{2^{n+2}-2}{2^{n+1}-2}x \to x$ and hence the series converges if $0 < x < 1$ and diverges for $x > 1$. For $x = 1$, $u_n = \frac{2^{n+1}-2}{2^{n+1}+1} \to 1$ and thus the series diverges.
(v). $u_{n+1}/u_n = 3$ or $1/9$ according as n is odd or even, so the ratio test is not applicable. However, since $\lim \sqrt[n]{u_n} = 1/\sqrt{3} < 1$ the root test is applicable, and the series converges.

9.8. (i). Let α be a constant such that $0 < \alpha < L$. Since $v_n \to L$, there exists some $N \in \mathcal{N}$ such that for $n \geq N$, $v_n > \alpha$, i.e., $P_n u_n - P_{n+1}u_{n+1} > \alpha u_{n+1}$. Thus, if $m \geq 1$, $P_N u_N - P_{N+m}u_{N+m} > \alpha(u_{N+1} + \cdots + u_{N+m}) = \alpha(U_{N+m} - U_N)$, i.e., $U_{N+m} < U_N + P_N u_N/\alpha$. Hence, if $n > N$, U_n is bounded above by $U_N + P_N u_N/\alpha$. Now let $M = \max\{U_1, \cdots, U_N, U_N + P_N u_N/\alpha\}$. Clearly, $\{U_n\}$ is bounded above by M. The result now follows from Theorem 9.4.
(ii). Since $L < 0$, there exists $N \in \mathcal{N}$ such that for $n \geq N$, $v_n < 0$, i.e., $P_n u_n < P_{n+1}u_{n+1}$. But this implies $P_N u_N < P_{N+1}u_{N+1} < \cdots < P_n u_n < \cdots$, i.e., $P_N u_N < P_n u_n$ for all $n \geq N+1$. Thus, we have $\sum_{n=N+1}^{\infty} u_n > P_N u_N \sum_{n=N+1}^{\infty} 1/P_n$.
(iii). For the p-series $\sum_{n=1}^{\infty} 1/n^p$, $p \geq 1$ with $P_n = 1$, we have $v_n = (1+1/n)^p - 1 \to 0$. The p-series diverges for $p = 1$, and converges for $p > 1$.

9.9. Let $P_n = n$ in Kummer's test, to obtain $v_n = n(u_n/u_{n+1}-1)-1$. By the ratio test $\sum_{n=0}^{\infty} u_n$ converges, or diverges according as $\lim_{n\to\infty} u_n/u_{n+1} > 1$, or < 1. This implies that $\lim_{n\to\infty} n(u_n/u_{n+1} - 1) = +\infty$, or $-\infty$ as $\lim_{n\to\infty} u_n/u_{n+1} > 1$, or < 1. Thus, $\sum_{n=0}^{\infty} u_n$ is also convergent or divergent according to Raabe-Duhamel's test. Now consider $\sum_{n=1}^{\infty} 1/n^2$ for which the ratio test fails, however, since $\lim_{n\to\infty} n(u_n/u_{n+1}-1) = \lim_{n\to\infty}(2n+1)/n = 2$, Raabe-Duhamel's test ensures its convergence. Similarly, for the series $\sum_{n=1}^{\infty} \frac{1\cdot3\cdots(2n-1)}{2\cdot4\cdots(2n)}$ the ratio test fails, whereas Raabe-Duhamel's test ensures its divergence.

9.10. Suppose $L > 1$ and choose $\epsilon > 0$ such that $a = L - \epsilon > 1$. Since $\lim_{n\to\infty} \left(n \ln \frac{u_n}{u_{n+1}} \right) = L$, there exists an $n_\epsilon \in \mathcal{N}$ such that $L - \epsilon < n \ln \frac{u_n}{u_{n+1}} < L + \epsilon$ for all $n \geq n_\epsilon$. Thus, $n \ln \frac{u_n}{u_{n+1}} > a$ for all $n \geq n_\epsilon$, which implies that $\frac{u_n}{u_{n+1}} > e^{\frac{a}{n}} > \left(1 + \frac{1}{n}\right)^a = \frac{(n+1)^a}{n^a} = \frac{v_n}{v_{n+1}}$ for all $n \geq n_\epsilon$, where $v_n = 1/n^a$. Therefore, $\frac{v_{n+1}}{v_n} > \frac{u_{n+1}}{u_n}$. Since $\sum v_n$ converges, from the ratio test $\lim \frac{v_{n+1}}{v_n} < 1$, which from the same test implies that $\sum u_n$ converges.

If $L < 1$ we choose $\epsilon > 0$ such that $a = L + \epsilon < 1$, and obtain $\frac{u_n}{u_{n+1}} < e^{\frac{a}{n}}$ for all $n \geq n_\epsilon$. This gives $\frac{u_{n+1}}{u_n} > e^{-\frac{a}{n}} > \left(1 - \frac{1}{n}\right)^a = \frac{(n-1)^a}{n^a} = \frac{v_{n+1}}{v_n}$ for all $n \geq n_\epsilon$, where $v_n = 1/(n-1)^a$. Since $\sum v_n$ diverges, from the ratio test $\lim \frac{v_{n+1}}{v_n} > 1$, which from the same test implies that $\sum u_n$ diverges.

Consider the series $\sum u_n$ where $u_n = 1/n^2$ for which the ratio test fails. To apply the logarithmic ratio test, it suffices to note that $\lim_{n\to\infty} \left(n \ln \frac{u_n}{u_{n+1}} \right) = $

$$\lim_{n\to\infty} \left(\ln \left(\frac{(n+1)^2}{n^2} \right)^n \right) = \lim_{n\to\infty} \left(\left(\ln \left(1 + \frac{1}{\frac{n^2}{2n+1}} \right)^{\frac{n^2}{2n+1}} \right)^{n\frac{2n+1}{n^2}} \right)$$

$$= \lim_{n\to\infty} \ln e^2 = 2.$$

Chapter 10

Infinite Series (Contd.)

In this chapter, some specific tests for the convergence of series of terms of arbitrary signs are provided. Some important results on rearrangements of terms of such series are also discussed.

A series of the form $\sum_{n=1}^{\infty}(-1)^{n+1}u_n$ where each $u_n > 0$ or < 0 is called an *alternating series*. For example, $1 - (1/2) + (1/3) - \cdots$ and $-1 + (1/2^2) - (1/3^2) + \cdots$ are alternative series. The following result is fundamental for the convergence of alternative series and one of the easiest to apply.

Theorem 10.1 (Alternative Series Test/Leibniz Criterion). If $\{u_n\}$ is a sequence of positive numbers such that

(i). $\{u_n\}$ is nonincreasing,

(ii). $\lim_{n\to\infty} u_n = 0$.

Then, the alternative series $\sum_{n=1}^{\infty}(-1)^{n+1}u_n$ converges to some $U \in \mathcal{R}$, and $|U_n - U| \leq u_{n+1}$, $n \in \mathcal{N}$ where $U_n = u_1 - u_2 + \cdots + (-1)^{n+1}u_n$.

Proof. Since

$$
\begin{aligned}
U_{2n} &= (u_1 - u_2) + (u_3 - u_4) + \cdots + (u_{2n-1} - u_{2n}) \\
&= u_1 - (u_2 - u_3) - (u_4 - u_5) - \cdots - (u_{2n-2} - u_{2n-1}) - u_{2n}
\end{aligned}
$$

from the hypothesis it follows that $\{U_{2n}\}$ is a positive monotonically nondecreasing sequence bounded above by u_1. Thus, $\{U_{2n}\}$ must converge to some U. Now from $U_{2n-1} = U_{2n} + u_{2n}$ it follows that $\lim U_{2n-1} = \lim U_{2n} + \lim u_{2n} = U + 0 = U$. Hence, for $\epsilon > 0 \, \exists \, N \in \mathcal{N}$ such that $|U_{2n} - U| < \epsilon$, $\forall n \geq N$ and $|U_{2n-1} - U| < \epsilon$, $\forall n \geq N$. This implies that $|U_n - U| < \epsilon$, $\forall n \geq N$. Hence, $\lim U_n = U$ and therefore $\sum_{n=1}^{\infty} u_n$ converges to U. Finally, we have

$$
\begin{aligned}
|U - U_n| &= \left| \sum_{k=n+1}^{\infty}(-1)^{k+1}u_k \right| \\
&= (u_{n+1} - u_{n+2}) + (u_{n+3} - u_{n+4}) + \cdots \\
&= u_{n+1} - (u_{n+2} - u_{n+3}) - (u_{n+4} - u_{n+5}) - \cdots < u_{n+1}.
\end{aligned}
$$

Example 10.1. From Theorem 10.1 it immediately follows that the series $\sum_{n=1}^{\infty}(-1)^{n+1}/(n-1)!$ converges to some $U \in \mathcal{R}$ (from elementary calculus we know $U = e^{-1} = 0.367879\cdots$), moreover,

$$\left| e^{-1} - \left(1 - \frac{1}{1!} + \frac{1}{2!} - \frac{1}{3!} + \frac{1}{4!} - \frac{1}{5!}\right)\right| \leq \frac{1}{6!},$$

i.e., $|e^{-1} - 0.366666| \leq 0.001389$.

Let $\sum_{n=1}^{\infty} u_n$ be a series of real numbers. If $\sum_{n=1}^{\infty} |u_n|$ converges, we say that $\sum_{n=1}^{\infty} u_n$ converges *absolutely*. For example, the series $\sum_{n=1}^{\infty}(-1)^{n+1}/n^2$ converges absolutely. If $\sum_{n=1}^{\infty} u_n$ converges but $\sum_{n=1}^{\infty} |u_n|$ diverges, we say that $\sum_{n=1}^{\infty} u_n$ converges *conditionally*. Since the series $\sum_{n=1}^{\infty} |u_n|$ consists of nonnegative terms, it either converges or diverges to $+\infty$. It is also clear that every known result about a series with nonnegative terms can be applied to the series $\sum_{n=1}^{\infty} |u_n|$. The following result ensures that the absolute convergence of a series implies its convergence.

Theorem 10.2. If $\sum_{n=1}^{\infty} u_n$ converges absolutely, then $\sum_{n=1}^{\infty} u_n$ converges.

Proof. If $\sum_{n=1}^{\infty} |u_n|$ converges, then by Theorem 9.3 for any given $\epsilon > 0$ there exists a number $N \in \mathcal{N}$ such that $|u_{n+1}| + |u_{n+2}| + \cdots + |u_{n+m}| < \epsilon$ for every $n \geq N$ and for each $m \in \mathcal{N}$. Since

$$|u_{n+1} + u_{n+2} + \cdots + u_{n+m}| \leq |u_{n+1}| + |u_{n+2}| + \cdots + |u_{n+m}|$$

again from Theorem 9.3 it follows that the series $\sum_{n=1}^{\infty} u_n$ converges.

The series $\sum_{n=1}^{\infty}(-1)^{n+1}/n$ converges conditionally, but not absolutely. Thus, the converse of Theorem 10.2 does not hold.

Theorem 10.3. (1). If $\sum_{n=1}^{\infty} u_n$ converges absolutely, then so do $\sum_{n=1}^{\infty} u_n^+$ and $\sum_{n=1}^{\infty} u_n^-$.

(2). If $\sum_{n=1}^{\infty} u_n$ converges conditionally, then $\sum_{n=1}^{\infty} u_n^+ = \sum_{n=1}^{\infty} u_n^- = \infty$.

Proof. (1). Since both $\sum_{n=1}^{\infty} |u_n|$ and $\sum_{n=1}^{\infty} u_n$ converge, and $u_n^+ = (|u_n| + u_n)/2$ and $u_n^- = (|u_n| - u_n)/2$ from Theorem 9.2 it follows that both

$$\sum_{n=1}^{\infty} u_n^+ = \frac{1}{2}\sum_{n=1}^{\infty} |u_n| + \frac{1}{2}\sum_{n=1}^{\infty} u_n$$

and

$$\sum_{n=1}^{\infty} u_n^- = \frac{1}{2}\sum_{n=1}^{\infty} |u_n| - \frac{1}{2}\sum_{n=1}^{\infty} u_n$$

converge.

(2). Suppose $\sum_{n=1}^{\infty} u_n^+$ converges. Then, from the convergence of $\sum_{n=1}^{\infty} u_n$ and $u_n^- = u_n^+ - u_n$ it follows that

$$\sum_{n=1}^{\infty} u_n^- = \sum_{n=1}^{\infty} u_n^+ - \sum_{n=1}^{\infty} u_n$$

also converges. But, then

$$\sum_{n=1}^{\infty} |u_n| = \sum_{n=1}^{\infty} u_n^+ + \sum_{n=1}^{\infty} u_n^-$$

converges, which is a contradiction.

A series $\sum_{n=1}^{\infty} v_n$ is called a *rearrangement* of $\sum_{n=1}^{\infty} u_n$ if there is a bijective mapping $\phi : \mathcal{N} \to \mathcal{N}$ such that $v_{\phi(n)} = u_n$, $n \in \mathcal{N}$. In our next result we shall show that any rearrangement of an absolutely convergent series converges to the same value as the original series.

Theorem 10.4. If $\sum_{n=1}^{\infty} u_n$ converges absolutely and $\sum_{n=1}^{\infty} v_n$ is any rearrangement of $\sum_{n=1}^{\infty} u_n$, then $\sum_{n=1}^{\infty} v_n$ converges. Moreover, if $\sum_{n=1}^{\infty} u_n = U$, then $\sum_{n=1}^{\infty} v_n = U$.

Proof. Let $\epsilon > 0$ be given. Since the series $\sum_{n=1}^{\infty} |u_n|$ converges, it follows from Problem 9.1 that there is a natural number N such that $\sum_{n=N+1}^{\infty} |u_n| < \epsilon/2$. Let $\phi : \mathcal{N} \to \mathcal{N}$ be the bijective mapping such that $v_{\phi(n)} = u_n$, $n \in \mathcal{N}$. Choose $k \in \mathcal{N}$ so large that $\{1, 2, \cdots, N\} \subseteq \{\phi(1), \phi(2), \cdots, \phi(k)\}$. Thus, if $j \geq k$, $\{u_1, u_2, \cdots, u_N\} \subseteq \{v_1, v_2, \cdots, v_j\}$, and

$$\left| \sum_{n=1}^{N} u_n - \sum_{n=1}^{j} v_n \right| \leq \sum_{n=N+1}^{\infty} |u_n| < \frac{\epsilon}{2}.$$

Therefore, if $j \geq k$, we have

$$\left| U - \sum_{n=1}^{j} v_n \right| \leq \left| U - \sum_{n=1}^{N} u_n \right| + \left| \sum_{n=1}^{N} u_n - \sum_{n=1}^{j} v_n \right|$$

$$\leq \left| \sum_{n=N+1}^{\infty} u_n \right| + \left| \sum_{n=1}^{N} u_n - \sum_{n=1}^{j} v_n \right| < \frac{\epsilon}{2} + \frac{\epsilon}{2} = \epsilon.$$

Hence, $\sum_{n=1}^{\infty} v_n = U$.

Our next result shows that for a conditionally convergent series, Theorem 10.4 is false.

Theorem 10.5. Let $\sum_{n=1}^{\infty} u_n$ be a conditionally convergent sequence, and let $a \leq b$ be any pair of real numbers. Then, there is a rearrangement $\sum_{n=1}^{\infty} v_n$ of $\sum_{n=1}^{\infty} u_n$ whose partial sums V_n satisfy

$$\liminf_{n \to \infty} V_n = a \quad \text{and} \quad \limsup_{n \to \infty} V_n = b. \tag{10.1}$$

In particular, given any real number $c \in \mathcal{R}$, there is a rearrangement of $\sum_{n=1}^{\infty} u_n$ which converges to c.

Proof. For simplicity we assume that both a and b are finite. Since each u_n^+ and $-u_n^-$ is either u_n or 0, it suffices to show that there are integers $0 < n_1 < m_1 < n_2 < m_2 < \cdots$ such that if $v_1 = u_1^+$, $v_2 = u_2^+, \cdots, v_{n_1} = u_{n_1}^+$, $v_{n_1+1} = -u_1^-, \cdots, v_{m_1} = -u_{m_1-n_1}^-$, $v_{m_1+1} = u_{n_1+1}^+, \cdots$, and $V_n = \sum_{k=1}^{h} v_k$, then (10.1) holds.

Since in view of Theorem 10.3, $\sum_{n=1}^{\infty} u_n^+ = \infty$, we can choose the least integer $n_1 \in \mathcal{N}$ such that

$$V_{n_1} = v_1 + v_2 + \cdots + v_{n_1} = u_1^+ + u_2^+ + \cdots + u_{n_1}^+ > b.$$

Since n_1 is least, $V_{n_1-1} \leq b$, and hence $V_{n_1} = V_{n_1-1} + v_{n_1} \leq b + v_{n_1}$. Similarly, since $\sum_{n=1}^{\infty} u_n^- = \infty$, we can choose the least integer $m_1 \in \mathcal{N}$ such that

$$V_{m_1} = v_1 + v_2 + \cdots + v_{m_1} = V_{n_1} - u_1^- u_2^- \cdots - u_{m_1-n_1}^- < a$$

and $V_{m_1} \geq a + v_{m_1}$. Since $-u_\ell^-$'s are nonpositive, it is clear that $V_{m_1} \leq V_\ell \leq V_{n_1}$ for $n_1 \leq \ell \leq m_1$. Therefore,

$$V_{n_1} > b \quad \text{and} \quad a + v_{m_1} \leq V_\ell \leq b + v_{n_1}$$

for all $n_1 \leq \ell \leq m_1$. By a similar argument, if $n_2 > m_1$ is the least integer such that $V_{n_2} > b$, then $V_{m_1} < a$ and $a + v_{m_1} \leq V_\ell \leq b + v_{n_2}$ for all $m_1 \leq \ell \leq n_2$. In particular, we have

$$b < \sup_{n_1 \leq \ell \leq n_2} V_\ell \leq b + \max\{v_{n_1}, v_{n_2}\} \leq b + \sup_{\ell \geq n_1} v_\ell.$$

In the same way, if $m_2 > n_2$ is the least integer such that $V_{m_2} < a$, then

$$a + \inf_{\ell \geq m_1} v_\ell \leq \inf_{m_1 \leq \ell \leq m_2} V_\ell < a.$$

Continuing this process, we find positive integers $n_1 < m_1 < n_2 < m_2 < \cdots$ such that for each $j \in \mathcal{N}$,

$$b < \sup_{n_j \leq \ell \leq n_{j+1}} V_\ell \leq b + \sup_{\ell \geq n_j} v_\ell \quad \text{and} \quad a + \inf_{\ell \geq m_j} v_\ell \leq \inf_{m_j \leq \ell \leq m_{j+1}} V_\ell < a.$$

The first of these inequalities implies that

$$b < \sup_{\ell \geq n_j} V_\ell \leq b + \sup_{\ell \geq n_j} v_\ell.$$

Taking the limit of this inequality as $j \to \infty$ and recalling that $v_n \to 0$ as $n \to \infty$, we get

$$b \leq \limsup_{n \to \infty} V_n \leq b + \limsup_{n \to \infty} v_n = b.$$

This proves the second equality of (10.1). A similar argument establishes the first equality of (10.1).

Problems

10.1. Determine whether each of the following series is absolutely convergent, conditionally convergent, or divergent.

(i). $u_n = \frac{(-1)^{n+1}}{\sin x/n}$, $x \in \mathcal{R}$.

(ii). $\sum_{n=1}^{\infty} \frac{(1-e^{-n})\ln n}{n}$.

(iii). $\sum_{n=1}^{\infty}(-1)^n \frac{n+2}{n(n+1)}$.

(iv). $\sum_{n=1}^{\infty}(-1)^n \frac{n \sin n}{2^n}$.

10.2. Suppose $\sum_{n=1}^{\infty} u_n$ converges absolutely and $U = \sum_{n=1}^{\infty} |u_n|$. Show that

(i). If there exists numbers $p \in (0,1)$ and $N \in \mathcal{N}$ such that $\sqrt[n]{|u_n|} \leq p$ for all $n > N$, then

$$0 \leq U - \sum_{k=1}^{n} |u_k| \leq \frac{p^{n+1}}{1-p}, \quad n \geq N.$$

(ii). If there exists numbers $p \in (0,1)$ and $N \in \mathcal{N}$ such that $|u_{n+1}|/|u_n| \leq p$, for all $n > N$, then

$$0 \leq U - \sum_{k=1}^{n} |u_k| \leq \frac{|u_N|p^{n+1-N}}{1-p}, \quad n \geq N.$$

10.3. Given two real sequences $\{u_n\}_{n=1}^{\infty}$ and $\{v_n\}_{n=1}^{\infty}$, let $V_n = \sum_{k=1}^{n} v_k$ and $S_n = \sum_{k=1}^{n} u_k v_k$ ($n \geq 1$). Show that for $n \geq 2$, $S_n = \sum_{k=1}^{n-1}(u_k - u_{k+1})V_k + u_n V_n$ (*Abel's formula*). Prove that if $\{V_n\}_{n=1}^{\infty}$ is bounded, $\lim_{n\to\infty} u_n = 0$ and $\sum_{n=1}^{\infty} |u_{n+1} - u_n|$ converges, then $\sum_{n=1}^{\infty} u_n v_n$ converges.

10.4. Show that $\sum_{n=1}^{\infty} \frac{\sin nx}{n}$ converges for every real x.

10.5. The Cauchy product of two series $\sum_{n=1}^{\infty} u_n$ and $\sum_{m=1}^{\infty} v_m$ is a new series $\sum_{k=1}^{\infty} w_k$, where $w_k = \sum_{\ell=1}^{k} u_\ell v_{k+1-\ell} = \sum_{\ell=1}^{k} u_{k+1-\ell} v_\ell$. Show that if $\sum_{n=1}^{\infty} u_n$ and $\sum_{m=1}^{\infty} v_m$ converge absolutely to U and V, respectively, then $\sum_{k=1}^{\infty} w_k$ converges absolutely to $W = UV$.

10.6. Show that $\sum_{n=0}^{\infty}(-1)^n/\sqrt{n+1}$ is a convergent series. Form the Cauchy product of this series with itself and show that the Cauchy product diverges.

Answers or Hints

10.1. (i). For any $x \in \mathcal{R} \exists N \in \mathcal{N}$ such that $\sin x/n > 0$, $\forall n \geq N$ and

$\sin x/n > \sin x/(n+1)$, $\forall\, n \geq N$. Also $\lim \sin x/n = 0$. Thus, Theorem 10.1 is applicable.

(ii). Let $u_n = (1/n)(1-e^{-n})\ln n \geq 0$. Since $\lim_{n\to\infty} e^{-n} = 0$, there exists a positive integer N such that $e^{-n} \leq 1/2$ for all $n \geq N$. Then, $u_n \geq \frac{1}{2}\frac{\ln n}{n} \geq \frac{1}{2}\frac{1}{n}$ if $n \geq \max(N,3)$. Since $\sum \frac{1}{n}$ diverges, by the comparison test, $\sum u_n$ also diverges.

(iii). Let $u_n = \frac{n+2}{n(n+1)}$. Then $\sum(-1)^n u_n$ is an alternating series. Since $\frac{u_{n+1}}{u_n} = \frac{n^2+3n}{n^2+4n+4} < 1$, $\{u_n\}_{n=1}^{\infty}$ is decreasing and obviously $\lim_{n\to\infty} u_n = 0$. Thus by the alternating series test, $\sum(-1)^n u_n$ converges. Now since $u_n = \frac{n+2}{n(n+1)} \geq \frac{1}{n}$ and series $\sum \frac{1}{n}$ diverges, by comparison test $\sum u_n$ diverges. Hence $\sum(-1)^n u_n$ is conditionally convergent.

(iv). Let $u_n = (-1)^n \frac{n \sin n}{2^n}$. Then, $|u_n| \leq \frac{n}{2^n} = v_n$. Since $\lim_{n\to\infty} \frac{v_{n+1}}{v_n} = \lim_{n\to\infty} \frac{n+1}{2n} = \frac{1}{2} < 1$, by the ratio test, $\sum v_n$ converges, and so by the comparison test, $\sum |u_n|$ converges. Hence $\sum u_n$ converges absolutely.

10.2. (i). For $n > N$, $|a_n| \leq p^n$. Now $0 \leq U - \sum_{k=1}^{n} |u_k| = \sum_{k=n+1}^{\infty} |u_k| \leq \sum_{k=n+1}^{\infty} p^k = p^{n+1}/(1-p)$.

(ii). For $n > N$, $|u_n| \leq p^{n-N}|u_N|$.

10.3. $S_n = u_1 v_1 + u_2 v_2 + \cdots + u_{n-1}v_{n-1} + u_n v_n = u_1 V_1 + u_2(V_2 - V_1) + \cdots + u_{n-1}(V_{n-1} - V_{n-2}) + u_n(V_n - V_{n-1}) = (u_1 - u_2)V_1 + (u_2 - u_3)V_2 + \cdots + (u_{n-1} - u_n)V_{n-1} + u_n V_n = \sum_{k=1}^{n-1}(u_k - u_{k+1})V_k + u_n V_n$. If $|V_n| \leq M$ for all n for some constant M, $\lim_{n\to\infty} u_n = 0$ and $\sum_{n=1}^{\infty} |u_{n+1} - u_n|$ converges, then $\lim_{n\to\infty} u_n V_n = 0$ and $|(u_k - u_{k+1})V_k| \leq M|u_k - u_{k+1}|$ for all k. Thus by the comparison test $\sum_{k=1}^{\infty} |(u_k - u_{k+1})V_k|$ converges and so $\lim_{n\to\infty} \sum_{k=1}^{n}(u_k - u_{k+1})V_k$ exists. Hence, $\lim_{n\to\infty} S_n$ exists, i.e., $\sum u_n v_n$ converges.

10.4. If $x = 2k\pi$ for an integer k, then $\sin nx = 0$ for all n and hence the series converges. If $x \neq 2k\pi$ for any integer k, then $\sin nx = [\cos(n - 1/2)x - \cos(n+1/2)x]/2\sin(x/2)$, $n \geq 1$ and hence $\sum_{j=1}^{n} \sin jx = [\cos(x/2) - \cos(n+1/2)x]/2\sin(x/2)$. Thus, $\left|\sum_{j=1}^{n} \sin jx\right| \leq 1/|\sin(x/2)|$, $n \geq 1$. Hence the partial sums of $\sum \sin nx$ are bounded. Next since $\lim_{n\to\infty} 1/n = 0$ and $\sum_{n=1}^{\infty}\left|\frac{1}{n+1} - \frac{1}{n}\right| = \sum_{n=1}^{\infty} \frac{1}{n(n+1)}$ converges, it follows from Problem 10.3 that $\sum_{n=1}^{\infty} \frac{\sin nx}{n}$ converges.

10.5. For any $n \in \mathcal{N}$, we have $|w_1| + |w_2| + \cdots + |w_n| \leq |u_1 v_1| + (|u_1 v_2| + |u_2 v_1|) + \cdots + (|u_1 v_n| + |u_2 v_{n-1}| + \cdots + |u_n v_1|) \leq (\sum_{k=1}^{n} |u_k|)(\sum_{j=1}^{n} |v_j|) \leq (\sum_{k=1}^{\infty} |u_k|)(\sum_{j=1}^{\infty} |v_j|) < \infty$.

10.6. Take $a_n = b_n = (-1)^n/\sqrt{n+1}$, $n \geq 0$. Then, $c_n = \sum_{k=0}^{n} a_k b_{n-k} = (-1)^n \sum_{k=0}^{n} 1/(\sqrt{k+1}\sqrt{n-k+1})$. Since $\sqrt{k+1}\sqrt{n-k+1} \leq (n+2)/2$, it follows that $|c_n| \geq \sum_{k=0}^{n} 2/(n+2) = 2(n+1)/(n+2) \geq 1$. Since $c_n \not\to 0$, we must have $\sum_{n=0}^{\infty} c_n$ divergent.

Chapter 11

Limits of Functions

The concept of limit of sequences introduced in Chapter 7 will be extended here to real functions. This extension is most important to the development of analysis. In fact, without understanding the limit of a function properly one cannot appreciate the later chapters.

Let $x_0 \in \mathcal{R}$, and let f be a real-valued function defined on some nbd of x_0, except possibly at x_0 itself. We say that $f(x)$ approaches to a limit ℓ as x approaches x_0 if given $\epsilon > 0$ there exists $\delta > 0$ such that

$$0 < |x - x_0| < \delta \;\Rightarrow\; |f(x) - \ell| < \epsilon. \tag{11.1}$$

In symbols we write this as $\lim_{x \to x_0} f(x) = \ell$, or $f(x) \to \ell$ as $x \to x_0$. Equivalently, $\lim_{x \to x_0} f(x) = \ell$ if given any nbd $I_\epsilon(\ell)$ there is a deleted nbd $I_\delta^*(x_0)$ of x_0 such that if $x \in I_\delta^*(x_0)$, then $f(x) \in I_\epsilon(\ell)$. Further, it is clear that (11.1) is the same as

$$0 < |y| < \delta \;\Rightarrow\; |f(y + x_0) - \ell| < \epsilon.$$

Thus, it follows that $\lim_{x \to x_0} f(x) = \lim_{x \to 0} f(x_0 + x) = \lim_{x \to 0} f(x_0 - x)$.

Example 11.1. We shall show that $\lim_{x \to x_0} 1/\sqrt{x} = 1/\sqrt{x_0}$, where $x_0 > 0$. For this, it suffices to note that

$$
\left| \frac{1}{\sqrt{x}} - \frac{1}{\sqrt{x_0}} \right| = \frac{|x - x_0|}{\sqrt{x}\sqrt{x_0}(\sqrt{x} + \sqrt{x_0})} < \frac{|x - x_0|}{\sqrt{(x_0/2)}\sqrt{x_0}(\sqrt{(x_0/2)} + \sqrt{x_0})}
$$

$$
= \frac{|x - x_0|}{(1/2 + 1/\sqrt{2})x_0^{3/2}} < \frac{1}{x_0^{3/2}}|x - x_0| < \epsilon
$$

provided $0 < |x - x_0| < \delta = \min\{x_0/2, x_0^{3/2}\epsilon\}$.

Our first result ensures that if a limit exists then it is unique. This result is analogous to Theorem 7.1.

Theorem 11.1. If $\lim_{x \to x_0} f(x)$ exists, then it is unique.

Proof. Suppose that $\lim_{x \to x_0} f(x) = \ell_1$ and $\lim_{x \to x_0} f(x) = \ell_2$. Given any $\epsilon > 0$ there exists positive numbers δ_1 and δ_2 such that $0 < |x - x_0| <$

$\delta_1 \Rightarrow |f(x) - \ell_1| < \epsilon/2$ and $0 < |x - x_0| < \delta_2 \Rightarrow |f(x) - \ell_2| < \epsilon/2$. Thus, if $\delta = \min\{\delta_1, \delta_2\}$, then $0 < |x - x_0| < \delta$ gives that

$$|\ell_1 - \ell_2| = |\ell_1 - f(x) + f(x) - \ell_2| \le |f(x) - \ell_1| + |f(x) - \ell_2| < \frac{\epsilon}{2} + \frac{\epsilon}{2} = \epsilon.$$

Hence, $\ell_1 = \ell_2$.

Now we shall prove the following useful result.

Theorem 11.2. $\lim_{x \to x_0} f(x) = \ell$ iff $f(x_n) \to \ell$ for every sequence $\{x_n\}$ on the domain of f with $x_n \to x_0$ and $x_n \ne x_0$, $n \in \mathcal{N}$.

Proof. Suppose $\lim_{x \to x_0} f(x) = \ell$. Then, given $\epsilon > 0$ there is a $\delta > 0$ such that (11.1) holds. Now since $x_n \to x_0$ there exists an $N \in \mathcal{N}$ such that $|x_n - x_0| < \delta$ for all $n \ge N$. Since $x_n \ne x_0$, it follows from (11.1) that $|f(x_n) - \ell| < \epsilon$ for all $n \ge N$. Thus, $f(x_n) \to \ell$ as $n \to \infty$.

Conversely, suppose $f(x_n) \to \ell$ as $n \to \infty$ for every sequence $\{x_n\}$ with $x_n \to x_0$ and $x_n \ne x_0$, $n \in \mathcal{N}$. If $\lim_{x \to x_0} f(x) \ne \ell$, then there is an $\epsilon > 0$ such that for each $\delta > 0$ there is an $x \in I_\delta^*(x_0)$ with $|f(x) - \ell| \ge \epsilon$. Now for each $n \in \mathcal{N}$ choose x_n in the domain of f such that $x_n \in I_{1/n}^*(x_0)$ and $|f(x_n) - \ell| \ge \epsilon$. Then, for each $n \in \mathcal{N}$ we have $0 < |x_n - x_0| < 1/n$ and $|f(x_n) - \ell| \ge \epsilon$. Clearly, $x_n \to x_0$ but $f(x_n) \nrightarrow \ell$.

We remark that one should be careful in using Theorem 11.2. In particular, it implies that if $\lim_{x \to \infty} f(x) = \ell$, then $\lim_{n \to \infty} f(n) = \ell$; however, its converse is not true. For example, $\lim_{n \to \infty} \sin n\pi = 0$, but $\lim_{x \to \infty} \sin \pi x$ does not exist. Further, Theorems 11.1 and 11.2 can be used to show that certain limits do not exist. We illustrate this in the following example.

Example 11.2. Consider the function

$$f(x) = \frac{x}{|x|} = \begin{cases} 1 & \text{if } x > 0 \\ -1 & \text{if } x < 0. \end{cases}$$

To show $\lim_{x \to 0} f(x)$ does not exist it suffices to consider the sequences $\{1/n\}$ and $\{-1/n\}$ which converge to zero, and note that $\{f(1/n)\} = \{1\} \to 1$ and $f(-1/n) = \{-1\} \to -1$.

Next, we state two results which are analogous to Theorems 7.4 and 7.5.

Theorem 11.3. Let the functions f, g, h be defined on a deleted nbd $I_\delta^*(x_0)$ of a point x_0.

(1). If $g(x) \le f(x) \le h(x)$, $x \in I_\delta^*(x_0)$ and $\lim_{x \to x_0} g(x) = \lim_{x \to x_0} h(x) = \ell$, then $\lim_{x \to x_0} f(x) = \ell$.

(2). If $|g(x)| \le M$, $x \in I_\delta^*(x_0)$ and $\lim_{x \to x_0} f(x) = 0$, then $\lim_{x \to x_0} f(x)g(x) = 0$.

Example 11.3. Since $-x^2 \leq x^2 \sin(1/x) \leq x^2$ on a deleted nbd of 0, and $\lim_{x \to 0} x^2 = \lim_{x \to 0} -x^2 = 0$, from Theorem 11.3 it follows that $\lim_{x \to 0} x^2 \sin(1/x) = 0$.

Theorem 11.4. Suppose $\lim_{x \to x_0} f(x) = \ell$ and $\lim_{x \to x_0} g(x) = m$ and c is a real number. Then,

(1). $\lim_{x \to x_0} (f(x) \pm g(x)) = \ell \pm m$.

(2). $\lim_{x \to x_0} cf(x) = c\ell$.

(3). $\lim_{x \to x_0} f(x)g(x) = \ell m$.

(4). $\lim_{x \to x_0} f(x)/g(x) = \ell/m$ provided $m \neq 0$.

Example 11.4. Consider the *rational function* $R(x) = P_n(x)/P_m(x)$, where $P_n(x) = \sum_{k=0}^{n} a_k x^k$ and $P_m(x) = \sum_{\ell=0}^{m} a_\ell x^\ell$ are *algebraic polynomials* of degree n and m, respectively. From Theorem 11.4 it follows that $\lim_{x \to x_0} R(x) = P_n(x_0)/P_m(x_0)$ provided $P_m(x_0) \neq 0$.

Now let f be defined on some interval of the form (b, ∞). We say that $f(x)$ approaches to a limit ℓ as x approaches ∞ if given $\epsilon > 0$ there exists $M > 0$ such that $x > M \Rightarrow |f(x) - \ell| < \epsilon$. In symbols we write this as $\lim_{x \to \infty} f(x) = \ell$, or $f(x) \to \ell$ as $x \to \infty$. Similarly, if f is defined on some interval of the form $(-\infty, a)$. We say that $f(x)$ approaches to a limit ℓ as x approaches $-\infty$ if given $\epsilon > 0$ there exists $M < 0$ such that $x < M \Rightarrow |f(x) - \ell| < \epsilon$. In symbols we write this as $\lim_{x \to -\infty} f(x) = \ell$, or $f(x) \to \ell$ as $x \to -\infty$.

Example 11.5. We shall show that $\lim_{x \to \infty} 1/x^2 = 0$. For this, given $\epsilon > 0$ we need to find $M > 0$ such that $|1/x^2 - 0| < \epsilon$, i.e., $1/x < \sqrt{\epsilon}$, and hence $M = 1/\sqrt{\epsilon}$.

We say $\lim_{x \to x_0} f(x) = +\infty$ if given any $M > 0$ there exists a $\delta > 0$ such that $0 < |x - x_0| < \delta \Rightarrow f(x) > M$. Similarly, $\lim_{x \to x_0} f(x) = -\infty$ if given any $M > 0$ there exists a $\delta > 0$ such that $0 < |x - x_0| < \delta \Rightarrow f(x) < -M$. It is clear that if $\lim_{x \to x_0} f(x) = +\infty$, or $-\infty$, then $\lim_{x \to x_0} 1/f(x) = 0$. Further, if $f(x) > 0$ for $0 < |x - x_0| < \delta$ and $\lim_{x \to x_0} 1/f(x) = 0$, then $\lim_{x \to x_0} f(x) = +\infty$.

Example 11.6. We shall show that $\lim_{x \to 0} \ln |x| = -\infty$. For this, given $M > 0$ we choose $\delta = e^{-M}$. Then, $0 < |x - 0| = |x| < \delta$ and $f(x) = \ln |x| < -M$. Similarly, it is easy to see that $\lim_{x \to 0} 1/|x| = \infty$.

Now we shall introduce one-sided limits. These limits are useful to find the behavior of a function particularly at the endpoints of the interval. We say that the function f approaches ℓ as x approaches x_0 from the right if given $\epsilon > 0$ there exists $\delta > 0$ such that $x \in (x_0, x_0 + \delta) \Rightarrow |f(x) - \ell| < \epsilon$. In this case, we write $\lim_{x \to x_0^+} f(x) = \ell$ or $f(x_0 + 0) = \ell$ and call it the *right-hand limit*. Similarly, we say that the function f approaches ℓ as x approaches x_0 from the left if given $\epsilon > 0$ there exists $\delta > 0$ such that $x \in (x_0 - \delta, x_0) \Rightarrow |f(x) - \ell| < \epsilon$. In this case we write $\lim_{x \to x_0^-} f(x) = \ell$, or $f(x_0 - 0) = \ell$ and call it the *left-hand limit*. Thus, $\lim_{x \to x_0^+} f(x)$ requires only the values of $f(x)$ for x to the right of x_0, while $\lim_{x \to x_0^-} f(x)$ involves only the values of $f(x)$ for x to the left of x_0. If $x_0 = 0$, the right- and left-hand limits are simply written as $f(+0)$ and $f(-0)$.

It is clear that $\lim_{x \to x_0} f(x) = \ell$ iff $\lim_{x \to x_0^+} f(x) = \lim_{x \to x_0^-} f(x) = \ell$. However, only one of the limits $\lim_{x \to x_0^+} f(x)$, $\lim_{x \to x_0^-} f(x)$ may exist, or both exist but may not be equal.

Example 11.7. For the function $f(x) = \sin(1/x)$ the right-hand limit at $x = 0$ does not exist. For this, in view of Theorem 11.2, it suffices to take the positive sequences $\{2/[(4n+1)\pi]\}$ and $\{2/[(4n+3)\pi]\}$ which converge to 0, and note that $\sin[(4n+1)\pi]/2 = 1$, whereas $\sin[(4n+3)\pi]/2 = -1$. For the function $f(x) = e^{-1/x}$ it is clear that $f(+0) = 0$, whereas $f(-0) = \infty$. For the function $f(x)$ considered in Example 11.2, we have $f(+0) = 1$, whereas $f(-0) = -1$. For the function $f(x) = x - [x]$ we have $f(1 + 0) = 0$, but $f(1 - 0) = 1$.

In our next result we shall prove that for monotone functions one-sided limits always exist.

Theorem 11.5. Let $f : (a, b) \to \mathcal{R}$.

(1). If f is monotone, then for each $x_0 \in (a, b)$, $f(x_0 + 0)$ and $f(x_0 - 0)$ exist.

(2). If f is nondecreasing, then $f(b - 0)$ $(f(a + 0))$ exists provided f is bounded above (below).

(3). If f is nonincreasing, then $f(b - 0)$ $(f(a + 0))$ exists provided f is bounded below (above).

Proof. (1). Assume that f is nondecreasing on (a, b). Since $x_0 \in (a, b)$ for every $x \in (a, x_0)$, $f(x) \leq f(x_0)$, and so $\sup_{x \in (a, x_0)} f(x)$, say, $\lambda \in \mathcal{R}$ exists. We shall show that $\lambda = f(x_0 - 0)$. Let $\epsilon > 0$. Since $\lambda - \epsilon < \lambda$ there exists $x_1 \in (a, x_0)$ such that $f(x_1) > \lambda - \epsilon$. Let $\delta = x_0 - x_1$. Now if $x \in (x_0 - \delta, x_0)$ then $f(x) \geq f(x_1)$, and therefore $|f(x) - \lambda| = \lambda - f(x) \leq \lambda - f(x_1) < \epsilon$. Hence, $f(x_0 - 0) = \lambda$. The proof of $f(x_0 + 0) = \inf_{x \in (x_0, b)} f(x)$ is similar. If f is nonincreasing on (a, b) we can apply the above arguments to the function $-f$ which will be nondecreasing.

(2). If f is bounded above and nondecreasing on (a, b), then $\sup_{x \in (a, b)} f(x)$ exists. Now following the argument exactly as in (1), we find $\sup_{x \in (a, b)} f(x) = f(b - 0)$. If f is bounded below and nondecreasing on (a, b), then we have $\inf_{x \in (a, b)} f(x) = f(a + 0)$.

(3). We can apply the arguments used in (2) to the function $-f$.

From Theorem 11.5 it follows that, if $f : (a, b) \to \mathcal{R}$ is nondecreasing, then for each $x_0 \in (a, b)$,

$$\lim_{x \to x_0^-} f(x) = \sup_{x \in (a, x_0)} f(x) \le f(x_0) \le \inf_{x \in (x_0, b)} f(x) = \lim_{x \to x_0^+} f(x), \quad (11.2)$$

whereas, if f is nonincreasing, then for each $x_0 \in (a, b)$,

$$\lim_{x \to x_0^-} f(x) = \inf_{x \in (a, x_0)} f(x) \ge f(x_0) \ge \sup_{x \in (x_0, b)} f(x) = \lim_{x \to x_0^+} f(x). \quad (11.3)$$

Thus, if f is monotone, then it is bounded at each point $x_0 \in (a, b)$; however, f may be unbounded on the interval (a, b), e.g., consider the nondecreasing function $f(x) = 1/(1 - x)$, $x \in (0, 1)$.

Now assume that the function $f : (a, b) \to \mathcal{R}$ is nondecreasing, and $a < x_1 < x_2 < b$. Let $x_0 \in (x_1, x_2)$. Since $x_0 \in (x_1, b)$, $f(x_0) \ge \inf_{x \in (x_1, b)} f(x)$, and since $x_0 \in (a, x_2)$, $f(x_0) \le \sup_{x \in (a, x_2)} f(x)$. Thus, from Theorem 11.5 it follows that

$$\lim_{x \to x_1^+} f(x) = \inf_{x \in (x_1, b)} f(x) \le f(x_0) \le \sup_{x \in (a, x_2)} f(x) = \lim_{x \to x_2^-} f(x),$$

i.e.,

$$\lim_{x \to x_1^+} f(x) \le \lim_{x \to x_2^-} f(x). \quad (11.4)$$

Similarly, we can show that if f is nonincreasing, then

$$\lim_{x \to x_1^+} f(x) \ge \lim_{x \to x_2^-} f(x). \quad (11.5)$$

The following result proves a very important property of monotone functions whose significance will be clear in Chapter 13.

Theorem 11.6. If $f : (a, b) \to \mathcal{R}$ is monotone, then $\lim_{x \to x_0} f(x)$ exists and is equal to $f(x_0)$ for all but countably many points $x_0 \in (a, b)$.

Proof. It suffices to assume that f is nondecreasing on (a, b). From (11.2) for every $x_0 \in (a, b)$, we have $\lim_{x \to x_0^-} f(x) \le f(x_0) \le \lim_{x \to x_0^+} f(x)$. Now define the set $D = \left\{ x_0 \in (a, b) : \lim_{x \to x_0^-} f(x) < \lim_{x \to x_0^+} f(x) \right\}$. It is clear that if $x_0 \notin D$, then $\lim_{x \to x_0^-} f(x) = f(x_0) = \lim_{x \to x_0^+} f(x)$, and hence $\lim_{x \to x_0} f(x)$ exists and equal to $f(x_0)$. We shall prove that the set D is countable. For this, we define a mapping $\phi : D \to \mathcal{Q}$ as follows: For each $x_0 \in D$ choose $q_{x_0} \in \mathcal{Q}$ such that $\lim_{x \to x_0^-} f(x) < q_{x_0} < \lim_{x \to x_0^+} f(x)$ and define $\phi(x_0) = q_{x_0}$. Now let $x_1, x_2 \in D$ with $x_1 < x_2$. Then, $\lim_{x \to x_1^-} f(x) < \lim_{x \to x_1^+} f(x)$ and $\lim_{x \to x_2^-} f(x) < \lim_{x \to x_2^+} f(x)$, which in view of (11.4) imply that

$$\lim_{x \to x_1^-} f(x) < \lim_{x \to x_1^+} f(x) \le \lim_{x \to x_2^-} f(x) < \lim_{x \to x_2^+} f(x),$$

and therefore

$$\phi(x_1) \;=\; q_{x_1} \;<\; \lim_{x\to x_1^+} f(x) \;\le\; \lim_{x\to x_2^-} f(x) \;<\; q_{x_2} \;=\; \phi(x_2).$$

Hence, $\phi : D \to \mathcal{Q}$ is a one-to-one function. Now since the set $\mathcal{Q}$ is countable, from Problem 5.2 it follows that the set D is also countable.

Finally, let f be a real-valued function defined on a deleted nbd $I_\delta^*(x_0)$ of x_0. We define

$$
\begin{aligned}
\limsup_{x\to x_0} f(x) &= \inf_{\delta>0} \sup_{0<|x-x_0|<\delta} f(x),\\
\liminf_{x\to x_0} f(x) &= \sup_{\delta>0} \inf_{0<|x-x_0|<\delta} f(x),\\
\limsup_{x\to x_0^+} f(x) &= \inf_{\delta>0} \sup_{0<x-x_0<\delta} f(x) = \overline{f(x_0+0)},\\
\liminf_{x\to x_0^+} f(x) &= \sup_{\delta>0} \inf_{0<x-x_0<\delta} f(x) = \underline{f(x_0+0)},\\
\limsup_{x\to x_0^-} f(x) &= \inf_{\delta>0} \sup_{0<x_0-x<\delta} f(x) = \overline{f(x_0-0)},\\
\liminf_{x\to x_0^-} f(x) &= \sup_{\delta>0} \inf_{0<x_0-x<\delta} f(x) = \underline{f(x_0-0)}.
\end{aligned}
$$

It is clear that

$$\liminf_{x\to x_0} f(x) \;\le\; \liminf_{x\to x_0^\pm} f(x) \;\le\; \limsup_{x\to x_0^\pm} f(x) \;\le\; \limsup_{x\to x_0} f(x).$$

Further, $\lim_{x\to x_0^+} f(x)$ exists (finite or infinite) iff $\overline{f(x_0+0)} = \underline{f(x_0+0)}$, $\lim_{x\to x_0^-} f(x)$ exists iff $\overline{f(x_0-0)} = \underline{f(x_0-0)}$, and $\lim_{x\to x_0} f(x)$ exists iff $\liminf_{x\to x_0} f(x) = \limsup_{x\to x_0} f(x)$.

Example 11.8. For the function $f(x) = [x] + x^2 \sin(1/x)$ it is clear that $\limsup_{x\to 0^+} f(x) = \liminf_{x\to 0^+} f(x) = 0$, and $\limsup_{x\to 0^-} f(x) = \liminf_{x\to 0^-} f(x) = -1$.

Problems

11.1. Use the $\epsilon - \delta$ definition of limits to prove the following.

(i). $\lim_{x\to 2}(x^2 + 1) = 5$.

(ii). $\lim_{x\to 2}(x^2 - 3x + 3) = 1$.

(iii). $\lim_{x\to -1} \frac{2x+1}{3+x} = -\frac{1}{2}$.

(iv). $\lim_{x\to 7^-} \frac{x^2|x-7|}{x-7} = -49$.

(v). $\lim_{x\to 0^+} \frac{1}{x} = \infty$.

(vi). $\lim_{x\to 1^+} \frac{x}{x^2-1} = \infty$.

11.2. Evaluate the following limits.

(i). $\lim_{x \to y} \frac{x^n - y^n}{x - y}$, $n \in \mathcal{N}$.

(ii). $\lim_{x \to 0} \frac{\sqrt[3]{(x+a)^2} - \sqrt[3]{a^2}}{x}$.

(iii). $\lim_{x \to -\infty} \frac{\sqrt{x^2 + 7}}{2x - 3}$.

(iv). $\lim_{x \to \infty} \left(1 + \frac{1}{x}\right)^x$.

(v). $\lim_{x \to 0} \frac{\sin x}{x}$.

(vi). $\lim_{x \to 0} \frac{a^x - 1}{x}$, $a > 0$.

11.3. Use Theorems 11.1 and 11.2 to show that

(i). For the function $f : \mathcal{R} \to \mathcal{R}$ defined by

$$f(x) = \begin{cases} 1 & \text{if } x \text{ is rational} \\ 0 & \text{if } x \text{ is irrational} \end{cases}$$

$\lim_{x \to x_0} f(x)$ does not exist at any $x_0 \in \mathcal{R}$.

(ii). $\lim_{x \to 0} \cos(1/x)$ does not exist.

11.4. For the function $f : \mathcal{R} \to \mathcal{R}$ defined by

$$f(x) = \begin{cases} x & \text{if } x \text{ is rational} \\ -x & \text{if } x \text{ is irrational} \end{cases}$$

show that $\lim_{x \to x_0} f(x)$ exists only when $x_0 = 0$.

11.5. Consider the *Dirichlet's function* f, defined on $(0,1)$ by

$$f(x) = \begin{cases} 0 & \text{if } x \text{ is irrational} \\ 1/q & \text{if } x = p/q; \end{cases}$$

here p, q are positive integers with no common factor. Show that $\lim_{x \to x_0} f(x) = 0$ for all $x_0 \in (0,1)$.

11.6. Show that if $\lim_{x \to x_0} f(x) = \ell$, then f is bounded on a deleted nbd of x_0. Also, if $\ell \neq 0$, then there exists a $x_0 \in \mathcal{R}^+$ such that $|f(x)| > c$ on a deleted nbd of x_0. However, the converse of each is not true.

11.7. Let the function f be defined on a deleted nbd of x_0. Show that $\lim_{x \to x_0} f(x)$ exists iff for given $\epsilon > 0$ there exists $\delta > 0$ such that $x_1, x_2 \in I_\delta^*(x_0) \Rightarrow |f(x_1) - f(x_2)| < \epsilon$.

11.8. Prove the following statements.

(i). If $\lim_{x \to x_0} f(x) = \ell$, then $\lim_{x \to x_0} |f(x)| = |\ell|$.

(ii). The converse of (i) is false; however, if $\lim_{x \to x_0} |f(x)| = 0$, then $\lim_{x \to x_0} f(x) = 0$.

11.9. Suppose that $f(x) \geq 0$ for all x on the domain of f and $\lim_{x \to x_0} f(x) = \ell > 0$. Show that $\lim_{x \to x_0} \sqrt{f(x)} = \sqrt{\ell}$.

11.10. Let $\lim_{x \to 0} f(x)/x = \ell$ and $c \neq 0$. Show that $\lim_{x \to 0} f(cx)/x = c\ell$. What happens if $c = 0$?

11.11. Suppose $\lim_{x \to x_0} f(x)$ and $\lim_{x \to x_0} g(x)$ do not exist. Show that $\lim_{x \to x_0} (f(x) \pm g(x))$ and $\lim_{x \to x_0} f(x)g(x)$ may or may not exist.

11.12. Prove the following.

(i).   If f is an even function, then $\lim_{x \to 0} f(x) = \ell$ iff $\lim_{x \to 0^+} f(x) = \ell$.

(ii).   If f is an odd function, then $\lim_{x \to 0} f(x) = \ell$ implies $\ell = 0$.

11.13. If $\lim_{x \to a} g(x) = \ell$ and $\lim_{y \to \ell} f(y) = f(\ell)$, then show that $\lim_{x \to a} f(g(x)) = f(\lim_{x \to a} g(x))$.

11.14. Let $A \subseteq \mathcal{R}$, $A \neq \emptyset$ and $f, g : A \to \mathcal{R}$ be two functions. If a is a limit point of the set A and $\lim_{x \to a} f(x) = \alpha$, $\lim_{x \to a} f(x) = \beta$ with $\alpha < \beta$, then show that there is a $\delta > 0$ such that $f(x) < g(x)$ for all $x \in A$ with $0 < |x - a| < \delta$.

Answers or Hints

11.1. (i). $|x^2 + 1 - 5| = |x^2 - 4| = |x + 2||x - 2| < 5|x - 2| < \epsilon$ if $0 < |x - 2| < \delta = \min\{1, \epsilon/5\}$.

(ii). $|x^2 - 3x + 3 - 1| = |x - 1||x - 2| \leq 2|x - 2| < \epsilon$ if $0 < |x - 2| < \delta = \min\{1, \epsilon/2\}$.

(iii). $\left|\frac{2x+1}{3+x} - \left(-\frac{1}{2}\right)\right| = \left|\frac{5}{2}\frac{x+1}{3+x}\right| < \frac{5}{2}|x + 1| < \epsilon$ if $0 < |x + 1| < \delta = \min\{1, 2\epsilon/5\}$.

(iv). Let $\epsilon > 0$. Then, $\left|\frac{x^2|x-7|}{x-7} + 49\right| = |-x^2 + 49| = |x + 7||x - 7| < 14|x - 7| < \epsilon$ if $-\delta < x - 7 < 0$ where $\delta = \min\{1, \epsilon/14\}$.

(v). Given $N > 0$ choose $\delta = 1/N$. Then $1/x > N$ if $0 < x < \delta$.

(vi). Let $M > 0$. Then, $\frac{x}{x^2-1} > \frac{(1/2)(x+1)}{(x-1)(x+1)} = \frac{1}{2(x-1)} > M$ if $0 < x - 1 < 1/2M$.

11.2. (i). ny^{n-1}; (ii). $2/(3a^{1/3})$; (iii). $-1/2$; (iv). e; (v). 1; (vi). $\ln a$.

11.3. (i). Consider the sequences $\{x_n\}$ and $\{y_n\}$ of rational and irrational numbers which converge to x_0. Then, $\{f(x_n)\} = \{1\} \to 1$ and $\{f(y_n)\} = \{0\} \to 0$.

(ii). Consider the function along the sequences $\{1/(2n\pi)\}$ and $\{1/[(2n+1)\pi]\}$ both of which tend to 0.

11.4. For $x_0 > 0$ and $\delta > 0$, let $x_1 \in \mathcal{Q}$, $x_2 \in \mathcal{R} - \mathcal{Q}$ and $x_1, x_2 \in (x_0, x_0 + \delta)$. Then, $|f(x_1) - f(x_2)| = |x_1 + x_2| > 2x_0$, and so $\lim_{x \to x_0} f(x)$ does not exist. The proof for $x_0 < 0$ is similar. If $x_0 = 0$, then for $\epsilon > 0 \, \exists \, \delta = \epsilon > 0$ such that $0 < |x - 0| < \epsilon \Rightarrow |f(x) - 0| = |x| < \epsilon$.

11.5. Let $\epsilon > 0$ and $n \in \mathcal{N}$ be so large that $1/n \le \epsilon$. The only numbers x for which $|f(x) - 0| > \epsilon$ could be are $1/2, 1/3, 2/3, 1/4, 3/4, 1/5, 2/5, 3/5, 4/5,$ $\cdots, 1/n, \cdots, (n-1)/n$. If x_0 is rational, then x_0 might be one of these numbers. These numbers are only finitely many, and hence there is one closest to x_0, i.e., $|p/q - x_0|$ is smallest for one p/q among these numbers. If x_0 is one of these numbers, then we consider values $|p/q - x_0|$ for $p/q \ne x_0$. This closest distance may be chosen as the δ. Now if $0 < |x - x_0| < \delta$, then x is not one of $1/2, \cdots, (n-1)/n$ and therefore $|f(x) - 0| < \epsilon$ is true.

11.6. Let $\epsilon = 1$ in the definition of limit to get $|f(x)| = |f(x) - \ell + \ell| \le |f(x) - \ell| + |\ell| < 1 + |\ell|$, $x \in I_\delta^*(x_0)$. If $\ell \ne 0$, then let $\epsilon = |\ell|/2$ to get $|\ell| - |f(x)| \le |f(x) - \ell| < |\ell|/2$ and hence $|f(x)| \ge |\ell|/2 = c$, $x \in I_\delta^*(x_0)$. The function $\cos(1/x)$ is bounded on a deleted nbd of 0, but $\lim_{x \to 0} \cos(1/x)$ does not exist. Clearly, $1/|x - 1| \ge 1/2$ on a deleted nbd of 1, but $\lim_{x \to 1} 1/(x - 1)$ does not exist.

11.7. If $\lim_{x \to x_0} f(x) = \ell$ in (11.1) let ϵ to be $\epsilon/2$. Then, for $x_1, x_2 \in I_\delta^*(x_0)$, we have $|f(x_1) - f(x_2)| \le |f(x_1) - \ell| + |f(x_2) - \ell| < (\epsilon/2) + (\epsilon/2) = \epsilon$. Conversely, let $\{x_n\}$, $x_n \ne x_0$ be an arbitrary sequence such that $x_n \to x_0$. Then, there exists an $N \in \mathcal{N}$ such that for all $n, m \ge N$, $x_n, x_m \in I_\delta^*(x_0) \Rightarrow |f(x_n) - f(x_m)| < \epsilon$, i.e., $\{f(x_n)\}$ is a Cauchy sequence, and thus has a limit. Now in view of Theorem 11.2, $\lim_{x \to x_0} f(x)$ exists provided for each sequence $\{x_n\}$, $x_n \ne x_0$ such that $x_n \to x_0$, $\{f(x_n)\}$ converges to the same limit. For this, let $\{y_n\}, \{z_n\}$ ($y_n, z_n \ne x_0$) be such that $y_n \to x_0$ and $z_n \to x_0$ but $f(y_n) \to \ell$ and $f(z_n) \to \ell'$. Now, $\ell = \ell'$ immediately follows from the inequality $|\ell - \ell'| \le |\ell - f(x_n)| + |f(x_n) - f(z_n)| + |f(z_n) - \ell'|$.

11.8. (i). Note that $||f(x)| - |\ell|| \le |f(x) - \ell|$.

(ii). For the function $f(x) = \begin{cases} -1 & \text{if } x < x_0 \\ 1 & \text{if } x \ge x_0 \end{cases}$ it is clear that $\lim_{x \to x_0} |f(x)| = 1$. Now if $x_1 < x_0$ and $x_2 > x_0$, then $|f(x_1) - f(x_2)| = 2$, which cannot be made less than ϵ. Thus, in view of Problem 11.4, $\lim_{x \to x_0} f(x)$ does not exist. To show the rest of the part note that $||f(x)| - 0| = |f(x) - 0|$.

11.9. In (11.1) choose ϵ to be $\epsilon\sqrt{\ell}$. Then, for $0 < |x - x_0| < \delta$ it follows that $|\sqrt{f(x)} - \sqrt{\ell}| = |f(x) - \ell|/(\sqrt{f(x)} + \sqrt{\ell}) \le |f(x) - \ell|/\sqrt{\ell} < \epsilon$.

11.10. $\lim_{x \to 0} f(cx)/x = \lim_{x \to 0} cf(cx)/(cx) = c \lim_{x \to 0} f(cx)/(cx) = c \lim_{y \to 0} f(y)/y = c\ell$. When $c = 0$, we have $f(cx)/x = f(0)/x$. Hence, as $x \to 0$, $f(cx)/x$ does not tend to any finite limit unless $f(0) = 0$.

11.11. Let $f(x) = \begin{cases} 1, & x < x_0 \\ -1, & x \geq x_0 \end{cases}$ and $g(x) = \begin{cases} 1, & x < x_0 \\ -1, & x \geq x_0. \end{cases}$ Then,

$f(x) + g(x) = \begin{cases} 2, & x < x_0 \\ -2, & x \geq x_0 \end{cases}$ and $f(x) - g(x) = 0$. Thus, $\lim_{x \to x_0^-} (f(x) +$

$g(x)) = 2 \neq \lim_{x \to x_0^+} = -2$ and $\lim_{x \to x_0} (f(x) - g(x)) = 0$.

11.12. Use the definitions of even and odd functions.

11.13. For $\epsilon > 0 \, \exists \, \delta_1 > 0$ such that $0 < |y - \ell| < \delta_1 \Rightarrow |f(y) - f(\ell)| < \epsilon$, and for $\epsilon = \delta_1 > 0 \, \exists \, \delta_2 > 0$ such that $0 < |x - a| < \delta_2 \Rightarrow |g(x) - \ell| < \delta_1$. Let $\delta = \min\{\delta_1, \delta_2\}$, then $0 < |x - a| < \delta \Rightarrow |g(x) - \ell| < \delta_1 \Rightarrow |f(g(x)) - f(\ell)| < \epsilon$.

11.14. Let $\varepsilon > 0$ be a real number such that $\beta - \alpha > 3\varepsilon$. Since, $\beta - \varepsilon - (\alpha + \varepsilon) = \beta - \alpha - 2\varepsilon > 3\varepsilon - 2\varepsilon = \varepsilon > 0$, we find that $\beta - \varepsilon > \alpha + \varepsilon$. Since, $\lim_{x \to a} f(x) = \alpha$, there is a $\delta_f > 0$ such that for all $x \in A$ with $0 < |x - a| < \delta_f$, we have $\alpha - \varepsilon < f(x) < \alpha + \varepsilon$. Similarly, for g there is a $\delta_g > 0$, such that for all $x \in A$ with $0 < |x - a| < \delta_g$, we have $\beta - \varepsilon < g(x) < \beta + \varepsilon$. Let $\delta = \min\{\delta_f, \delta_g\}$ and consider $V = (\delta - a, \delta + a)$. If $x \in V \cap A$, $x \neq a$ we have for $0 < |x - a| < \delta$ that $f(x) < \alpha + \varepsilon < \beta - \varepsilon < g(x)$.

Chapter 12

Continuous Functions

We begin this chapter with several equivalent definitions of continuity of a function at a point which is closely connected with the concept of limits. Then, we will show that continuous functions over closed and bounded intervals possess several remarkable properties.

A function f, defined on some nbd of a point x_0, is said to be continuous at x_0 iff either any of the following holds:

(D1). $\lim_{x \to x_0} f(x) = f(x_0)$.

(D2). $f(x_0 + 0) = f(x_0 - 0) = f(x_0)$.

(D3). $\limsup_{x \to x_0^+} f(x) = \liminf_{x \to x_0^+} f(x) = \limsup_{x \to x_0^-} f(x)$
$= \liminf_{x \to x_0^-} f(x) = f(x_0)$.

(D4). Given $\epsilon > 0$ there exists a $\delta > 0$ such that

$$|x - x_0| < \delta \Rightarrow |f(x) - f(x_0)| < \epsilon. \qquad (12.1)$$

(D5). For every sequence $\{x_n\}$ defined in the nbd of x_0 with $x_n \to x_0$ the sequence $\{f(x_n)\}$ converges to $f(x_0)$.

(D6). Given $\epsilon > 0$ there exists a $\delta > 0$ such that

$$x_1, \ x_2 \in (x_0 - \delta, x_0 + \delta) \Rightarrow |f(x_1) - f(x_2)| < \epsilon.$$

From the definitions of limit, left-hand and right-hand limits, lim sup and lim inf, Theorem 11.2 and Problem 11.7 it immediately follows that all the conditions (D1) to (D6) are equivalent. If f is not continuous at x_0, it is called *discontinuous at x_0*.

A function f, defined on some nbd of a point x_0 of the form $[x_0, x_0 + \delta)$ $((x_0 - \delta, x_0])$, is called *right-continuous (left-continuous)* at x_0 iff $f(x_0 + 0) = f(x_0)$ $(f(x_0 - 0) = f(x_0))$. Thus, a function is continuous at x_0 iff it is left-continuous as well as right-continuous.

A function f, defined on an interval I, is said to be *continuous on I* if it is continuous at every interior point of I and is right-(left-)continuous at the left (right) endpoint of I, provided it exists. Now $C(I)$ represents the class of all continuous functions on the interval I.

A function f, is said to be *continuous on a set S* if for given $\epsilon > 0$ and $x_0 \in S$ there exists $\delta > 0$ such that

$$x \in S \;\wedge\; |x - x_0| < \delta \;\Rightarrow\; |f(x) - f(x_0)| \;<\; \epsilon. \tag{12.2}$$

Example 12.1. From Example 11.1 it follows that the function $1/\sqrt{x}$ is continuous at any point $x_0 > 0$. From Example 11.2 it is clear that the function $x/|x|$ is discontinuous at $x = 0$; here the function is also not defined at $x = 0$. From Example 11.3 we find that the function $f(x) = x^2 \sin(1/x)$, $x \neq 0$ is continuous at $x = 0$ only if $f(0) = 0$. From Example 11.4 we can conclude that the rational function $R(x) = P_n(x)/P_m(x)$ is continuous at any point $x_0 \in \mathcal{R}$ provided $P_m(x_0) \neq 0$.

Our first result provides a necessary and sufficient condition for the continuity of a function defined on $\mathcal{R}$.

Theorem 12.1. A function $f : \mathcal{R} \to \mathcal{R}$ is continuous on $\mathcal{R}$ iff for every open set $S \subseteq \mathcal{R}$, $f^{-1}(S)$ is open in $\mathcal{R}$.

Proof. Let f be continuous on $\mathcal{R}$. If $S = \emptyset$, then $f^{-1}(S) = \emptyset$ is open. Now let $x_0 \in f^{-1}(S) \Rightarrow f(x_0) \in S \Rightarrow$ for some $\epsilon > 0$, $(f(x_0) - \epsilon, f(x_0) + \epsilon) \subseteq S$. Since f is continuous at $x_0 \;\exists\, \delta > 0$ such that $|x - x_0| < \delta \Rightarrow |f(x) - f(x_0)| < \epsilon$. Hence, $x \in (x_0 - \delta, x_0 + \delta) \Rightarrow f(x) \in (f(x_0) - \epsilon, f(x_0) + \epsilon) \subseteq S \Rightarrow x \in f^{-1}(S)$, i.e., $x_0 \in (x_0 - \delta, x_0 + \delta) \subseteq f^{-1}(S)$. Thus, $f^{-1}(S)$ is open.

Conversely, assume that $f^{-1}(S)$ is open whenever S is open. Let $x_0 \in \mathcal{R}$ be arbitrary, and choose $\epsilon > 0$. Let $S = (f(x_0) - \epsilon, f(x_0) + \epsilon)$. Clearly, S is open, and this in turn implies that $f^{-1}(S)$ is open. Thus, since $x_0 \in f^{-1}(S)$ there exists a $\delta > 0$ such that $(x_0 - \delta, x_0 + \delta) \subseteq f^{-1}(S)$. Hence, for any $x \in (x_0 - \delta, x_0 + \delta)$, we have $f(x) \in (f(x_0) - \epsilon, f(x_0) + \epsilon)$.

Theorem 12.1 can be restated as follows: A function $f : \mathcal{R} \to \mathcal{R}$ is continuous on $\mathcal{R}$ iff for every closed set $S \subseteq \mathcal{R}$, $f^{-1}(S)$ is closed in $\mathcal{R}$. For this, we note that if S is closed, then $\mathcal{R} - S$ is open. Thus, if f is continuous, then $f^{-1}(\mathcal{R} - S)$, i.e., $\mathcal{R} - f^{-1}(S)$ is open. Hence, $f^{-1}(S)$ is closed. Conversely, if S and $f^{-1}(S)$ are closed, then $\mathcal{R} - S$ and $\mathcal{R} - f^{-1}(S)$, i.e., $f^{-1}(\mathcal{R} - S)$ are open, and so f is continuous.

Now we state two results which respectively follow from Theorem 11.4 and Problem 11.13.

Theorem 12.2. If f and g are continuous at x_0, then $f \pm g$, $f \cdot g$ are continuous at x_0, and f/g is continuous at x_0 provided $g(x_0) \neq 0$.

Theorem 12.3. If g is continuous at x_0 and f is continuous at $g(x_0)$, then $f \circ g(x) = f(g(x))$ is continuous at x_0.

In the following theorems we shall prove several properties of continuous functions.

Theorem 12.4 (Extreme-Value Theorem/Weierstrass's Theorem).

If f is continuous on a closed bounded (compact) interval $[a, b]$, then f is bounded on $[a, b]$. Further, if $M = \sup_{x \in [a,b]} f(x)$ and $m = \inf_{x \in [a,b]} f(x)$, then there exists points x_M, $x_m \in [a, b]$ such that $f(x_M) = M$ and $f(x_m) = m$.

Proof. If f is not bounded on $[a, b]$, then there exists $x_n \in [a, b]$ such that $|f(x_n)| > n$, $n \in \mathcal{N}$. Since $[a, b]$ is bounded, the sequence $\{x_n\}$ is bounded. Thus, by Theorem 4.1 the sequence $\{x_n\}$ has a subsequence $\{x_{n_k}\}$ which converges to some $x_0 \in \mathcal{R}$. However, since $[a, b]$ is closed $x_0 \in [a, b]$. Now since f is continuous at x_0, $f(x_{n_k})$ must converge to $f(x_0)$. But this contradicts the fact that $|f(x_{n_k})| > n_k > k$ for all $k \in \mathcal{N}$. Hence, f is bounded on $[a, b]$.

Now since f is bounded on $[a, b]$, $M = \sup_{x \in [a,b]} f(x)$ and $m = \inf_{x \in [a,b]} f(x)$ are finite numbers. To show that there is an $x_M \in [a, b]$ such that $f(x_M) = M$, we assume the contrary that $f(x) < M$ for all $x \in [a, b]$. Then, the function $g(x) = 1/(M - f(x))$ is continuous, and hence bounded on $[a, b]$, i.e., there is some $K > 0$ such that $|g(x)| = g(x) \leq K$. But, this implies that $f(x) \leq M - (1/K)$ for all $x \in [a, b]$, and therefore $M = \sup_{x \in [a,b]} f(x) \leq M - (1/K)$, which is a contradiction. A similar argument proves that there exists an $x_m \in [a, b]$ such that $f(x_m) = m$.

From Theorem 12.4 it immediately follows that any continuous periodic function is bounded on the real line. The functions $f(x) = 1/x$ over the open interval $(0, 1)$, and $f(x) = x$ over the closed interval $[0, \infty)$ are continuous, but unbounded. For the function $f(x) = x^2$ over the open interval $(1, 2)$, $M = 4$, $m = 1$, but there are no points x_M, $x_m \in (1, 2)$ with $f(x_M) = 4$ and $f(x_m) = 1$. Thus, Theorem 12.4 holds only if the interval $[a, b]$ is closed and bounded. The numbers M and m in Theorem 12.4 are called the *maximum* and *minimum* values of f on $[a, b]$.

Theorem 12.5 (Sign Preserving Property).

If f is continuous on a set S, and at $x_0 \in S$, $f(x_0) > 0$, then there are positive constants ϵ and δ such that $S \wedge |x - x_0| < \delta$ implies $f(x) > \epsilon$.

Proof. In (12.2) let $\epsilon = f(x_0)/2$, to obtain $-f(x_0)/2 < f(x) - f(x_0) < f(x_0)/2$. Thus, $f(x) > f(x_0)/2 = \epsilon$ for all $S \wedge |x - x_0| < \delta$.

Theorem 12.6 (Intermediate-Value Theorem).

Let f be continuous on a closed bounded interval $[a, b]$. If $f(a) \neq f(b)$ and y_0 is a real number between $f(a)$ and $f(b)$, then there is an $x_0 \in [a, b]$ such that $f(x_0) = y_0$.

Proof. Let $f(a) \leq y_0 \leq f(b)$. The proof for the case $f(b) \leq y_0 \leq f(a)$

is similar. If $y_0 = f(a)$ or $y_0 = f(b)$, we can choose $x_0 = a$ or $x_0 = b$. Thus, we assume that $f(a) < y_0 < f(b)$. Define the continuous function $g(x) = f(x) - y_0$, $x \in [a, b]$. Since $g(a) < 0$ and $g(b) > 0$ in view of Theorem 12.5 there exists positive constants δ_1 and δ_2 such that $g(x) < 0$, $x \in [a, a+\delta_1)$ and $g(x) > 0$, $x \in (b - \delta_2, b]$. Consider the set $S = \{x \in [a, b] : g(x) < 0\}$. Since $a \in S$ and $S \subseteq [a, b]$, S is a nonempty bounded subset of $\mathcal{R}$, and hence by the completeness axiom $x_0 = \sup S$ is a finite real number. Clearly, $x_0 \in (a, b)$. We shall show that $g(x_0) = 0$. For this, since $x_0 = \sup S$, for every $x \in (x_0, b]$, $g(x) \geq 0$. Hence, by the continuity of g at x_0, $\lim_{x \to x_0^+} g(x) = g(x_0) \geq 0$. On the other hand, let $\{x_n\} \subseteq [a, b]$ be a sequence such that $x_n < x_0$, $x_n \to x_0$ and $g(x_n) < 0$, $n \in \mathcal{N}$. The existence of such a sequence is guaranteed, otherwise the set S would have an upper bound less than x_0. Again by the continuity of g at x_0, $g(x_n) \to g(x_0)$, and hence, $g(x_0) \leq 0$. Thus, it follows that $g(x_0) = 0$, and consequently, $f(x_0) = y_0$.

In view of the intermediate-value theorem, geometrically the graph of a continuous function on $[a, b]$ is without breaks. It can be traced by a pencil on a piece of paper without raising the pencil from the paper. The converse of Theorem 12.6 is not true. For this, it suffices to consider the discontinuous function $f(x) = \sin 1/x$ when $x \neq 0$, and $f(0) = 0$. Clearly, $f : \mathcal{R} \to [-1, 1]$, and for each $y_0 \in [-1, 1]$ there are infinite numbers $x_0 \in \mathcal{R}$ such that $f(x_0) = y_0$. We also remark that in Theorem 12.6 the continuity of f on a closed interval is essential. In fact, the function $f(x) = [x]$, $x \in [0, 1]$ is continuous on $[0, 1)$ and $f(0) = 0$, $f(1) = 1$, but it does not pass through any value between 0 and 1 as x passes from 0 to 1.

Corollary 12.1. Let f be a nonconstant, continuous function on a closed bounded interval $[a, b]$, and let $M = \sup_{x \in [a,b]} f(x)$ and $m = \inf_{x \in [a,b]} f(x)$. Then, the range of f is the closed bounded interval $[m, M]$.

Proof. From Theorem 12.4 there exists points x_M, $x_m \in [a, b]$ such that $f(x_M) = M$ and $f(x_m) = m$. Since f is nonconstant, $m < M$. Now by Theorem 12.6 for every $y_0 \in (m, M)$ there exists an $x_0 \in (x_m, x_M)$ with $f(x_0) = y_0$. Thus, the range of f is $[m, M]$.

Corollary 12.2. Let f be a continuous and strictly monotone function on a closed bounded interval $[a, b]$. If y_0 is a real number between $f(a)$ and $f(b)$, then there is a unique $x_0 \in [a, b]$ such that $f(x_0) = y_0$.

Corollary 12.3. Let f be a continuous function on a closed bounded interval $[a, b]$, and let $f(a)f(b) < 0$. Then, there exists at least one $x_0 \in (a, b)$ such that $f(x_0) = 0$.

As an application of Corollary 12.3 we shall show that every polynomial $P_n(x) = p_0 x^n + p_1 x^{n-1} + \cdots + p_{n-1} x + p_n$ of odd degree has at least one real zero. For definiteness, we assume that $p_0 > 0$. Clearly, P_n is continuous on $\mathcal{R}$.

Also, $\lim_{x \to \infty} P_n(x) = \infty$, and $\lim_{x \to -\infty} P_n(x) = -\infty$. Thus, there exists real numbers a and b such that $P_n(a) < 0 < P_n(b)$. Now by Corollary 12.3 there exists an $x_0 \in (a, b)$ such that $P_n(x_0) = 0$.

Theorem 12.7 (Fixed Point Property). If f is continuous on $[a, b]$, and $f(x) \in [a, b]$, $x \in [a, b]$ then f has a fixed point, i.e., there exists an $x_0 \in [a, b]$ such that $f(x_0) = x_0$.

Proof. If $f(a) = a$ or $f(b) = b$ then we are done, hence we assume that $f(a) > a$ and $f(b) < b$. Let $g(x) = f(x) - x$. Clearly, g is continuous on $[a, b]$, and $g(a) > 0$, $g(b) < 0$. Hence, by Corollary 12.3 there exists an $x_0 \in (a, b)$ such that $g(x_0) = 0$, i.e., $f(x_0) = x_0$.

For example, the function $f(x) = x^2 - x + 1$ maps the interval $[0, 1]$ into itself, and $x^2 - x + 1 = x$ at $x = 1$. Similarly, the function $\sin x$ maps the interval $[-1, 1]$ into itself, and $\sin x = x$ at $x = 0$.

Finally, as an application of the intermediate-value theorem we shall prove the following result.

Theorem 12.8. Let f be a strictly monotone and continuous function on $[a, b]$ and g be the inverse of f. Then, g is one-to-one and continuous on $[f(a), f(b)]$.

Proof. Assume that f is strictly increasing. Then, from the property (M3) in Chapter 6, the inverse function $f^{-1} \equiv g : [f(a), f(b)] \to [a, b]$ exists and is one-to-one. Thus, it suffices to show that g is continuous on $[f(a), f(b)]$. Let $y_0 \in [f(a), f(b))$ and define $x_0 = g(y_0)$, then $x_0 \in [a, b]$ and $0 < \epsilon \le b - x_0$. Then, f is continuous on $[x_0, x_0 + \epsilon]$. Thus, in view of the intermediate-value theorem $[f(x_0), f(x_0 + \epsilon)) \subseteq f([x_0, x_0 + \epsilon))$. Take $\delta = f(x_0 + \epsilon) - f(x_0)$. Then, $\delta > 0$ and for each $y \in (y_0, y_0 + \delta) = (f(x_0), f(x_0 + \epsilon))$ $(y_0 = f(x_0))$ there exists $x \in [x_0, x_0 + \epsilon)$ such that $f(x) = y$ which implies that $0 < g(y) - g(y_0) = x - x_0 < \epsilon$, and hence $|g(y) - g(y_0)| < \epsilon$ for all $y_0 < y < y_0 + \delta$. Thus, g is right-continuous at $y_0 \in [f(a), f(b))$. Similarly, g is left-continuous on $(f(a), f(b)]$. Therefore, g is continuous on $[f(a), f(b)]$. The proof when f is strictly decreasing is similar.

Remark 12.1. There are functions which are nowhere continuous. For example, consider the function

$$f(x) = \begin{cases} -1, & \text{if } x \text{ is rational} \\ 1, & \text{if } x \text{ is irrational.} \end{cases} \tag{12.3}$$

This function is not continuous at any point x_0 because we can choose a sequence $\{x_n\}$ converging to x_0 such that the odd terms of $\{x_n\}$ are rational and the even terms irrational. Then, $\{f(x_n)\} = \{(-1)^n\}$, which diverges. Hence, in view of (D5) this function is discontinuous at x_0. Clearly, the absolute value of this function is everywhere continuous. Another interesting example is the function

$$f(x) = \begin{cases} 0, & \text{if } x \text{ is irrational or } 0 \\ 1/n, & \text{if } x = m/n \text{ in lowest terms,} \end{cases} \qquad (12.4)$$

which is continuous at the irrationals and discontinuous at the nonzero rationals.

Problems

12.1. Given

$$f(x) = \begin{cases} x^2 + 5 & \text{for} & x < 0 \\ ax + b & \text{for} & 0 < x < 2 \\ c & \text{for} & x = 2 \\ 3x - 5 & \text{for} & x > 2. \end{cases}$$

For what values of a, b, c is f continuous?

12.2. Let

$$f(x) = \begin{cases} x^2 & \text{for} & x < -1 \\ 0 & \text{for} & x = -1 \\ 2x & \text{for} & -1 < x < 0 \\ 0 & \text{for} & x = 0 \\ \sin(\pi/x) & \text{for} & 0 < x \le 1 \\ -1 + 1/x & \text{for} & 1 < x < \infty. \end{cases}$$

For any $a \in \mathcal{R}$ find $\lim_{x \to a^-} f(x)$, $\lim_{x \to a^+} f(x)$ if possible. State the points of continuity of f. What are $\lim_{x \to -\infty} f(x)$ and $\lim_{x \to \infty} f(x)$?

12.3. Suppose $a < c < b$ and $f_1 : (a, c) \to \mathcal{R}$ and $f_2 : (c, b) \to \mathcal{R}$ are continuous functions. Show that f_1 and f_2 can be extended to a continuous function $f : (a, b) \to \mathcal{R}$ iff $\lim_{x \to c^-} f_1(x) = \lim_{x \to c^+} f_2(x)$.

12.4. Show that if f is continuous on $\mathcal{R}$ and is zero on a dense set S in $\mathcal{R}$, then f is identically zero.

12.5. Prove the following.

(i). If f is continuous at x_0, then so is $|f|$.

(ii). If f and g are continuous, then so are $\max(f, g)$ and $\min(f, g)$.

(iii). Every continuous function f can be written as $f = g - h$, where g and h are nonnegative and continuous functions.

12.6. Let $f : \mathcal{R} \to \mathcal{R}^+$ be continuous, and $\lim_{x \to \infty} f(x) = 0 = \lim_{x \to -\infty} f(x)$. Prove that there exists a number $x_0 \in \mathcal{R}$ such that $f(x_0) \ge f(x)$ for all $x \in \mathcal{R}$. Give an example to show that we cannot replace $\mathcal{R}^+$ by $\mathcal{R}$ in the hypothesis.

12.7. Prove that if g is continuous at 0, $g(0) = 0$, and for some $\delta > 0$, $|f(x)| \le |g(x)|$ for all $|x| < \delta$, then f is continuous at 0.

12.8. (i). A function f is said to be *Lipschitz continuous* on an interval I if there is a constant $k > 0$ such that for all x, $y \in I$,

$$|f(x) - f(y)| \le k|x - y|.$$

Prove that Lipschitz continuity implies that f is continuous on I.

(ii). Show that $\sin x$ and $\cos x$ are continuous on $\mathcal{R}$.

12.9. (i). Prove that there exists a number $x_0 \in [a, b] \subseteq \mathcal{R}$ such that

$$\frac{7 + 2x + 18x^2}{1 + x^2 + 3x^6} \le \frac{7 + 2x_0 + 18x_0^2}{1 + x_0^2 + 3x_0^6}$$

for all $x \in [a, b]$.

(ii). Let f be defined on $\mathcal{R}$ as $f(x) = (x - 1)(x - 2)(x - 3)(x - 4)$. Show that there exists a number $x_0 \in \mathcal{R}$ such that $f(x_0) \le f(x)$ for all $x \in \mathcal{R}$.

12.10. Find $M = \sup_{x \in A} f(x)$ and $m = \inf_{x \in A} f(x)$ (if they exist) for the following functions f defined on the indicated domain A, and then find points (if possible) x_1, $x_2 \in A$ such that $f(x_1) = M$, $f(x_2) = m$.

(i). $f(x) = 5 - 2|x - 9|$, $A = [-7, 12]$.
(ii). $f(x) = \frac{1}{1+x^2}$, $A = (-7, 1)$.
(iii). $f(x) = \frac{x}{1+x^2}$, $A = (-1, \infty)$.

12.11. (i). Let f be a continuous function on $[a, b]$ and suppose that $f(a) \ne f(b)$. Prove that there is a number $x_0 \in (a, b)$ such that

$$f(x_0) = \frac{1}{5}f(a) + \frac{4}{5}f(b).$$

(ii). Prove that there is a negative real number x_0 such that

$$x_0^{179} + \frac{163}{1 + x_0^2 + \sin^2 x_0} = 119.$$

12.12. Let f be monotone and $a \in \mathcal{R}$. Show that $\lim_{x \to a} f(x)$ exists iff f is continuous at a.

12.13. Show that if f is one-to-one and continuous on $[a, b]$, then f is strictly monotone on $[a, b]$. Hence, deduce that

(i). In Theorem 12.8 "strictly monotone" can be replaced by "one-to-one."

(ii). If in Theorem 12.8, f is strictly increasing (decreasing) on $[a, b]$, then f^{-1} is strictly increasing (decreasing) on $[f(a), f(b)]$ $([f(b), f(a)])$.

12.14. A function $f : \mathcal{R} \to \mathcal{R}$ is called *multiplicative* if it satisfies $f(x+y) = f(x)f(y)$ for all $x, \ y \in \mathcal{R}$. If f is continuous show that either $f(x) \equiv 0$, or there exists an $a > 0$ such that $f(x) = a^x, \ x \in \mathcal{R}$.

12.15. A function $f : \mathcal{R} \to \mathcal{R}$ is called *additive* if it satisfies $f(x + y) = f(x) + f(y)$ for all $x, \ y \in \mathcal{R}$. Prove that f is continuous on $\mathcal{R}$ if it is continuous at 0. Moreover, show that $f(x) = xf(1)$ for all $x \in \mathcal{R}$.

12.16. A function $f : \mathcal{R} \to \mathcal{R}$ is called *subadditive* if it satisfies $f(x+y) \leq f(x) + f(y)$ for all $x, \ y \in \mathcal{R}$. If the inequality is reversed, then f is called *superadditive*. Prove that if $f(0) = 0$ and f is continuous at 0, then it is continuous on $\mathcal{R}$.

12.17. A function $f : I \to \mathcal{R}$ is called *upper semicontinuous* at a point $x_0 \in I^i$ iff $\overline{f(x_0 + 0)} \leq f(x_0)$; f is *lower semicontinuous* at x_0 iff $\underline{f(x_0 + 0)} \geq f(x_0)$; f is *semicontinuous* at x_0 iff f is either upper semicontinuous at x_0 or lower semicontinuous at x_0. Find the values of x_0 at which each of the following functions is upper semicontinuous.

(i). $f(x) = [x]$.

(ii). $f(x) = x - [x]$.

(iii). $f(x) = \begin{cases} 0 \ \text{ if } \ x \in \mathcal{Q} \\ 1 \ \text{ if } \ x \in \mathcal{R} - \mathcal{Q}. \end{cases}$

(iv). $f(x) = x^2 + 4$.

12.18. Let f be a continuous function on $[0, 1]$, and $\|f\|$ be the maximum value of $|f|$ on $[0, 1]$.

(i).   Prove that for any number c, $\|cf\| = |c|\|f\|$.

(ii).   If g is also continuous on $[0, 1]$, then $\|f + g\| \leq \|f\| + \|g\|$. Give an example where $\|f + g\| \neq \|f\| + \|g\|$.

(iii). If h is also continuous on $[0, 1]$, then $\|f - h\| \leq \|f - g\| + \|g - h\|$.

Answers or Hints

12.1. $a = -2, b = 5, c = 1$.

12.2. $\lim_{x \to a^-} f(x) = \lim_{x \to a^+} f(x) = \begin{cases} a^2 \ \text{ if } \ a < -1 \\ 2a \ \text{ if } \ -1 < a < 0 \\ \sin(\pi/a) \ \text{ if } \ 0 < a < 1 \\ -1 + 1/a \ \text{ if } \ 1 < a < \infty, \end{cases}$

$\lim_{x \to a^-} f(x) = \begin{cases} 1, & a = -1 \\ 0, & a = 0 \\ 0, & a = 1, \end{cases}$ $\lim_{x \to a^+} f(x) = \begin{cases} 0, & a = -1 \\ \text{not exists}, & a = 0 \\ 0, & a = 1. \end{cases}$

Hence, f is continuous everywhere except -1 and 0. $\lim_{x\to-\infty} f(x) = \infty$ and $\lim_{x\to\infty} f(x) = -1$.

12.3. If such an extension f exists, then the equality of the limits follows from the definition of continuity. Conversely, define $f : (a,b) \to \mathcal{R}$ as

$$f(x) = \begin{cases} f_1(x), & x \in (a,c) \\ \lim_{x\to c^-} f_1(x) = \lim_{x\to c^+} f_2(x), & x = c \\ f_2(x), & x \in (c,b) \end{cases}$$

and show f is continuous on (a,b).

12.4. For any $x \in \mathcal{R}$ there exists a sequence $\{x_n\}$ in S such that $x_n \to x$, and $f(x_n) = 0, \forall n \in \mathcal{N} \Rightarrow \{f(x_n)\} \to f(x) = 0$.

12.5. (i). Given $\epsilon > 0$, there exists a $\delta > 0$ such that $|f(x) - f(x_0)| < \epsilon$ if $|x - x_0| < \delta$. Since $||f(x)| - |f(x_0)|| \le |f(x) - f(x_0)|$, we have $||f(x)| - |f(x_0)|| < \epsilon$ if $|x - x_0| < \delta$.

(ii). Since $\max(f,g) = (1/2)(f + g + |f - g|)$ and $\min(f,g) = (1/2)(f + g - |f - g|)$ the result follows from (i).

(iii). Let $g = \max(f,0)$ and $h = -\min(f,0)$. Then, $f = g - h$ and the result follows from (ii).

12.6. Since $\lim_{x\to\infty} f(x) = 0 = \lim_{x\to-\infty} f(x)$, there exists a real number M such that

$$f(x) < f(0) \ (> 0) \quad \text{for all} \quad |x| > M. \tag{12.5}$$

Since f is continuous on $[-M, M]$ there exists a $x_0 \in [-M, M]$ such that $f(x_0) \ge f(x)$ for all $x \in [-M, M]$. In view of (12.5), we now have $f(x_0) \ge f(x)$ for all $x \in \mathcal{R}$. The condition $f : \mathcal{R} \to \mathcal{R}^+$ is essential. For this consider the function $f(x) = \begin{cases} -1/(x+1), & x \ge 0 \\ 1/(x-1), & x \le 0. \end{cases}$ Then, f is continuous on $\mathcal{R}$ and $f(x_0) < f(2x_0)$, $x_0 \in \mathcal{R} - \{0\}$. Hence, there does not exist a $x_0 \in \mathcal{R}$ such that $f(x_0) \ge f(x)$, $x \in \mathcal{R}$.

12.7. Since g is continuous at 0 and $g(0) = 0$, given $\epsilon > 0$ there exists a $\delta > 0$ such that if $|x| < \delta$, then $|f(x) - 0| \le |g(x) - g(0)| < \epsilon$, and therefore, $\lim_{x\to0} f(x) = 0$. Now since $|f(x)| \le |g(x)|$ for all $x \in (x - \delta, x + \delta)$, we have $|f(0)| \le 0$, i.e., $f(0) = 0$. Hence, $\lim_{x\to0} f(x) = f(0) = 0$. Thus, f is continuous at $x = 0$.

12.8. (i). Let $x_0 \in I$. Given $\epsilon > 0$, choose $\delta = \epsilon/k$. Then, if $|x - x_0| < \delta$, we have $|f(x) - f(x_0)| \le k|x - x_0| < k\delta = \epsilon$.

(ii). Since for all $x, y \in \mathcal{R}$

$$
\begin{aligned}
|\sin x - \sin y| &= 2\left|\cos \tfrac{x+y}{2}\right|\left|\sin \tfrac{x-y}{2}\right| \le |x - y|, \\
|\cos x - \cos y| &= 2\left|\sin \tfrac{x+y}{2}\right|\left|\sin \tfrac{x-y}{2}\right| \le |x - y|
\end{aligned}
$$

the continuity of functions $\sin x$ and $\cos x$ on $\mathcal{R}$ follows from (i).

12.9. (i). Define $f(x) = (7 + 2x + 18x^2)/(1 + x^2 + 3x^6)$. Clearly, f is continuous on $[a, b]$. Thus, the extremum value theorem implies that there exists a point $x_0 \in [a, b]$ such that $f(x) \leq f(x_0)$ for all $x \in [a, b]$.

(ii). Since f is a polynomial, it is continuous on $\mathcal{R}$. Thus, the extremum value theorem implies that there exists a point $x_0 \in [1, 4]$ such that $f(x) \geq f(x_0)$. Clearly, $f(x_0) < 0$. Now since for all $x \notin [1, 4]$, $f(x) > 0$ it follows that $f(x_0) \leq f(x)$ for all $x \in \mathcal{R}$.

12.10. (i). $\sup_A f(x) = 5 - 2|12 - 9| = 5 - 6 = -1 = f(12) = f(6); \inf_A f(x) = f(-7) = 5 - 2(16) = -27$.

(ii). $\inf_A f(x) = f(-7) = 1/50$, $-7 \notin A$; $\sup_A f(x) = f(0) = 1$.

(iii). $\inf_A f(x) = f(-1) = -1/2$, $-1 \notin A$; $\sup_A f(x) = f(1) = 1/2$.

12.11. (i). Assume that $f(a) < f(b)$. Then, $f(a) = \frac{1}{5}f(a) + \frac{4}{5}f(a) < \frac{1}{5}f(a) + \frac{4}{5}f(b) < f(b)$. Now the intermediate-value theorem implies that there exists a $x_0 \in (a, b)$ such that $f(x_0) = \frac{1}{5}f(a) + \frac{4}{5}f(b)$.

(ii). Define $f(x) = x^{179} + 163/(1 + x^2 + \sin^2 x)$, $x \in \mathcal{R}$. Clearly, f is continuous on $[-2, 0]$ and $f(0) = 163 > 119 > f(-2)\ (< 0)$. Now intermediate-value theorem implies that there exists a $x_0 \in (-2, 0)$ such that $f(x_0) = 119$.

12.12. Suppose f is monotone increasing, $\lim_{x \to a^-} f(x)$, $\lim_{x \to a^+} f(x)$ exist and $\lim_{x \to a^-} f(x) \leq f(a) \leq \lim_{x \to a^+} f(x)$. Thus, if $\lim_{x \to a} f(x)$ exists, $\lim_{x \to a^-} f(x) = f(a) = \lim_{x \to a^+} f(x)$, and hence f is continuous at a.

12.13. Since f is one-to-one, $f(a) \neq f(b)$. We assume that $f(a) < f(b)$. The arguments for the case $f(a) > f(b)$ can be repeated to the function $-f$. If f is not strictly increasing on $[a, b]$, then there exists $x_1, x_2 \in [a, b]$ such that $x_1 < x_2$ and $f(x_1) > f(x_2)$. Now there are two cases to consider. (1) $f(x_1) > f(b)$. Let $k \in (f(b), f(x_1))$. Then, $k \in (f(a), f(x_1))$ and by Theorem 12.6 there exists $s_1 \in (a, x_1)$ and $s_2 \in (x_1, b)$ such that $f(s_1) = f(s_2) = k$. But, $s_1 \neq s_2$ contradicts the fact that f is one-to-one. (2) $f(x_1) < f(b)$. Then, $f(x_2) < f(b)$. If $k \in (f(x_2), f(x_1))$, then $k \in (f(x_2), f(b))$ and by Theorem 12.6 there exists $s_1 \in (x_1, x_2)$ and $s_2 \in (x_2, b)$ such that $f(s_1) = f(s_2) = k$. But, again $s_1 \neq s_2$ contradicts the fact that f is one-to-one.

(i). See the hypotheses of Theorem 12.8.

(ii). From the conclusion of Theorem 12.8, the function f^{-1} is strictly monotone.

12.14. Clearly, $f(1) = f(1/2)f(1/2) \geq 0$. If $f(1) = 0$, then $f(x) = f(x - 1)f(1) = 0$, $x \in \mathcal{R}$. If $f(1) \neq 0$, let $f(1) = a$, then $a > 0$. By successive computation it follows that $f(r) = a^r$, $r \in \mathcal{Q}$. Now since $f(x)$ and a^x are continuous at every point $x \in \mathcal{R}$, we have $\lim_{r \to x} f(r) = f(x) = \lim_{r \to x} a^r = a^x$, i.e., $f(x) = a^x$, $x \in \mathcal{R}$.

12.15. By induction, we have for any $z \in \mathcal{R}$ and $m \in \mathcal{N}$, $f((m + 1)z) = f(mz) + f(z) = (m + 1)f(z)$. Also, $f(0) = f(0 + 0) = f(0) + f(0)$, and hence $f(0) = 0$. Thus, $0 = f(mz - mz) = f(mz) + f(-mz)$, which implies

$f(-mz) = -f(mz)$. Therefore,

$$f(mz) = mf(z) \quad \text{for all} \quad m \in \mathcal{Z}. \tag{12.6}$$

Now for any $z_0 \in \mathcal{R}$ and $n \in \mathcal{N}$ let $z = z_0/n$, then by (12.6) it follows that

$$f(z_0) = f\left(n\frac{z_0}{n}\right) = nf\left(\frac{z_0}{n}\right),$$

and hence $f(z_0/n) = (1/n)f(z_0)$. Therefore,

$$f\left(\frac{m}{n}z_0\right) = mf\left(\frac{z_0}{n}\right) = \frac{m}{n}f(z_0) \quad \text{for} \quad m \in \mathcal{Z}, \ n \in \mathcal{N}.$$

Take $z_0 = 1$, we see that $f(r) = rf(1)$ for $r \in \mathcal{Q}$. Now since f is continuous at 0, given $\epsilon > 0$ there exists a $\delta > 0$ such that $|f(x) - 0| < \epsilon$ if $|x - 0| < \delta$. But, this implies for $x_0 \in \mathcal{R}$, $|f(x - x_0)| = |f(x) - f(x_0)| < \epsilon$ if $|x - x_0| < \delta$. Hence, f is continuous at any $x_0 \in \mathcal{R}$. Finally, for any $x \in \mathcal{R}$ we can find a sequence $\{x_n\} \subset \mathcal{Q}$ such that $x_n \to x$, and then $f(x) = \lim_{n\to\infty} f(x_n) = \lim_{n\to\infty} f(1)x_n = xf(1)$.

12.16. Modify the answer of Problem 12.15.

12.17. (i). $\mathcal{R}$.
(ii). $\mathcal{R} - \mathcal{Z}$.
(iii). $\mathcal{R} - \mathcal{Q}$.
(iv). $\mathcal{R}$.

12.18. (i). $\|cf\| = \max_{x\in[0,1]} |cf(x)| = |c| \max_{x\in[0,1]} |f(x)| = |c|\|f\|$.
(ii). $\|f + g\| = \max_{x\in[0,1]} |f(x) + g(x)| \le \max_{x\in[0,1]}\{|f(x)| + |g(x)|\} \le \max_{x\in[0,1]} |f(x)| + \max_{x\in[0,1]} |g(x)| = \|f\| + \|g\|$. As an example, let $f = 1$, $g = -1$.
(iii). $\|f - h\| = \|(f - g) + (g - h)\| \le \|f - g\| + \|g - h\|$.

Chapter 13

Discontinuous Functions

In this chapter, we will first classify different types of discontinuities, and then prove some interesting results. In real-world applications discontinuous functions are at least as important as continuous functions.

Recall that if f is not continuous at x_0, it is called *discontinuous at x_0*. Thus, from (D3) in Chapter 12, f is discontinuous at x_0 when the five numbers $\overline{f(x_0 + 0)}$, $\underline{f(x_0 + 0)}$, $\overline{f(x_0 - 0)}$, $\underline{f(x_0 - 0)}$ and $f(x_0)$ are not equal. Since this can happen in several different ways, we have several different types of discontinuities which we classify as follows:

(T1). *Removable discontinuity.* If $\lim_{x \to x_0} f(x)$ exists, but $f(x_0)$ has a different value or is undefined, then f is said to have a removable discontinuity at x_0. In such a case by defining $f(x_0) = \lim_{x \to x_0} f(x)$ the function f can be made continuous at x_0.

Example 13.1. Let the function f be defined on $[0, 1]$ as follows: $f(x) = x$ for $0 \le x < 1/2$, $f(x) = 0$ for $x = 1/2$, $f(x) = 1 - x$ for $1/2 < x \le 1$. Clearly, $\lim_{x \to 1/2} f(x) = 1/2$, but since $f(1/2) = 0$ this function has a removable singularity at $x = 1/2$. Similarly, since the function $f(x) = \lim_{n \to \infty} \cos^n \pi x$, $x \in (-1, 1)$ has the value 0 for all $x \in (-1, 0) \cup (0, 1)$ and $f(0) = 1$ it has a removable singularity at $x = 0$.

(T2). *Discontinuity of the first kind.* If the limits $f(x_0 + 0)$ and $f(x_0 - 0)$ both exist but have different values, then f is said to have a discontinuity of the first kind, or a *ordinary discontinuity*, or a *jump discontinuity* at x_0. If $f(x_0 - 0) = f(x_0)$, while $f(x_0 + 0) \ne f(x_0)$, the point x_0 is a point of discontinuity of the first kind on the right. Similarly, if $f(x_0 + 0) = f(x_0)$, while $f(x_0 - 0) \ne f(x_0)$, the point x_0 is a point of discontinuity of the first kind on the left.

Example 13.2. Consider $f(x) = (x)$, where (x) denotes the positive or negative excess of x over the nearest integer, and when x is mid-way between two consecutive integers $(x) = 0$. Thus, if n, $n+1$ are two consecutive integers, then $f(x) = x - n$ for $n \le x < n + 1/2$, $f(x) = x - (n+1) = -(n+1-x)$ for $n + 1/2 < x \le n + 1$, $f(x) = 0$ for $x = n + 1/2$. Clearly, $f(n + 1/2 + 0) = -1/2$, $f(n + 1/2 - 0) = 1/2$, $f(n + 1/2) = 0$. Therefore, $f(x)$ has an ordinary discontinuity at $x = n + 1/2$, where n is an integer. Similarly, since for the greatest-integer function $f(x) = [x]$, $f(n+0) = n$, $f(n-0) = n-1$ and $f(n) = n$ it has an ordinary discontinuity on the left for all integer values of x, but is continuous on the right at these points.

(T3). *Discontinuity of the second kind.* The function f is said to have a discontinuity of the second kind at x_0 if none of $f(x_0+0)$ and $f(x_0-0)$ exist. If $f(x_0-0)$ exists and $f(x_0-0) = f(x_0)$, while $f(x_0+0)$ does not exist, the point x_0 is a point of discontinuity of the second kind on the right. Similarly, if $f(x_0+0)$ exists and $f(x_0+0) = f(x_0)$, while $f(x_0-0)$ does not exist, the point x_0 is a point of discontinuity of the second kind on the left.

Example 13.3. The function $f(x) = \sin(1/x)$ is continuous for all $x \neq 0$. At $x = 0$, we have $\limsup_{x\to 0^+} f(x) = 1$, $\liminf_{x\to 0^+} f(x) = -1$, $\limsup_{x\to 0^-} f(x) = 1$, $\liminf_{x\to 0^-} f(x) = -1$, and hence none of $f(+0)$ and $f(-0)$ exist. Therefore, $f(x)$ has a discontinuity of the second kind at $x = 0$. It is clear that the function is not defined at $x = 0$, but this does not affect the continuity of $f(x)$ as long as the right and left limits do not exist.

(T4). *Mixed discontinuity.* The function f is called mixed discontinuous at x_0 if it has a discontinuity of the second kind on one side of x_0, and on the other side either it is continuous or has a discontinuity of the first kind.

Example 13.4. For the function $f(x) = e^{1/x}\sin(1/x)$, $x \neq 0$ we have $\limsup_{x\to 0^+} f(x) = \infty$, $\liminf_{x\to 0^+} f(x) = -\infty$, $\limsup_{x\to 0^-} f(x) = 0$, $\liminf_{x\to 0^-} f(x) = 0$. Thus, this function has a discontinuity of the first kind from the left, and a discontinuity of the second kind from the right at $x = 0$, i.e., a mixed discontinuity at $x = 0$.

(T5). *Infinite discontinuity.* If f is discontinuous at x_0 and at least one of $\overline{f(x_0+0)}$, $f(x_0+0)$, $\overline{f(x_0-0)}$, $f(x_0-0)$ is $+\infty$ or $-\infty$, i.e., f is unbounded in every nbd of x_0, then f is said to have an infinite discontinuity at x_0.

Example 13.5. For the function $f(x) = 1/(x - x_0)$, $x \neq x_0$ we have $f(x_0+0) = +\infty$, $f(x_0-0) = -\infty$, and hence it has an infinite discontinuity of the first kind at x_0. Note that according to our definition the function $f(x) = 1/|x - x_0|$, $x \neq x_0$ and $f(x_0) = +\infty$ is continuous.

In what follows except removable and first kind discontinuities, all other types will be called discontinuities of the second kind.

Theorems 11.5 and 11.6 tell us that a monotone function on (a, b) cannot have discontinuities of the second kind, and are removable and first kind discontinuities are at most countable. In the following results we shall show that an arbitrary function f can have only countably many removable and first kind discontinuities. Hence, if f has no discontinuities of the second kind, then f is continuous at all but countably many points.

Theorem 13.1. An arbitrary function $f : \mathcal{R} \to \mathcal{R}$ can have at most countably many removable discontinuities.

Proof. Let S be the set of all $x_0 \in \mathcal{R}$ such that $\lim_{x \to x_0} f(x)$ exists, but is different from $f(x_0)$. For each $x_0 \in S$, we define $\ell(x_0) = \lim_{x \to x_0} f(x)$ and $g(x_0) = |f(x_0) - \ell(x_0)| > 0$. Let r_1, r_2, $r_3 \in \mathcal{Q}$ be such that $r_1 < x_0 < r_2$ and for each $x \in [r_1, x_0) \cup (x_0, r_2]$, $|f(x) - \ell(x_0)| < (1/3)g(x_0)$, and $|f(x_0) - r_3| < (1/3)g(x_0)$. The existence of such rational numbers is guaranteed from the existence of $\lim_{x \to x_0} f(x)$. Now consider the function $\phi : S \to \mathcal{Q} \times \mathcal{Q} \times \mathcal{Q}$ defined by $\phi(x_0) = (r_1, r_2, r_3)$. Clearly, it suffices to show that ϕ is one-to-one. Suppose there exists an $x_1 \in S$ with $\phi(x_1) = (r_1, r_2, \bar{r}_3)$ and $x_1 \neq x_0$. By the definition of ϕ, we have $r_1 < x_1 < r_2$ and $|f(x_1) - \bar{r}_3| < (1/3)g(x_1)$. Now since $x \in [r_1, x_0) \cup (x_0, r_2]$ implies $|f(x) - \ell(x_0)| < (1/3)g(x_0)$, and $x_1 \in (r_1, x_0) \cup (x_0, r_2)$ it follows that $|f(x_1) - \ell(x_0)| < (1/3)g(x_0)$ and $|\ell(x_1) - \ell(x_0)| \leq (1/3)g(x_0)$. Thus, we have

$$\begin{aligned} |f(x_1) - \bar{r}_3| < \tfrac{1}{3}g(x_1) &= \tfrac{1}{3}|f(x_1) - \ell(x_1)| \\ &\leq \tfrac{1}{3}(|f(x_1) - \ell(x_0)| + |\ell(x_0) - \ell(x_1)|) \\ &< \tfrac{1}{3}\left(\tfrac{1}{3}g(x_0) + \tfrac{1}{3}g(x_0)\right) = \tfrac{2}{9}g(x_0), \end{aligned}$$

and hence

$$|\ell(x_0) - \bar{r}_3| \leq |\ell(x_0) - f(x_1)| + |f(x_1) - \bar{r}_3| < \frac{1}{3}g(x_0) + \frac{2}{9}g(x_0) = \frac{5}{9}g(x_0).$$

From these inequalities it follows that

$$\begin{aligned} |\bar{r}_3 - r_3| &\geq |\ell(x_0) - f(x_0)| - |\ell(x_0) - \bar{r}_3| - |r_3 - f(x_0)| \\ &> g(x_0) - \tfrac{5}{9}g(x_0) - \tfrac{1}{3}g(x_0) = \tfrac{1}{9}g(x_0) > 0. \end{aligned}$$

Hence, $r_3 \neq \bar{r}_3$, and therefore $\phi(x_0) \neq \phi(x_1)$.

Theorem 13.2 (Froda's Theorem). An arbitrary function $f : \mathcal{R} \to \mathcal{R}$ can have at most countably many discontinuities of the first kind.

Proof. Let S be the set of all $x_0 \in \mathcal{R}$ such that $f(x_0 + 0)$ and $f(x_0 - 0)$ exist, but are different. For each $x_0 \in S$, we define $g(x_0) = |f(x_0 + 0) - f(x_0 - 0)| > 0$. Let r_1, $r_2 \in \mathcal{Q}$ be such that $r_1 < x_0 < r_2$ and for each $x \in [r_1, x_0)$, $|f(x) - f(x_0 - 0)| < (1/6)g(x_0)$ and for each $x \in (x_0, r_2]$, $|f(x) - f(x_0 + 0)| < (1/6)g(x_0)$. The existence of such rational numbers is guaranteed from the existence of $f(x_0 - 0)$ and $f(x_0 + 0)$. Then, we have

$$\begin{aligned} &|f(r_1) - f(r_2)| \\ &\leq |f(r_1) - f(x_0 - 0)| + |f(x_0 - 0) - f(x_0 + 0)| + |f(x_0 + 0) - f(r_2)| \\ &< \tfrac{1}{6}g(x_0) + g(x_0) + \tfrac{1}{6}g(x_0) = \tfrac{4}{3}g(x_0) \end{aligned}$$

and

$$|f(r_2) - f(r_1)|$$
$$\geq |f(x_0 + 0) - f(x_0 - 0)| - |f(x_0 + 0) - f(r_2)| - |f(r_1) - f(x_0 - 0)|$$
$$> g(x_0) - \tfrac{1}{6}g(x_0) - \tfrac{1}{6}g(x_0) = \tfrac{2}{3}g(x_0).$$

Thus, it follows that

$$\frac{2}{3}g(x_0) < |f(r_1) - f(r_2)| < \frac{4}{3}g(x_0).$$

Now consider the function $\phi : S \to \mathcal{Q} \times \mathcal{Q}$ defined by $\phi(x_0) = (r_1, r_2)$. Clearly, it suffices to show that ϕ is one-to-one. Suppose there exists an $x_1 \in S$, $x_1 \neq x_0$ with $\phi(x_1) = (r_1, r_2) = \phi(x_0)$, then either $x_1 \in (r_1, x_0)$ or $x_1 \in (x_0, r_2)$. We assume that $x_1 \in (r_1, x_0)$. The arguments for the other case are similar. Since $x \in [r_1, x_0)$ implies $|f(x) - f(x_0 - 0)| < (1/6)g(x_0)$ it follows that $|f(x_1 - 0) - f(x_0 - 0)| \leq (1/6)g(x_0)$ and $|f(x_1 + 0) - f(x_0 - 0)| \leq (1/6)g(x_0)$. Hence, we have

$$0 < g(x_1) \leq |f(x_1 + 0) - f(x_0 - 0)| + |f(x_0 - 0) - f(x_1 - 0)| \leq \frac{1}{3}g(x_0).$$

However, since $\phi(x_1) = (r_1, r_2)$,

$$\frac{2}{3}g(x_1) < |f(r_1) - f(r_2)| < \frac{4}{3}g(x_1).$$

Thus, we have

$$\frac{2}{3}g(x_0) < |f(r_1) - f(r_2)| < \frac{4}{3}g(x_1) \leq \frac{4}{9}g(x_0),$$

which is a contradiction, and hence $\phi(x_1) = \phi(x_0)$ holds only if $x_1 = x_0$.

Finally, in this chapter we shall prove the following interesting result.

Theorem 13.3. If $f : (a, b) \to \mathcal{R}$ has a discontinuity of the second kind at $x_0 \in (a, b)$, then in every nbd of x_0 the function f makes an infinite number of oscillations.

Proof. We shall show that f makes an infinite number of oscillations in the right nbd of x_0. The arguments in the left nbd are similar. Since x_0 is a discontinuity of the second kind, $\overline{f(x_0 + 0)} \neq \underline{f(x_0 + 0)}$. Now for a given $\epsilon > 0$ in any right nbd I of x_0 there are an infinite number of points at which $f(x) > \overline{f(x_0 + 0)} - \epsilon$; for if there are only a finite number of such points, we can take I so small as to exclude all these points, and then $f(x) \leq \overline{f(x_0 + 0)} - \epsilon$ in I, i.e., the upper limit of $f(x)$ at x_0 on the right is not $\overline{f(x_0 + 0)}$. Similarly, it can be shown that $f(x) < \underline{f(x_0 + 0)} + \epsilon$ at an infinite number of points in I. Thus, the function f makes an infinite number of oscillations in any nbd of x_0 on the right. The amplitude of oscillation is finite if both the limits $\overline{f(x_0+0)}$, $\underline{f(x_0+0)}$ are finite, and is infinite if one or both of these limits are infinite.

Problems

13.1. Show that the following functions have discontinuity of the first kind at $x = 0$.

(i). $f(x) = \begin{cases} \frac{x}{|x|}, & x \neq 0 \\ 1, & x = 0. \end{cases}$

(ii). $f(x) = \begin{cases} 0, & x = 0 \\ \frac{1}{1-e^{1/x}}, & x \neq 0. \end{cases}$

(iii). $f(x) = \begin{cases} 1, & x = 0 \\ \frac{2}{\pi} \lim_{n\to\infty} \tan^{-1} nx, & x \neq 0. \end{cases}$

13.2. Show that the following functions have discontinuity of the second kind at $x = 0$.

(i). $f(x) = \begin{cases} 0, & x = 0 \\ e^{-1/x}, & x \neq 0. \end{cases}$

(ii). $f(x) = \begin{cases} x, & x \leq 0 \\ \sin \frac{1}{x}, & x > 0. \end{cases}$

13.3. Show that

(i). The function f considered in Problem 11.3(i) is discontinuous at every point $x \in \mathcal{R}$.

(ii). The function f considered in Problem 11.4 is continuous only at $x = 0$.

(iii). The function f considered in Problem 11.5 is continuous at every irrational point in $(0, 1)$, but discontinuous at every rational point in $(0, 1)$.

13.4. For a given function $f : [-n, n] \to \mathcal{R}$, $n \in \mathcal{N}$ let S_n be the set of all removable and first kind discontinuities. Consider the function

$$F_n(x) = \begin{cases} \sup_{y<x} f(y) & \text{if } x \notin S_n \\ \sup_{y<x} f(y) + 1 & \text{if } x \in S_n. \end{cases}$$

Show that

(i). F_n has the same removable and first kind discontinuities as the function f.

(ii). The set S_n is countable.

(iii). An arbitrary function $f : \mathcal{R} \to \mathcal{R}$ can have a countable number of removable and first kind discontinuities.

13.5. A function f is called *piecewise continuous* (*sectionally continuous*) on an interval (a, b) if there are finitely many points $a = x_0 < x_1 < \cdots < x_n = b$ such that

(a). f is continuous on each subinterval $x_0 < x < x_1, x_1 < x < x_2, \cdots,$ $x_{n-1} < x < x_n$, and

(b). f has discontinuities of the first kind at the points $x_0, x_1, \cdots, x_n$.

The function $f(x)$ need not be defined at the points $x_0, x_1, \cdots, x_n$.

Show that the following functions are piecewise continuous.

(i). $f(x) = \begin{cases} x^2, & 0 < x \le 1 \\ -1, & 1 < x < 2 \\ x - 1, & 2 < x \le 3. \end{cases}$

(ii). $f(x) = x^2 \sin 1/x, \ 0 < x < 1.$

(iii). $f(x) = \sqrt{x} \cos(1/x^2), \ 0 < x < 1.$

(iv). $f(x) = \begin{cases} \sqrt{1 - x^2} & \text{for} \quad 0 < x < 1 \\ x & \text{for} \quad 1 < x < 2. \end{cases}$

13.6. A function $f : [a, b] \to \mathcal{R}$ is said to satisfy the *intermediate-value property* on $[a, b]$ if for every $x_1, x_2 \in [a, b]$ with $x_1 < x_2$ and for every y_0 between $f(x_1)$ and $f(x_2)$ there is a $x_0 \in (x_1, x_2)$ such that $f(x_0) = y_0$. Show that if $f : [a, b] \to \mathcal{R}$ satisfies the intermediate-value property, then

(i). f need not be continuous on $[a, b]$.

(ii). f has no removable or first kind discontinuity on $[a, b]$.

(iii). If in addition f is one-to-one, then it is continuous on $[a, b]$.

13.7. A set $S \subseteq \mathcal{R}$ is said to be of the *first category* if S is expressible as a countable union of nowhere dense sets. If a set is not of the first category, it is called of the *second category*. Show that

(i).   If each of the countable number of sets $S_n, \ n = 1, 2, \cdots$ is of the first category, then $\cup_{n=1}^{\infty} S_n$ is of the first category.

(ii). $\mathcal{R}$ is of the second category.

(iii). $\mathcal{R} \backslash \mathcal{Q}$ is of the second category.

(iv). $\mathcal{R} \backslash \mathcal{Q}$ cannot be written as $\cup_{n=1}^{\infty} S_n$, where each S_n is a closed subset of $\mathcal{R}$.

13.8. Let $f : \mathcal{R} \to \mathcal{R}$, $x_0 \in \mathcal{R}$ and $\delta > 0$. For $x \in [x_0 - \delta, x_0 + \delta]$, let $\phi_f(x_0, \delta) = \sup_x f(x) - \inf_x f(x)$. It is clear that if $0 < \delta_1 \le \delta_2$, then $\phi_f(x_0, \delta_1) \le \phi_f(x_0, \delta_2)$. The *oscillation* of f at x_0 is defined by $w_f(x_0) = \inf_{\delta > 0} \phi_f(x_0, \delta)$. It is clear that $w_f(x_0)$ is nonnegative, and bounded iff f is bounded at x_0. Show that

(i). f is discontinuous at x_0 iff $w_f(x_0) > 0$.

(ii). For each $k > 0$ the set $S_k = \{x : \omega_f(x) \geq k\}$ is closed.

(iii). The set of all points of discontinuity of f is the union of a countable number of closed sets.

(iv). There is no function $f : \mathcal{R} \to \mathcal{R}$ which is continuous at every rational point and discontinuous at every irrational point.

Answers or Hints

13.1. (i). See Example 11.2, $f(0+) = 1$, $f(-0) = -1$.

(ii). $f(0 + h) = 1/(1 - e^{1/h})$, and hence $f(+0) = \lim_{h \to 0} f(0 + h) = \lim_{h \to 0} 1/(1 - e^{1/h}) = 1/(1 - \infty) = 0$; $f(0 - h) = 1/(1 - e^{-1/h})$, and hence $f(-0) = \lim_{h \to 0} f(0 - h) = \lim_{h \to 0} 1/(1 - e^{-1/h}) = 1/(1 - 0) = 1$.

(iii). $f(+0) = \lim_{h \to 0} f(0 + h) = \lim_{h \to 0} (2/\pi) \lim_{n \to \infty} \tan^{-1} nh = (2/\pi) (\pi/2) = 1$; $f(-0) = \lim_{h \to 0} f(0 - h) = \lim_{h \to 0} (2/\pi) \lim_{n \to \infty} \tan^{-1} n(-h) = (2/\pi) \times (-\pi/2) = -1$.

13.2 (i). $f(+0) = \lim_{h \to 0} f(0 + h) = \lim_{h \to 0} e^{-1/h} = 0$; $f(-0) = \lim_{h \to 0} f(0 - h) = \lim_{h \to 0} e^{1/h} = \infty$.

(ii). $f(-0) = \lim_{h \to 0} f(0 - h) = \lim_{h \to 0}(-h) = 0$; $f(+0) = \lim_{h \to 0} f(0 + h) = \lim_{h \to 0} \sin(1/h)$ does not exist.

13.3. (i). See Problem 11.3(i).
(ii). See Problem 11.4.
(iii). See Problem 11.5.

13.4. (i). Note that F_n is monotone increasing.
(ii). Follows from Theorem 11.6 and (i).
(iii). Consider $S = \cup_{n=1}^{\infty} S_n$ and use (P3) in Chapter 5.

13.5. Verify directly.

13.6. (i). Consider the function $f(x) = \sin 1/x$ with $f(0) = 0$ defined on $[-1, 1]$. This function satisfies the intermediate-value property on $[-1, 1]$, but is discontinuous at $x = 0$.

(ii). Suppose f has a removable or first kind discontinuity at $x_0 \in [a, b]$, then either $f(x_0 + 0) \neq f(x_0)$ or/and $f(x_0 - 0) \neq f(x_0)$. Assume that $f(x_0 + 0) = \ell < f(x_0)$. The arguments for the other cases are similar. Let $\alpha = (1/2)(f(x_0) - \ell) > 0$. Now there exists a $\delta > 0$ such that if $x \in (x_0, x_0 + \delta)$ then $|f(x) - \ell| < \alpha$, i.e., $f(x) < \ell + \alpha$. On the other hand, $f(x_0) = \ell + 2\alpha > \ell + \alpha$. Fix $x_1 \in (x_0, x_0 + \delta)$. Then, $f(x_1) < \ell + \alpha$, and if we let $y \in (\ell + \alpha, f(x_0)) \subset (f(x_1), f(x_0))$, then since for each $x \in (x_0, x_1)$, $f(x) < \ell + \alpha$ there is no $x \in (x_0, x_1)$ with $f(x) = y$, i.e., f does not satisfy intermediate-value property.

(iii). From part (ii), the function f has no removable or first kind discontinuity on $[a, b]$. Now in the answer to Problem 12.13, we used only the assumption that f is one-to-one and satisfies the intermediate-value property on $[a, b]$. Thus, f is strictly monotone on $[a, b]$. Hence, from Theorem 11.6, f has no discontinuity of the second kind also.

13.7. (i). Let $S_n = \cup_{m=1}^{\infty} A_m^n$, where each A_m^n is nowhere dense. Then, $\cup_{n=1}^{\infty} S_n = \cup_{n=1}^{\infty} \cup_{m=1}^{\infty} A_m^n$.

(ii). Let $S \subseteq \mathcal{R}$ be an arbitrary set of the first category. Then, $S = \cup_{m=1}^{\infty} A_m$, where each set A_m is nowhere dense. Since A_1 is nowhere dense, there exists a closed interval I_1 of length less than 1 such that $I_1 \cap A_1 = \emptyset$. Since A_2 is nowhere dense, there is a closed interval $I_2 \subset I_1$ of length less than $1/2$ such that $A_2 \cap I_2 = \emptyset$. Continuing this process we get a sequence of closed intervals $\{I_m\}$ such that for each $m \in \mathcal{N}$, $I_{m+1} \subset I_m$, the length of I_m is less than $1/m$, and $I_m \cap A_m = \emptyset$. Thus, by the nested interval property $\cap_{m=1}^{\infty} I_m = \{x_0\}$, $x_0 \in \mathcal{R}$. Since $x_0 \in I_m$ for each $m \in \mathcal{N}$ and $I_m \cap A_m = \emptyset$, it follows that $x_0 \notin A_m$ for each m, and so $x_0 \notin S$. Hence, $S \neq \mathcal{R}$.

(iii). Since the set $\mathcal{Q}$ is nowhere dense, it is of the first category. If $\mathcal{R} \backslash \mathcal{Q}$ is also of the first category, then $\mathcal{R} = \mathcal{Q} \cup \mathcal{R} \backslash \mathcal{Q}$ must also be of the first category.

(iv). Suppose $\mathcal{R} \backslash \mathcal{Q} = \cup_{n=1}^{\infty} S_n$, where each S_n is a closed subset of $\mathcal{R}$. Since each S_n contains only irrationals, each S_n contains no nonempty open interval. Thus, from Problem 4.8, each S_n must be nowhere dense. This implies $\mathcal{R} \backslash \mathcal{Q}$ is of the first category. But, this contradicts (iii).

13.8. (i). $\omega_f(x_0)$ is the difference between the greatest and the least of $\overline{f(x_0 + 0)}$, $\underline{f(x_0 + 0)}$, $\overline{f(x_0 - 0)}$, $\underline{f(x_0 - 0)}$ and $f(x_0)$.

(ii). Suppose x_1 is a limit point of S_k. Then, for any $\delta > 0$, $(x_1 - \delta, x_1 + \delta)$ contains a point $x_0 \in S_k$. Let $\delta_1 > 0$ be so small that $[x_0 - \delta_1, x_0 + \delta_1] \subset [x_1 - \delta, x_1 + \delta]$. Now $\omega_f(x_0) \geq k$, and hence $\phi_f(x_0, \delta_1) \geq k$, but this implies that $\phi_f(x_1, \delta) \geq k$. However, since $\delta > 0$ is arbitrary, we must have $\omega_f(x_1) \geq k$, i.e., $x_1 \in S_k$.

(iii). At a point of discontinuity x, $\omega_f(x) > 0$. Thus, the set of points of discontinuity of f can be expressed as $\cup_{n=1}^{\infty} \{x : \omega_f(x) \geq 1/n\}$.

(iv). In view of (iii) it suffices to show that the set of irrational points cannot be expressed as a countable union of closed sets. But, this is precisely what we have shown in Problem 13.7(iv).

Chapter 14

Uniform and Absolute Continuities and Functions of Bounded Variation

In this chapter, we will introduce uniform and absolute continuities. Uniform continuity is stronger than the continuity, whereas absolute continuity is stronger than the uniform continuity. Absolute continuity plays an important role in Lebesgue's theory of differentiation and integration. Here, we will also discuss functions of bounded variation.

Let $f : I \to \mathcal{R}$ and $x_0 \in I$. In our definition of continuity of f at x_0 given in (D4) in Chapter 12, the number δ generally depends not only on ϵ but also on x_0, i.e., $\delta = \delta(x_0)$. If the infimum of the set of values of $\delta(x_0)$ for all points x_0 in I is different from zero, then f is said to be uniformly continuous in I, i.e., in (D4) for a given $\epsilon > 0$ the same δ works for all $x_0 \in I$. We formalize it as follows: A function $f : I \to \mathcal{R}$ is said to be *uniformly continuous* iff for a given $\epsilon > 0$ there exists a $\delta > 0$ such that for all $x_1, x_2 \in I$,

$$|x_1 - x_2| < \delta \implies |f(x_1) - f(x_2)| < \epsilon. \tag{14.1}$$

Example 14.1. The function $f(x) = x^2$ is uniformly continuous on $(-1, 2)$. Indeed for a given $\epsilon > 0$ choose $\delta = \epsilon/4$ so that $x_1, x_2 \in (-1, 2)$ and $|x_1 - x_2| < \delta$ imply $|x_1^2 - x_2^2| = |x_1 - x_2||x_1 + x_2| < 4|x_1 - x_2| < 4\delta = \epsilon$.

It is clear that if f is uniformly continuous on I, then it is continuous on I. Indeed, if we let $x_1 = x \in I$, and $x_2 = x_0 \in I$ fixed but arbitrary, then (14.1) is the same as (12.1). However, the following examples demonstrate that the converse is not true.

Example 14.2. The function $f(x) = 1/x$ is continuous on $(0, 1)$, but not uniformly continuous. To show this, let $\delta > 0$ be arbitrary. Then, there exists an $m \in \mathcal{N}$ such that $1/n < \delta$ for all $n \geq m$. Let $x_1 = 1/m$, $x_2 = 1/2m$, so that $x_1, x_2 \in (0, 1)$ and $|x_1 - x_2| = 1/2m < \delta$, but $|f(x_1) - f(x_2)| = m$ cannot be smaller than every $\epsilon > 0$.

Example 14.3. The function $f(x) = x^2$ is continuous on $\mathcal{R}$, but not uniformly continuous. To show this, let $\delta > 0$ be arbitrary. Then, there exists an $m \in \mathcal{N}$ such that $1/m < \delta$. Let $x_1 = m$, $x_2 = m + (1/m)$, so that $x_1, x_2 \in \mathcal{R}$

and $|x_1 - x_2| = 1/m < \delta$, but $|f(x_1) - f(x_2)| \geq 2$ cannot be smaller than every $\epsilon > 0$.

In the following results we shall provide sufficient conditions so that the continuity of f on I implies its uniform continuity on I.

Theorem 14.1. If f is continuous on a closed bounded interval $I = [a, b]$, then it is uniformly continuous on I.

Proof. Assume that f is not uniformly continuous on I, then for $\epsilon > 0$ there exists a pair of sequences $\{x_n\}$, $\{y_n\}$ in I such that $|x_n - y_n| < 1/n \Rightarrow |f(x_n) - f(y_n)| \geq \epsilon$. By Theorem 8.3, there exists a subsequence $\{x_{n_k}\}$ of $\{x_n\}$ which converges to, say, $x_0 \in I$. Since f is continuous, we have $\lim_{k\to\infty} f(x_{n_k}) = f(x_0)$. Further, since

$$|y_{n_k} - x_0| \leq |y_{n_k} - x_{n_k}| + |x_{n_k} - x_0| \leq \frac{1}{n_k} + |x_{n_k} - x_0|$$

the subsequence $\{y_{n_k}\}$ of $\{y_n\}$ also converges to x_0, and therefore $\lim_{k\to\infty} f(y_{n_k}) = f(x_0)$. Hence, $\lim_{k\to\infty} |f(x_{n_k}) - f(y_{n_k})| = 0$, which is a contradiction to $|f(x_{n_k}) - f(y_{n_k})| \geq \epsilon$.

Theorem 14.2. If f is continuous on a bounded interval $I = (a, b)$, then it is uniformly continuous on I iff $f(a + 0)$ and $f(b - 0)$ exist.

Proof. If $f(a+0)$ and $f(b-0)$ exist, we define the function $g : [a, b] \to \mathcal{R}$ as follows

$$g(x) = \begin{cases} f(a + 0) & \text{if } x = a \\ f(x) & \text{if } a < x < b \\ f(b - 0) & \text{if } x = b. \end{cases}$$

Clearly, g is continuous on the closed bounded interval $[a, b]$, and hence from Theorem 14.1 it is uniformly continuous on $[a, b]$. This in turn implies that g, and therefore, f is uniformly continuous on (a, b).

Conversely, suppose that $f(a+0)$ does not exist. Then, there is a sequence $\{x_n\}$ in (a, b) such that $x_n \to a$, but the sequence $\{f(x_n)\}$ does not converge, and hence it is not a Cauchy sequence. Thus, there exists an $\epsilon > 0$ such that for all large $i, j \in \mathcal{N}$, $|f(x_i) - f(x_j)| \geq \epsilon$. Now, since $\{x_n\}$ is a Cauchy sequence, $\lim_{i,j\to\infty} |x_i - x_j| = 0$, and therefore we can find a pair of points $x_i, x_j \in (a, b)$ which are arbitrarily close, but $|f(x_i) - f(x_j)| \geq \epsilon$. However, this contradicts the defintion of uniform continuity of f on (a, b). The arguments for the case when $f(b - 0)$ fails to exist is similar.

Example 14.4. Since $\lim_{x\to a^+} x^2 = a^2$, $\lim_{x\to b^-} x^2 = b^2$, the function $f(x) = x^2$ is uniformly continuous on every bounded interval (a, b). However, since $\lim_{x\to 0^+} 1/x$ does not exist the function $f(x) = 1/x$ is not uniformly continuous on $(0, 1)$.

Theorem 14.3. If f is continuous on $I = [a, \infty)$ and $\lim_{x\to\infty} f(x)$ exists, then it is uniformly continuous on I.

Proof. Let $\epsilon > 0$ be given. If $\lim_{x\to\infty} f(x) = \ell$, then there exists an $M > a$ such that $x > M \Rightarrow |f(x) - \ell| < \epsilon/2$. Since f is continuous at $x = M$, there exists a $\delta_1 > 0$ such that $|x - M| < \delta_1 \Rightarrow |f(x) - f(M)| < \epsilon/2$. Now since in view of Theorem 14.1, f is uniformly continuous on $[a, M]$, there exists a $\delta_2 > 0$ such that $x_1, x_2 \in [a, M]$ with $|x_1 - x_2| < \delta_2 \Rightarrow |f(x_1) - f(x_2)| < \epsilon$. Let $\delta = \min\{\delta_1, \delta_2\}$. Suppose that $y, z \in [a, \infty)$ be such that $|y - z| < \delta$. To show $|f(y) - f(z)| < \epsilon$, we consider the following three cases:

If $y, z > M$, then we have $|f(y) - f(z)| \le |f(y) - \ell| + |\ell - f(z)| < (\epsilon/2) + (\epsilon/2) = \epsilon$.

If $y, z \le M$, then since $|y - z| < \delta \le \delta_2$, we have $|f(y) - f(z)| < \epsilon$.

If $y \le M$ and $z > M$, then since $|y - z| < \delta \Rightarrow |y - M| < \delta \le \delta_1$ and $|z - M| < \delta \le \delta_1$, we have $|f(y) - f(z)| \le |f(y) - f(M)| + |f(M) - f(z)| < (\epsilon/2) + (\epsilon/2) = \epsilon$.

The converse of Theorem 14.3 does not hold. For example, the function $f(x) = x$ is uniformly continuous on $[0, \infty)$, but $\lim_{x\to\infty} f(x)$ does not exist. We also note that in Theorem 14.3 the interval $[a, \infty)$ cannot be replaced by (a, ∞). For example, the function $f(x) = 1/x$ is continuous on $(0, \infty)$ and $\lim_{x\to\infty} f(x) = 0$, but f is not uniformly continuous on $(0, \infty)$.

Theorem 14.4. If f is continuous on $I = (-\infty, b]$ and $\lim_{x\to-\infty} f(x)$ exists, then it is uniformly continuous on I.

Proof. The proof is similar to that of Theorem 14.3.

Combining Theorems 14.3 and 14.4, we have the following result.

Theorem 14.5. If f is continuous on $\mathcal{R}$ and $\lim_{x\to-\infty} f(x)$, $\lim_{x\to\infty} f(x)$ exist, then it is uniformly continuous on $\mathcal{R}$.

A collection of points $P = \{x_0, x_1, \cdots, x_n\}$ satisfying $a = x_0 < x_1 < \cdots < x_n = b$ is called a partition of the interval $[a, b]$. A function $f : [a, b] \to \mathcal{R}$ is said to be of *bounded variation* if the total variation $V_f(a, b)$ of f defined by

$$V_f(a, b) = \sup \left\{ \sum_{k=1}^{n} |f(x_k) - f(x_{k-1})| : P \text{ is a partition of } [a, b] \text{ with } n \in \mathcal{N} \right\}$$

is finite. The following example suggests that a continuous function need not be of bounded variation.

Example 14.5. For the continuous function

$$f(x) = \begin{cases} x \sin(1/x), & 0 < x \le 2/\pi \\ 0, & x = 0 \end{cases}$$

we choose the partition

$$x_0 = 0, x_1 = \frac{2}{(2n+1)\pi}, x_2 = \frac{2}{2n\pi}, \cdots, x_{2n} = \frac{2}{2\pi}, x_{2n+1} = \frac{2}{\pi}.$$

Then, we have

$$\begin{aligned}
\sum_{k=1}^{2n+1} |f(x_k) - f(x_{k-1})| &= \sum_{k=1}^{2n+1} |x_k \sin x_k - x_{k-1} \sin x_{k-1}| \\
&= \sum_{k=1}^{n} |x_{2k-1} \sin x_{2k-1}| + |x_{2n+1} \sin x_{2n+1}| \\
&= \frac{4}{\pi} \sum_{k=1}^{n} \frac{1}{2n-2k+3} + \frac{2}{\pi},
\end{aligned}$$

which diverges to ∞.

The following result provides an easily verifiable condition so that a given function is of bounded variation.

Theorem 14.6. If $f : [a, b] \to \mathcal{R}$ is monotone, then f is of bounded variation. However, the converse is not true.

Proof. Assume that $a = x_0 < x_1 < x_2 < \cdots < x_n = b$ is a partition, and f is increasing, then it follows that

$$\sum_{k=1}^{n} |f(x_k) - f(x_{k-1})| = \sum_{k=1}^{n} [f(x_k) - f(x_{k-1})] = f(b) - f(a).$$

For converse, note that the function $f(x) = x^2$, $x \in [-1, 1]$ is of bounded variation, but it is not monotone.

Theorem 14.7. If f is of bounded variation on both $[a, b]$ and $[b, c]$, then f is of bounded variation on $[a, c]$ and $V_f(a, c) = V_f(a, b) + V_f(b, c)$.

Proof. Let $a = x_0 < x_1 < \cdots < x_n = c$ be a partition of $[a, c]$. Suppose $b = x_r$ for some $r \in \{1, 2, \cdots, n-1\}$. Then, we have

$$\sum_{k=1}^{n} |f(x_k) - f(x_{k-1})| = \sum_{k=1}^{r} |f(x_k) - f(x_{k-1})| + \sum_{k=r+1}^{n} |f(x_k) - f(x_{k-1})|$$
$$\leq V_f(a, b) + V_f(b, c).$$

Suppose $b \neq x_r$ for $r \in \{1, 2, \cdots, n-1\}$, i.e., $x_{s-1} < b < x_s$ for some $s \in \{1, 2, \cdots, n\}$. Let

$$y_k = \begin{cases} x_k, & 0 \leq k \leq s-1 \\ b, & k = s \\ x_{k-1}, & s+1 \leq k \leq n+1 \end{cases}$$

and note that from above, we have

$$\sum_{k=1}^{n} |f(x_k) - f(x_{k-1})| \leq \sum_{k=1}^{n+1} |f(y_k) - f(y_{k-1})| \leq V_f(a, b) + V_f(b, c).$$

Thus, it follows that

$$V_f(a,c) \leq V_f(a,b) + V_f(b,c). \tag{14.2}$$

Now let $\epsilon > 0$ be given. Then, there exists partitions $a = x_0' < x_1' < \cdots < x_p' = b$ of $[a,b]$ and $b = y_0' < y_1' < \cdots < y_q' = c$ of $[b,c]$ such that

$$\sum_{i=1}^{p} |f(x_i') - f(x_{i-1}')| > V_f(a,b) - \frac{\epsilon}{2}$$

and

$$\sum_{j=1}^{q} |f(y_j') - f(y_{j-1}')| > V_f(b,c) - \frac{\epsilon}{2}.$$

Combining all points x_i', y_j' we get a partition of $[a,c]$ (we call these points z_k for simplicity), so that

$$\sum_{k=1}^{p+q} |f(z_k) - f(z_{k-1})| = \sum_{i=1}^{p} |f(x_i') - f(x_{i-1}')| + \sum_{j=1}^{q} |f(y_j') - f(y_{j-1}')|$$
$$> V_f(a,b) + V_f(b,c) - \epsilon.$$

Since $\epsilon > 0$ is arbitrary from the above inequality it follows that

$$V_f(a,c) \geq V_f(a,b) + V_f(b,c). \tag{14.3}$$

The required result now follows from (14.2) and (14.3).

Our next result provides necessary and sufficient conditions for a function to be of bounded variation.

Theorem 14.8 (Jordan Decomposition Theorem). A function $f : [a,b] \to \mathcal{R}$ is of bounded variation iff f can be written as the difference of two nondecreasing functions.

Proof. Suppose $f(x) = f_1(x) - f_2(x)$, $x \in [a,b]$ where f_1 and f_2 are nondecreasing functions, then from Theorem 14.6 both of these functions are of bounded variation, and now from Problem 14.11 it follows that the function f is of bounded variation.

Conversely, suppose that f is of bounded variation. Let $f_1(x) = V_f(a,x)$. Clearly, Theorem 14.7 guarantees that f_1 is nondecreasing. Let $f_2(x) = f_1(x) - f(x)$ (so that $f(x) = f_1(x) - f_2(x)$). We need to show that f_2 is nondecreasing. For this, let $a \leq x < y \leq b$. Now from Theorem 14.7, we have

$$f_1(y) - f_1(x) = V_f(a,y) - V_f(a,x) = V_f(x,y)$$
$$\geq |f(y) - f(x)| \geq f(y) - f(x).$$

Thus, it follows that

$$
\begin{aligned}
f_2(y) \;=\; f_1(y) - f(y) \;&=\; [f_1(y) - f_1(x)] + [f_1(x) - f(y)] \\
&\geq\; [f(y) - f(x)] + [f_1(x) - f(y)] \\
&=\; f_1(x) - f(x) \;=\; f_2(x).
\end{aligned}
$$

A function $f : [a,b] \to \mathcal{R}$ is said to be *absolutely continuous* iff for any given $\epsilon > 0$ there exists a $\delta > 0$ such that $\sum_{k=1}^{n} |f(b_k) - f(a_k)| < \epsilon$ for every n nonoverlapping intervals (a_k, b_k) in $[a,b]$ with $\sum_{k=1}^{n} (b_k - a_k) < \delta$. It is clear that for $n = 1$ absolute continuity is the same as the uniform continuity. We also remark that Lipschitz continuity implies absolute continuity, however, the converse is not true. For this, we note that the function $f(x) = \sqrt{x}$, $x \in [0,1]$ is absolutely continuous, but not Lipschitz. Finally, in this chapter we shall prove the following result.

Theorem 14.9. If f is absolutely continuous on $[a,b]$, then f is of bounded variation on $[a,b]$.

Proof. Since f is absolutely continuous, given $\epsilon > 0$ there exists $\delta > 0$ such that $\sum_{k=1}^{n} |f(b_k) - f(a_k)| < \epsilon$ for nonoverlapping intervals (a_k, b_k) in $[a,b]$ with $\sum_{k=1}^{n} (b_k - a_k) < \delta$. Thus, if $[c,d]$ is any interval of length less than δ then $V_f(c,d) \leq \epsilon$. Choose $m \in \mathcal{N}$ such that $(b-a)/m < \delta$ and let $x_i = a + i(b-a)/m$ for $0 \leq i \leq m$. Then, from Theorem 14.7, we have

$$
V_f(a,b) \;=\; \sum_{i=1}^{m} V_f(x_{i-1}, x_i) \;\leq\; \sum_{i=1}^{m} \epsilon \;=\; m\epsilon.
$$

Remark 14.1. There are functions which are continuous but not absolutely continuous. For this, Cantor's function $c : [0,1] \to [0,1]$ is one of the well-known examples. This function is constructed as follows: Express x in base 3, if x contains a 1, replace every digit after the first 1 by 0, replace all 2's with 1's, and interpret the result as a binary number. The resulting function $c(x)$ is the Cantor function. This function $c(x)$ is also a classical example which is of bounded variation but not absolutely continuous.

Problems

14.1. Show that the following functions are continuous, but not uniformly continuous on the given interval.

(i). $f(x) = \sin 1/x$, $x \in (0,1)$.

(ii). $f(x) = \cos 1/x$, $x \in (0,1]$.

14.2. Show that Lipschitz continuity implies uniform continuity. In particular, show that functions $\sin x$ and $\cos x$ are uniformly continuous on $\mathcal{R}$.

14.3. Let f and g be uniformly continuous functions on an interval I. Show that

(i). $f + g$ and $f - g$ are uniformly continuous on I.

(ii). fg is uniformly continuous on I provided I is compact.

(iii). f/g is uniformly continuous on I provided I is compact and g has no zeros in I.

(iv). Give examples to show that the conclusion of (ii) and (iii) may fail if I is not compact.

14.4. Suppose that f is uniformly continuous on an interval I, g is uniformly continuous on an interval J, and $g(x) \in I$ for every $x \in J$. Show that $f \circ g$ is uniformly continuous on J.

14.5. Let f be uniformly continuous on (a, b). Show that f is bounded on (a, b).

14.6. Let f be monotone increasing, continuous, and bounded on an interval I. Show that f is uniformly continuous on I.

14.7. Let f be continuous and periodic on $\mathcal{R}$. Show that f is uniformly continuous on $\mathcal{R}$.

14.8. (i). Show that if f is uniformly continuous on an interval I and $\{x_n\} \subset (a, b)$ is a Cauchy sequence, then $\{f(x_n)\}$ is a Cauchy sequence.

(ii). Show that if f is continuous but not uniformly continuous on (a, b), then there is a Cauchy sequence $\{x_n\} \subset (a, b)$ for which $\{f(x_n)\}$ is not a Cauchy sequence.

14.9. Show that if $f : [a, b] \to \mathcal{R}$ is of bounded variation, then it is bounded. However, the converse is not true.

14.10. Show that if a function f is Lipschitz continuous, then f is of bounded variation, but the converse is not true.

14.11. Let f and g be of bounded variation on $[a, b]$. Show that $f + g$, $f - g$, fg, and f/g provided g is bounded away from zero, are also of bounded variation on $[a, b]$.

14.12. Let f and g be absolutely continuous on $[a, b]$. Show that $f + g$, $f - g$, fg, and f/g provided g is bounded away from zero, are also absolutely continuous on $[a, b]$.

Answers or Hints

14.1. (i). Let m be an odd positive integer, $x_1 = 2/m\pi, x_2 = 2/3m\pi$ so that $|x_1 - x_2| = (4/3m\pi) < \delta$, but $|\sin\frac{n\pi}{2} - \sin\frac{3n\pi}{2}| = 2$.
(ii). $|f(1/n\pi) - f(1/(n+1)\pi)| = 2, \ n = 1, 2, \cdots$.

14.2. See Problem 12.8.

14.3. (i). Use the definition.
(ii). Use the definition and the fact that uniform continuity implies continuity and I is compact, so that f and g are bounded.
(iii). There exists numbers $M, m > 0$ such that $\sup_{x \in I} |f(x)| < M$, $\sup_{x \in I} |g(x)| < M$, and $\inf_{x \in I} |g(x)| > m$ for all $x \in I$. For a given $\epsilon > 0$ there exists $\delta_1, \delta_2 > 0$ such that for all $x_1, x_2 \in I$, we have $|x_1 - x_2| < \delta_1 \Rightarrow$ $|f(x_1) - f(x_2)| < \epsilon m^2/2M$ and $|x_1 - x_2| < \delta_2 \Rightarrow |g(x_1) - g(x_2)| < \epsilon m^2/2M$. Let $\delta = \min\{\delta_1, \delta_2\}$. For $|x_1 - x_2| < \delta$, it follows that

$$\left|\frac{f}{g}(x_1) - \frac{f}{g}(x_2)\right| \leq \frac{|g(x_2)|}{|g(x_1)g(x_2)|}|f(x_1) - f(x_2)| + \frac{|f(x_2)|}{|g(x_1)g(x_2)|}|g(x_2) - g(x_1)|$$

$$< \frac{M}{m^2}\frac{\epsilon m^2}{2M} + \frac{M}{m^2}\frac{\epsilon m^2}{2M} = \epsilon.$$

(iv). For (ii), notice that $f(x) = g(x) = x, \ x \in \mathcal{R}$ are uniformly continuous, but $(fg)(x) = x^2$ is not uniformly continuous on $\mathcal{R}$. Similarly, the functions $f(x) = x, g(x) = \sin x$ are uniformly continuous on $\mathcal{R}$, but $(fg)(x) = x\sin x$ is not uniformly continuous on $\mathcal{R}$. For (iv) functions $f(x) = 1, g(x) = x$ are uniformly continuous on $(0,1)$, but the function $(f/g)(x) = 1/x$ is not uniformly continuous on $(0,1)$.

14.4. Since g is uniformly continuous on J it follows that for each $\epsilon > 0$ there exists a $\delta_1 > 0$ such that for all $x_1, x_2 \in J$, we have $|x_1 - x_2| < \delta_1 \Rightarrow$ $|g(x_1) - g(x_2)| < \epsilon$. Now, since f is uniformly continuous on I it follows that for each $\epsilon > 0$ there is $\delta_2 > 0$ such that for all $y_1, y_2 \in I$, we have $|y_1 - y_2| < \delta_1 \Rightarrow |f(y_1) - f(y_2)| < \epsilon$. Thus, for all $x_1, x_2 \in J$, with $|x_1 - x_2| < \delta_1$, we have $|g(x_1) - g(x_2)| < \epsilon$, and for $\delta_2 = \epsilon$, it results $|f(g(x_1)) - f(g(x_2))| < \epsilon$, since $g(x_1), g(x_2) \in I$.

14.5. Suppose that f is not bounded above. We fix an element $x_0 \in (a, b)$. It follows that there is a point $x_1 \in (a, b)$ such that $f(x_0) + \epsilon < f(x_1)$. From the same reason, there is an element $x_2 \in (a, b)$ such that $f(x_1) + \epsilon < f(x_2)$, and so on. This way we construct a sequence $\{x_i\}$ such that for $i \neq j$ we have $|f(x_i) - f(x_j)| > \epsilon$. If $\delta = \max\{|x_i - x_j| : i \neq j\}$, then for $\epsilon > 0$ there exists a $\delta > 0$ and $x_i \neq x_j$ such that $|f(x_i) - f(x_j)| > \epsilon$, which contradicts the definition of uniformly continuity of f.

14.6. If $I = [a, b]$ the result follows from Theorem 14.1. If I is an arbitrary interval and f is a monotone increasing bounded function, then there are $a, b \in I$ such that $m = f(a)$ and $M = f(b)$ such that $m \leq f(x) \leq M$, for all $x \in I$. Let $g : [a, b] \to \mathcal{R}, g(x) = f(x)$, then from Theorem 14.1 the function g is uniformly continuous on $[a, b]$, but this implies that f is uniformly continuous on I.

14.7. Let f be periodic of period π, so that $f(x + \pi) = f(x)$. Since f is continuous on the closed interval $[0, \pi]$ it is uniformly continuous on $[0, \pi]$. Thus, for a given $\epsilon > 0$ there exists a $\delta > 0$ such that for all $x_1, x_2 \in [0, \pi]$, we have $|x_1 - x_2| < \delta \Rightarrow |f(x_1) - f(x_2)| < \epsilon$. If x_1, x_2 are real numbers such that $|x_1 - x_2| < \delta$, we can find an integer k such that $x_1 + k\pi$ and $x_2 + k\pi$ are in $[0, \pi]$ and so we have $|(x_1 + k\pi) - (x_2 + k\pi)| = |x_1 - x_2| < \delta$ implies $|f(x_1 + k\pi) - f(x_2 + k\pi)| = |f(x_1) - f(x_2)| < \epsilon$. Therefore, for a given $\epsilon > 0$ there exists a $\delta > 0$ such that for all $x_1, x_2 \in \mathcal{R}$, we have $|x_1 - x_2| < \delta \Rightarrow |f(x_1) - f(x_2)| < \epsilon$.

14.8. (i). Let ϵ, δ be as in (14.1). Since $\{x_n\}$ is Cauchy, choose $N \in \mathcal{N}$ such that $n, m \geq N$ implies $|x_n - x_m| < \epsilon$. Then, $n, m \geq N$ implies $|f(x_n) - f(x_m)| < \epsilon$.

(ii). Suppose that for each Cauchy sequence $\{x_n\} \subset (a, b)$ the sequence $\{f(x_n)\}$ is Cauchy. Then, for a given $\epsilon > 0$ there exists an N such that for $m, n \geq N$, we have $|x_n - x_m| < \epsilon$. But this implies that $|f(x_n) - f(x_m)| < \epsilon$, which contradicts the definition. For example, let $f(x) = 1/x$, $x \in (0, 1)$. This function is continuous on $(0, 1)$, but it is not uniformly continuous. Let $x_n = 1/n$ be a Cauchy sequence and let $\epsilon, \delta > 0$ be as in the definition of uniformly continuous. Since f is not uniformly continuous, there exists m, n natural numbers such that for $|1/m - 1/n| < \delta$, we have $|f(1/m) - f(1/n)| = |m - n| > \epsilon$, which shows that $\{f(x_n)\}$ is not a Cauchy sequence.

14.9. Assume that $a \leq x \leq b$ is a partition, then since f is of bounded variation, $|f(x) - f(a)| + |f(b) - f(x)| \leq M$, which implies $|f(x)| \leq M + |f(a)|$. For the converse, consider the bounded function $f(x) = \begin{cases} \sin(1/x), & x \neq 0 \\ 0, & x = 0. \end{cases}$

For $x_k = \begin{cases} 0, & k = 0 \\ 2/[(n - k)\pi], & 0 < k < n, \end{cases}$ we have $\sum_{k=1}^{n} |f(x_k) - f(x_{k-1})| = 2n \to \infty$.

14.10. Since $f : [a, b] \to \mathcal{R}$ is Lipschitz continuous, there exists a constant $C > 0$ such that $|f(x) - f(y)| \leq C|x - y|$ for all x, y. Therefore, $V_f(a, b)$ is finite. For the converse, we note that the function $f(x) = \sqrt{x}, x \in [0, 1]$ is not Lipschitz continuous, but $V_f(0, 1)$ is finite ($\sqrt{x}$ is bounded and monotone).

14.11. We shall show that fg is of bounded variation. The proof of other cases is similar. From Problem 14.9, there exists constants P and Q such that $|f(x)| \leq P$ and $|g(x)| \leq Q$ for $x \in [a, b]$. Thus, we have
$$|f(x_k)g(x_k) - f(x_{k-1})g(x_{k-1})|$$
$$= |f(x_k)g(x_k) - f(x_{k-1})g(x_k) + f(x_{k-1})g(x_k) - f(x_{k-1})g(x_{k-1})|$$
$$\leq |g(x_k)||f(x_k) - f(x_{k-1})| + |f(x_{k-1})||g(x_k) - g(x_{k-1})|$$
$$\leq Q|f(x_k) - f(x_{k-1})| + P|g(x_k) - g(x_{k-1})|$$

and hence
$$\sum_{k=1}^{n} |f(x_k)g(x_k) - f(x_{k-1})g(x_{k-1})|$$
$$\leq Q \sum_{k=1}^{n} |f(x_k) - f(x_{k-1})| + P \sum_{k=1}^{n} |g(x_k) - g(x_{k-1})|$$
$$\leq QV_f(a,b) + PV_g(a,b),$$
since f and g are of bounded variation.

14.12. Similar to that of Problems 14.3 and 14.11.

Chapter 15

Differentiable Functions

In this chapter, we will address differentiability of functions, which is a stronger concept than the continuity. For differentiable functions we will prove some major results which have a wide range of applications.

A function f, defined on some nbd of a point x_0, is said to be *differentiable* at x_0 if

$$\lim_{x \to x_0} \frac{f(x) - f(x_0)}{x - x_0}$$

exists, and this limit is called the *derivative* of f at x_0. In symbols this derivative is denoted as $f'(x_0)$, $Df(x_0)$, or $df(x_0)/dx$. Thus, we have

$$f'(x_0) = \lim_{x \to x_0} \frac{f(x) - f(x_0)}{x - x_0} = \lim_{h \to 0} \frac{f(x_0 + h) - f(x_0)}{h}.$$

It follows that f is differentiable at x_0 iff for every sequence $\{x_n\}$ with $x_n \neq x_0$ converging to x_0, the sequence $\{(f(x_n) - f(x_0))/(x_n - x_0)\}$ converges.

Graphically, $f'(x_0)$ means the gradient of the curve $y = f(x)$ at the point $(x_0, f(x_0))$. Quantitatively, $f'(x_0)$ represents the rate of change of the function f at x_0.

A function f is said to be differentiable from the left or right at a point x_0 according as

$$\lim_{h \to 0^-} \frac{f(x_0 + h) - f(x_0)}{h} \quad \text{or} \quad \lim_{h \to 0^+} \frac{f(x_0 + h) - f(x_0)}{h}$$

exists. In symbols, these limits are denoted by $f'(x_0^-)$ and $f'(x_0^+)$ and are called derivatives of f from the left and from the right at x_0. It is clear that $f'(x_0)$ exists iff $f'(x_0^-) = f'(x_0^+) = f'(x_0)$. The function f is said to be differentiable on an interval I, if it is differentiable at every interior point of I, and right and left differentiable at left and right endpoints of I provided these points belong to I.

Example 15.1. Consider the continuous function $f(x) = x^2 \sin(1/x)$, $x \neq 0$, $f(0) = 0$. Clearly, this function is differentiable at all $x \neq 0$, and

$$f'(x) = -\cos \frac{1}{x} + 2x \sin \frac{1}{x}, \quad x \neq 0.$$

Further, since
$$\lim_{h \to 0} \frac{f(h) - f(0)}{h} = \lim_{h \to 0} h \sin \frac{1}{h} = 0$$
it is also differentiable at $x = 0$, and $f'(0) = 0$. Clearly, f' is not continuous at $x = 0$.

Example 15.2. Since for the continuous function $f(x) = x \sin(1/x)$, $x \neq 0$, $f(0) = 0$,
$$\lim_{h \to 0} \frac{f(h) - f(0)}{h} = \lim_{h \to 0} \sin \frac{1}{h}$$
does not exist, it is not differentiable at $x = 0$. Similarly, since for the continuous function $g(x) = |x|$,
$$\lim_{h \to 0^-} \frac{g(h) - g(0)}{h} = \frac{|h|}{h} = \frac{-h}{h} = -1$$
and
$$\lim_{h \to 0^+} \frac{g(h) - g(0)}{h} = \frac{|h|}{h} = \frac{h}{h} = 1$$
it is not differentiable at $x = 0$.

From Examples 15.1 and 15.2 it is clear that continuity at a point of a function need not imply its differentiability. However, the converse is always true, i.e., differentiability is a stronger concept than that of continuity. This we shall prove in the following result.

Theorem 15.1. If f is differentiable at x_0, then f is continuous at x_0.

Proof. Since f is differentiable at x_0, we have
$$\lim_{x \to x_0} [f(x) - f(x_0)] = \lim_{x \to x_0} \frac{f(x) - f(x_0)}{x - x_0} (x - x_0)$$
$$= \lim_{x \to x_0} \frac{f(x) - f(x_0)}{x - x_0} \lim_{x \to x_0} (x - x_0)$$
$$= f'(x_0) \cdot 0 = 0,$$
i.e., $\lim_{x \to x_0} [f(x) - f(x_0)] = 0$, or $\lim_{x \to x_0} f(x) = f(x_0)$. Thus, f is continuous at x_0.

It is clear that if f is differentiable from the left or right at a point x_0, then accordingly it is continuous from the left or right at x_0.

Now we state the following familiar results.

Theorem 15.2. If f and g are differentiable at x_0, then $f \pm g$, $f \cdot g$ are differentiable at x_0, and f/g is differentiable at x_0 provided $g(x_0) \neq 0$, and
$$(f \pm g)'(x_0) = f'(x_0) \pm g'(x_0)$$
$$(fg)'(x_0) = f'(x_0)g(x_0) + f(x_0)g'(x_0)$$
$$\left(\frac{f}{g}\right)'(x_0) = \frac{f'(x_0)g(x_0) - f(x_0)g'(x_0)}{(g(x_0))^2}.$$

Proof. The proof is based on forming appropriate difference quotients. For example,

$$\left(\frac{f}{g}\right)'(x_0) = \lim_{x \to x_0} \frac{\left(\frac{f}{g}\right)(x) - \left(\frac{f}{g}\right)(x_0)}{(x - x_0)}$$

$$= \lim_{x \to x_0} \frac{f(x)g(x_0) - f(x_0)g(x_0) + f(x_0)g(x_0) - f(x_0)g(x)}{(x - x_0)g(x)g(x_0)}$$

$$= \lim_{x \to x_0} \frac{1}{g(x)} \lim_{x \to x_0} \frac{f(x) - f(x_0)}{(x - x_0)}$$
$$- \frac{f(x_0)}{g(x_0)} \lim_{x \to x_0} \frac{1}{g(x)} \lim_{x \to x_0} \frac{g(x) - g(x_0)}{(x - x_0)}$$

$$= \frac{f'(x_0)g(x_0) - f(x_0)g'(x_0)}{(g(x_0))^2}.$$

Theorem 15.3 (Chain Rule). If g is differentiable at x_0 and f is differentiable at $g(x_0)$, then $f \circ g(x) = f(g(x))$ is differentiable at x_0, and

$$(f \circ g)'(x_0) = f'(g(x_0))g'(x_0).$$

Proof. Since f is differentiable at $g(x_0)$, Problem 15.1 gives

$$f(t) - f(g(x_0)) = [f'(g(x_0)) + E(t)][t - g(x_0)],$$

where $\lim_{t \to g(x_0)} E(t) = E(g(x_0)) = 0$. Let $t = g(x)$, to obtain

$$\frac{f(g(x)) - f(g(x_0))}{x - x_0} = [f'(g(x_0)) + E(g(x))]\frac{g(x) - g(x_0)}{x - x_0}.$$

Since g is continuous at x_0 it follows that $\lim_{x \to x_0} E(g(x)) = E(g(x_0)) = 0$. Thus, we have

$$(f \circ g)'(x_0) = \lim_{x \to x_0} \frac{f(g(x)) - f(g(x_0))}{x - x_0} = f'(g(x_0))g'(x_0).$$

In particular, if f is differentiable at x_0 and is one-to-one on some nbd of x_0, then the inverse of f is differentiable at $f(x_0)$ and $(f^{-1})'(f(x_0)) = 1/f'(x_0)$ provided $f'(x_0) \neq 0$.

In the following theorems we shall prove several important properties of differentiable functions.

Theorem 15.4 (Rolle's Theorem). If f is continuous on a closed bounded interval $[a, b]$, differentiable on (a, b) and $f(a) = f(b)$, then there exists a point $x_0 \in (a, b)$ such that $f'(x_0) = 0$.

Proof. Since f is continuous on $[a, b]$, it attains its supremum M and infimum m on $[a, b]$. If $M = m$, then f is a constant function, and so $f'(x_0) = 0$ for all $x_0 \in [a, b]$. If $M \neq m$, then $f(a)$, $f(b)$ differ from at least one of M and m. If $M \neq f(a)$, then there exists an $x_0 \in (a, b)$ such that $f(x_0) = M$.

We shall show that $f'(x_0) = 0$. Suppose $f'(x_0) > 0$. Then, there exists $\delta_1 > 0$ such that

$$0 < h < \delta_1 \;\Rightarrow\; \left| \frac{f(x_0+h) - f(x_0)}{h} - f'(x_0) \right| < f'(x_0) \;\Rightarrow\; f(x_0) < f(x_0 + h).$$

But, this contradicts our assumption that $M = f(x_0)$. The argument for the case $f'(x_0) < 0$ is similar. The case $m \neq f(a)$ can be treated analogously.

Geometrically, Rolle's theorem demonstrates that, if the smooth curve $y = f(x)$, $x \in [a, b]$ satisfies $f(a) = f(b)$, then at some point on the curve the tangent will be parallel to the x-axis. We emphasize that in this theorem we need f' to exist on the open interval (a, b) and not on the closed interval $[a, b]$. This enlarges the class of functions to which this result can be applied. For example, the function $f(x) = \sqrt{1 - x^2}$ is continuous on $[-1, 1]$, differentiable on $(-1, 1)$ but has no derivatives at $x = \pm 1$, and $f(-1) = f(1) = 0$. For this function, clearly $f'(x) = -x/\sqrt{1 - x^2}$ vanishes at $x = 0$. We also observe that if f' fails to exist even at a single point in (a, b), then Rolle's theorem may not hold. For example, the function $g(x) = 1 - |x|$ is continuous on $[-1, 1]$, differentiable on $(-1, 0) \cup (0, 1)$ but has no derivatives at $x = 0$, $x = \pm 1$, and $g(-1) = g(1) = 0$. For this function there is no x_0 in $(-1, 1)$ for which $g'(x_0) = 0$.

Theorem 15.5 (Mean-Value Theorem). If f is continuous on a closed bounded interval $[a, b]$, differentiable on (a, b), then there exists a point $x_0 \in (a, b)$ such that

$$f'(x_0) = \frac{f(b) - f(a)}{b - a}. \tag{15.1}$$

Proof. We define an auxiliary function $g : [a, b] \to \mathcal{R}$ as

$$g(x) = f(x) - f(a) - \frac{f(b) - f(a)}{b - a}(x - a).$$

Clearly, g is continuous on $[a, b]$, differentiable on (a, b) and $g(a) = g(b) = 0$. Hence, g satisfies all the conditions of Rolle's theorem, and therefore, there exists an $x_0 \in (a, b)$ such that $g'(x_0) = 0$. Since,

$$g'(x) = f'(x) - \frac{f(b) - f(a)}{b - a}$$

the desired result follows.

In Theorem 15.5 if $b = a + h$, $h > 0$, then (15.1) can be written as

$$f(a + h) = f(a) + hf'(a + \theta h), \tag{15.2}$$

where $0 < \theta < 1$. This relation is often referred to as the formula for finite increments.

Corollary 15.1. If f is continuous on $[a, b]$, differentiable on (a, b), and $f'(x) = 0$ on (a, b), then f must be a constant function on $[a, b]$.

Corollary 15.2. If f and g are continuous on $[a, b]$, differentiable on (a, b), and $f'(x) = g'(x)$ on (a, b), then $f - g$ must be a constant on $[a, b]$.

Corollary 15.3. If f is continuous on an interval I, differentiable on I^i, then f is nondecreasing or nonincreasing on I according as $f'(x) \geq 0$ or $f'(x) \leq 0$ on I^i.

Proof. If x_1, $x_2 \in I$ with $x_1 < x_2$, then from Theorem 15.5, we have $f(x_2) - f(x_1) = f'(x_0)(x_2 - x_1)$ for some $x_0 \in (x_1, x_2)$. But, $f'(x_0) \geq 0$ or $f'(x_0) \leq 0$ by the hypothesis, and hence $f(x_2) \geq f(x_1)$ or $f(x_2) \leq f(x_1)$.

Theorem 15.6 (Cauchy's Mean-Value Theorem). If f and g are continuous on a closed bounded interval $[a, b]$, differentiable on (a, b), $g(b) \neq g(a)$, and f' and g' do not vanish at the same point x, then there exists a point $x_0 \in (a, b)$ such that

$$\frac{f'(x_0)}{g'(x_0)} = \frac{f(b) - f(a)}{g(b) - g(a)}. \tag{15.3}$$

Proof. It suffices to check that the function $h : [a, b] \to \mathcal{R}$ defined as

$$h(x) = f(x)[g(b) - g(a)] - g(x)[f(b) - f(a)]$$

has the property that $h(a) = h(b) = f(a)g(b) - g(a)f(b)$ and satisfies the hypotheses of Rolle's theorem.

The motivation of Theorem 15.6 comes by considering a smooth curve in parametric representation. Further, it is clear that (15.3) reduces to (15.1) when $g(x) = x$. We note that in Theorem 15.6 the assumption that f' and g' do not vanish at the same point x is essential. Indeed, for $f(x) = x^2$, $g(x) = x^3$, $x \in [-1, 1]$ we notice that there is no $x_0 \in (-1, 1)$ for which the equality (15.3) holds.

Theorem 15.7 (Darboux Theorem). If f is differentiable on a closed bounded interval $[a, b]$ and $f'(a) \neq f'(b)$, then for each y_0' lying between $f'(a)$ and $f'(b)$ there exists a point $x_0 \in (a, b)$ such that $f'(x_0) = y_0'$.

Proof. We assume that $f'(a) > f'(b)$. The arguments for the case $f'(a) < f'(b)$ are similar. We define $g(x) = f(x) - y_0'x$, $x \in [a, b]$. Clearly, $g'(a) > 0$ and $g'(b) < 0$. Since g is right differentiable at a, and left differentiable at b, there exists δ_1, $\delta_2 > 0$ such that

$$0 < h < \delta_1 \Rightarrow \left| \frac{g(a + h) - g(a)}{h} - g'(a) \right| < g'(a) \Rightarrow g(a) < g(a + h)$$

and

$$0 < h < \delta_2 \;\Rightarrow\; \left| \frac{g(b-h) - g(b)}{-h} - g'(b) \right| \;<\; -g'(b) \;\Rightarrow\; g(b) \;<\; g(b-h).$$

Now since g is differentiable on $[a,b]$, it is continuous on $[a,b]$, and hence it attains its supremum on $[a,b]$. The above inequalities show that the preceding supremum is not attained at a or b. Hence, g attains its supremum at some point $x_0 \in (a,b)$. We shall show that $g'(x_0) = 0$, i.e., $f'(x_0) = y_0'$. Suppose that $g'(x_0) > 0$. The argument for the case $g'(x_0) < 0$ is similar. Since g is differentiable at x_0, there exists a $\delta_3 > 0$ such that

$$0 < h < \delta_3 \;\Rightarrow\; \left| \frac{g(x_0+h) - g(x_0)}{h} - g'(x_0) \right| < g'(x_0) \;\Rightarrow\; g(x_0) < g(x_0+h).$$

Therefore, $g(x_0) \neq \sup_{x \in [a,b]} g(x)$, and hence $g'(x_0) = 0$.

The Darboux theorem shows that f' although need not be continuous (see Example 15.1) satisfies the intermediate-value property, and hence in view of Problem 13.6(ii), f' has no removable or first kind discontinuities. Thus, if $\lim_{x \to x_0} f'(x)$, $\lim_{x \to x_0^-} f'(x)$ and $f'(x_0)$ exist, then f' is continuous at x_0.

Remark 15.1. There are functions which are continuous everywhere, but nowhere differentiable. In fact, Karl Weierstrass in the year 1872 surprised the mathematical community by giving the following such function

$$W(x) \;=\; \sum_{k=0}^{\infty} a^k \cos\left(b^k \pi x\right),$$

where a is a real number with $0 < a < 1$, b is an odd integer, and $ab > 1 + 3\pi/2$.

Problems

15.1. Let f be differentiable at x_0. Show that

$$f(x) \;=\; f(x_0) + [f'(x_0) + E(x)](x - x_0),$$

where the function E is defined in an nbd of x_0, and $\lim_{x \to x_0} E(x) = E(x_0) = 0$.

15.2. Let

$$f(x) \;=\; \begin{cases} x+2, & x < -1 \\ x^2, & -1 \le x \le 0 \\ 4x - 3, & 0 < x \le 2 \\ 1 + x^2, & x > 2. \end{cases}$$

Compute one-sided derivatives $f'(x^+)$, $f'(x^-)$ and find where f is differentiable.

15.3. Determine whether the following functions are continuous at 0 or differentiable at 0.

(i). $f(x) = \begin{cases} x & \text{if } x \text{ is rational} \\ 0 & \text{if } x \text{ is irrational.} \end{cases}$

(ii). $g(x) = \begin{cases} x^n \sin \frac{1}{x} & \text{if } x \neq 0 \\ 0 & \text{if } x = 0, \quad n = 0, 1, 2, \cdots. \end{cases}$

(iii). $\max(h, k)$ where h and k are functions differentiable at 0 and $h(0) \neq k(0)$.

15.4. Prove that if $f : \mathcal{R} \to \mathcal{R}$ is a differentiable even (odd) function, then f' is an odd (even) function.

15.5. Let f and g be functions defined on $(-1, 1)$ such that $f(x) = xg(x)$ for all $x \in (-1, 1)$. Prove that g is continuous at 0 iff f is differentiable at 0 and $f'(0) = g(0)$.

15.6. Let g and h be functions defined on an open interval I and $x_0 \in I$. Define f on I by

$$f(x) = \begin{cases} g(x) & \text{if } x \leq x_0, \\ h(x) & \text{if } x > x_0. \end{cases}$$

Prove that if f, g, h are differentiable at x_0, then $g(x_0) = h(x_0)$ and $g'(x_0) = h'(x_0) = f'(x_0)$.

15.7. A function $f : I \to \mathcal{R}$ is said to have a *relative (local) maximum* at a point $x = c \in I$ iff there exists a $\delta > 0$ such that $f(c) \geq f(x)$ for every $x \in I_\delta(c) \cap I$. Similarly, the function is said to have a *relative (local) minimum* at a point $x = c \in I$ iff there exists a $\delta > 0$ such that $f(c) \leq f(x)$ for every $x \in I_\delta(c) \cap I$. The function f is said to have a *relative (local) extremum* at c iff f has either a relative maximum or minimum at c. Show that

(i). If f has a relative extremum at $c \in I^i$, then either $f'(c) = 0$ or $f'(c)$ does not exist.

(ii). Give an example to show that part (i) is not applicable if c is an endpoint of I.

(iii). Give an example of a function f which has a relative extremum at $c \in I^i$, but $f'(c)$ does not exist.

(iv). Give an example of a function to show that the converse of part (i) is not true.

15.8. If $f'(x_0)$ exists and is not zero, show that

$$\lim_{h \to 0} \frac{f(x_0 + h) - 2f(x_0) + f(x_0 - h)}{h} \quad \text{and} \quad \lim_{h \to 0} \frac{f(x_0 + h) - f(x_0)}{f(x_0 - h) - f(x_0)}$$

exist, and find their values. Do either or both of the limits exist if $f(x) = |x|$ and $x_0 = 0$?

15.9. Show that the equation $x^3 + ax + b = 0$, where $a > 0$ has exactly one real root.

15.10. Use Rolle's theorem to show that if

$$\frac{p_0}{n+1} + \frac{p_1}{n} + \cdots + \frac{p_{n-1}}{2} + p_n = 0,$$

then the equation $P_n(x) = p_0 x^n + p_1 x^{n-1} + \cdots + p_{n-1}x + p_n = 0$ has at least one root in $[0, 1]$. In particular, show that the equation $4x^3 + 3x^2 - 6x + 1 = 0$ has a root in $[0, 1]$.

15.11. Let f and g be continuous on $[a, b]$ and differentiable on (a, b). If $g'(x) \neq 0$ for all $x \in (a, b)$ prove that there exists an $x_0 \in (a, b)$ such that

$$\frac{f'(x_0)}{g'(x_0)} = \frac{f(x_0) - f(a)}{g(b) - g(x_0)}.$$

15.12. Suppose that $f(x)g'(x) - f'(x)g(x) \neq 0$ in an interval I. Show that between two consecutive solutions of $f(x) = 0$ there is exactly one solution of $g(x) = 0$.

15.13. Show that if f' exists and is bounded on some interval I, then f is uniformly continuous on I. In particular, show that $f(x) = \tan^{-1} x$, $x \in \mathcal{R}$ is uniformly continuous.

15.14. Let

$$f(x) = \begin{cases} x^2 \sin \frac{1}{x} + \frac{2}{\pi}x, & x \neq 0 \\ 0, & x = 0. \end{cases}$$

Show that f is not an increasing function on $[0, h]$ for any positive h, even though $f'(0) = 2/\pi > 0$.

15.15. Show that if f is differentiable on $(0, 1)$ and continuous on $[0, 1]$, $f(0) = 0$ and f' is strictly increasing on $(0, 1)$, then $f(x)/x$ is strictly increasing on $(0, 1]$.

15.16. Let f be differentiable on $(0, \infty)$ and $\lim_{x \to \infty} f'(x) = 0$. Show that $\lim_{x \to \infty}[f(x + x_0) - f(x)] = 0$, where x_0 is a positive number.

15.17. Let f be differentiable in (x_0, ∞) for some x_0. Suppose that $\lim_{x \to \infty} f(x) = A$ and $\lim_{x \to \infty} f'(x) = B$ where A and B are finite. Show that $B = 0$.

15.18. A function $f(x)$ is said to be *piecewise smooth* (*sectionally smooth*) on an interval (a, b) if both $f(x)$ and $f'(x)$ are piecewise continuous on (a, b). Show that the following functions are piecewise smooth.

(i). $f(x) = \begin{cases} x + 1, & -1 < x < 0 \\ \sin x, & 0 < x < \pi/2. \end{cases}$

(ii). $f(x) = \begin{cases} (1 - e^x)/x, & x \in (0,1) \\ 0, & x = 0. \end{cases}$

Answers or Hints

15.1. The required function E is defined as

$E(x) = \begin{cases} \frac{f(x)-f(x_0)}{x-x_0} - f'(x_0), & x \neq x_0 \\ 0, & x = x_0. \end{cases}$

15.2. $f'(x^+) = f'(x^-) = \begin{cases} 1, & x < -1 \\ 2x, & -1 < x < 0 \\ 4, & 0 < x \leq 2 \\ 2x, & x > 2. \end{cases}$ At $x = -1$, $f'(-1^-) =$

1, $f'(-1^+) = -2$. At $x = 0$, $f'(0^-) = 0$, $f'(0^+)$ does not exist. At $x = 2$, $f'(2^-) = 4$, $f'(2^+) = 4$. So, f is differentiable everywhere except at -1 and 0.

15.3 (i). $\lim_{x \to 0} f(x) = 0$ implies that f is continuous at 0; $\lim_{n \to \infty} \frac{f(\sqrt{2}/n)}{\sqrt{2}/n} = 0$, $\lim_{n \to \infty} \frac{f(1/n)}{1/n} = 1$ and hence f is not differentiable at 0.
(ii). For $n = 0$, g is not continuous at 0 and hence not differentiable; for $n = 1$, g is continuous at 0 but not differentiable; for $n \geq 2$, g is continous as well as differentiable at 0.
(iii). Assume that $h(0) > k(0)$. Then, in view of the continuity of h and k there exists a $\delta > 0$ such that $k(x) < k(0) + (h(0) - k(0))/2 = h(0) - (h(0) - k(0))/2 < h(x)$ for all $|x| < \delta$. Hence, $\lim_{x \to 0} \max(h(x), k(x)) = \lim_{x \to 0} h(x) = h(0) = \max(h(0), k(0))$, and therefore $\max(h, k)$ is continuous at 0. Next note that
$\lim_{x \to 0} \frac{\max(h(x),k(x))-\max(h(0),k(0))}{x-0} = \lim_{x \to 0} \frac{h(x)-h(0)}{x-0} = h'(0)$
and hence $\max(h, k)$ is differentiable at 0.

15.4. If $f(x) = f(-x)$, then $f'(x) = -f'(-x)$.

15.5. Let g be continuous at 0. Then $f'(0) = \lim_{x \to 0} \frac{f(x)-f(0)}{x-0} = \lim_{x \to 0} \frac{f(x)}{x} = \lim_{x \to 0} g(x) = g(0)$ and hence f is differentiable at 0 and $f'(0) = g(0)$. Now let f be differentiable at 0 and $f'(0) = g(0)$. Then $\lim_{x \to 0} g(x) = \lim_{x \to 0} \frac{f(x)}{x} = \lim_{x \to 0} \frac{f(x)-f(0)}{x-0} = f'(0) = g(0)$ and hence g is continuous at 0.

15.6. Since f, g, h are differentiable at x_0, f, g, h are continuous at x_0, and $\lim_{x \to x_0} (f(x)-f(x_0))/(x-x_0)$, $\lim_{x \to x_0} (g(x)-g(x_0))/(x-x_0)$, $\lim_{x \to x_0} (h(x) - h(x_0))/(x - x_0)$ exist. Thus, $g(x_0) = \lim_{x \to x_0} g(x) = \lim_{x \to x_0^-} g(x) = \lim_{x \to x_0^-} f(x) = \lim_{x \to x_0^+} f(x)$ and $h(x_0) = \lim_{x \to x_0} h(x) = \lim_{x \to x_0^+} h(x) = \lim_{x \to x_0^+} f(x) = \lim_{x \to x_0^-} f(x)$, thus $g(x_0) = h(x_0) = f(x_0)$. Now since

$g(x_0) = f(x_0) = h(x_0)$ we have $g'(x_0) = \lim_{x \to x_0^-} (g(x) - g(x_0))/(x - x_0) = \lim_{x \to x_0^-} (f(x) - f(x_0))/(x - x_0)$ and $h'(x_0) = \lim_{x \to x_0^+} (h(x) - h(x_0))/(x - x_0) = \lim_{x \to x_0^+} (f(x) - f(x_0))/(x - x_0)$, and hence from the existence of $f'(x_0)$ it follows that $g'(x_0) = f'(x_0) = h'(x_0)$.

15.7. (i). Suppose that f has a relative maximum at c. Then, there is a $\delta > 0$ such that for every $x \in I_\delta(c) \subset I$, $f(x) \le f(c)$. Hence,

$$\frac{f(x) - f(c)}{x - c} \begin{cases} \le 0 & \text{if } c < x < c + \delta \\ \ge 0 & \text{if } c - \delta < x < c. \end{cases}$$

Now if $f'(c)$ exists, then necessarily

$$0 \le \lim_{x \to c^-} \frac{f(x) - f(c)}{x - c} = f'(c) = \lim_{x \to c^+} \frac{f(x) - f(c)}{x - c} \le 0$$

and hence $f'(c) = 0$.

(ii). Consider $f(x) = x$, $x \in [0, 1]$. This function has extremas at $x = 0$ and 1, but $f'(0^+) = f'(1^-) = 1$.

(iii). Consider $f(x) = |x|$, $x \in [-1, 1]$. This function has relative minimum at $x = 0$, where it fails to be differentiable.

(iv). Consider $f(x) = x^3$, $x \in [-1, 1]$. This function does not have a relative extrema at $x = 0$, but $f'(0) = 0$.

15.8. $A = \lim_{h \to 0} \frac{f(x_0 + h) - 2f(x_0) + f(x_0 - h)}{h} = \lim_{h \to 0} \frac{f'(x_0 + h) - f'(x_0 - h)}{1} = 0$. $B = \lim_{h \to 0} \frac{f(x_0 + h) - f(x_0)}{f(x_0 - h) - f(x_0)} = -\lim_{h \to 0} \frac{[f(x_0 + h) - f(x_0)]/h}{[f(x_0 - h) - f(x_0)]/-h} = -\frac{f'(x_0)}{f'(x_0)} = -1$. When $f(x) = |x|$, then $f(h) = h$, $f(-h) = |-h| = h$ $(h > 0)$, $f(0) = 0$. Therefore, $A = \lim_{h \to 0} \frac{h - 0 + h}{h} = 2$ and $B = \lim_{h \to 0} \frac{h - 0}{h - 0} = 1$. Note that $f(x) = |x|$ is not differentiable at 0 and hence the limits are not the same.

15.9. Consider the function $f(x) = x^3 + ax + b$. Since for all $x \in \mathcal{R}$, $f'(x) = 3x^2 + a > 0$ $(a > 0)$ the function $f(x)$ is increasing for all $x \in \mathcal{R}$. Therefore, $f(x)$ can cross the x-axis at most once.

15.10. Consider the function $f(x) = \frac{p_0 x^{n+1}}{n+1} + \frac{p_1 x^n}{n} + \cdots + \frac{p_{n-1} x^2}{2} + p_n x$. Obviously, $f(0) = f(1) = 0$ and f is continuous on $[0, 1]$, differentiable on $(0, 1)$. Therefore, by Rolle's theorem $f'(x) = P_n(x)$ must vanish at least once in $(0, 1)$. Consider the function $f(x) = x^4 + x^3 - 3x^2 + x$.

15.11. Let $F(x) = f(x)g(b) + g(x)f(a) - f(x)g(x)$. Then, $F(a) = f(a)g(b) = F(b)$. Thus, Rolle's theorem implies the existence of an $x_0 \in (a, b)$ such that $F'(x_0) = 0$, i.e., $f'(x_0)[g(b) - g(x_0)] = g'(x_0)[f(x_0) - f(a)]$. Since $g'(x) \ne 0$ for all $x \in (a, b)$, we have $g(b) \ne g(x_0)$. Hence, $f'(x_0)/g'(x_0) = (f(x_0) - f(a))/(g(b) - g(x_0))$.

15.12. Let a and b $(a < b)$ be two consecutive solutions of the equation $f(x) = 0$. If $g(x) \ne 0$ in $[a, b]$, then for the function $\phi(x) = f(x)/g(x)$, we have $\phi(a) = \phi(b) = 0$ and $\phi'(x) = (g(x)f'(x) - f(x)g'(x))/g^2(x) \ne 0$ for all $x \in$

$[a, b]$, but this contradicts Rolle's theorem. Therefore, $g(x)$ must vanish at least once in $[a, b]$. $g(a) \neq 0$ since otherwise $f(a)g'(a) - g(a)f'(a) = 0$. Similarly, $g(b) \neq 0$. Now let (if possible) c and d be two solutions of $g(x) = 0$, such that $a < c < d < b$. Consider the function $\psi(x) = g(x)/f(x)$ in the interval $[c, d]$. Since $\psi(c) = \psi(d) = 0$ and $\psi'(x) = (g'(x)f(x) - g(x)f'(x))/f^2(x) \neq 0$ once again we have a contradiction to Rolle's theorem. Therefore, $g(x)$ cannot have more than one solution in (a, b).

15.13. Assume that $|f'(x)| \leq M$, $x \in I$. Then, for any $x_1, x_2 \in I$ with $x_1 < x_2$ the mean-value theorem implies that there exists $x_0 \in (x_1, x_2)$ such that $|f(x_2) - f(x_1)| \leq M|x_2 - x_1|$. Now see Problem 14.2. Since $|(\tan^{-1} x)'| = 1/(1 + x^2) \leq 1$, $\tan^{-1} x$ is uniformly continuous on $\mathcal{R}$.

15.14. There exists an $n \in \mathcal{N}$ such that $0 < 1/(2n\pi + \pi/2) < 1/(2n\pi - \pi/2) < h$. Since

$$f\left(\frac{1}{(2n\pi - \pi/2)}\right) - f\left(\frac{1}{(2n\pi + \pi/2)}\right)$$

$$= -\frac{1}{(2n\pi - \pi/2)^2} + \frac{2/\pi}{(2n\pi - \pi/2)} - \frac{1}{(2n\pi + \pi/2)^2} - \frac{2/\pi}{(2n\pi + \pi/2)}$$

$$= \frac{\left[-\left(2n+\frac{1}{2}\right)^2 + 2\left(2n-\frac{1}{2}\right)\left(2n+\frac{1}{2}\right)^2 - \left(2n-\frac{1}{2}\right)^2 - 2\left(2n+\frac{1}{2}\right)\left(2n-\frac{1}{2}\right)^2\right]}{\pi^2\left(2n-\frac{1}{2}\right)^2\left(2n+\frac{1}{2}\right)^2}$$

$$= -\frac{1}{\pi^2\left(2n-\frac{1}{2}\right)^2\left(2n+\frac{1}{2}\right)^2} < 0,$$

which implies $f(1/(2n\pi + \pi/2)) > f(1/2n\pi - \pi/2))$. Thus, f is not an increasing function on $[0, h]$ for any positive h.

15.15. Clearly, $f(x)/x$ is differentiable on $(0, 1)$ and continuous on $(0, 1]$, and $(f(x)/x)' = (f'(x) - f(x)/x)/x$ for $x \in (0, 1)$. Let $0 < a < b \leq 1$ be given. Then, using the mean-value theorem twice and the fact that f' is strictly increasing on $(0, 1)$, we get

$$\frac{f(b)}{b} - \frac{f(a)}{a} = \left(f'(c) - \frac{f(c)}{c}\right)(b - a)/c \quad \text{for some} \quad c \in (a, b)$$

$$= \left(f'(c) - \frac{f(0) + f'(d)(c - 0)}{c}\right)(b - a)/c \quad \text{for some} \quad d \in (0, c)$$

$$= (f'(c) - f'(d))(b - a)/c > 0.$$

Hence, $f(x)/x$ is strictly increasing on $(0, 1]$.

15.16. Let $\epsilon > 0$ be given. Since $f'(x) \to 0$ as $x \to \infty$, there exists a $M > 0$ such that $|f'(x)| < \epsilon/x_0$ for all $x > M$. Thus, when $x > M$ by the mean-value theorem, we have $|f(x + x_0) - f(x)| = |f'(\xi)||x + x_0 - x|$, $x < \xi < x + x_0 < (\epsilon/x_0)x_0 = \epsilon$. Thus, $\lim_{x \to \infty}[f(x + x_0) - f(x)] = 0$.

15.17. Suppose $\lim_{x \to \infty} f'(x) = B > 0$. Then there exists an $N > x_0$ such that $|f'(x) - B| < B/2$, $x > N$. Thus, in particular, for $x > N$, $f'(x) > B/2 > 0$. Now by the mean-value theorem, for $N < x < y$, there exists a $\xi \in (x, y)$ such that $f(y) = f(x) + f'(\xi)(y - x) > f(x) + (B/2)(y - x)$. Thus, $\lim_{y \to \infty} f(y) \geq \lim_{y \to \infty}[f(x) + (B/2)(y - x)] = \infty$, which is a contradiction.

Chapter 16

Higher Order Differentiable Functions

The derivative of a function f is again a function f'. The function f' may be differentiable, and its derivative we denote by f'' and call it the second derivative of f. In this manner, the function f may be differentiable $n \geq 1$ times, and we denote its nth derivative by $f^{(n)}$. In order for $f^{(n)}(x_0)$ to exist, $f^{(n-1)}(x)$ must exist in an nbd of x_0 (or in a one-sided nbd, if x_0 is an endpoint of the interval on which f is defined), and $f^{(n-1)}$ must be differentiable at x_0. Since $f^{(n-1)}$ must exist in an nbd of x_0, $f^{(n-2)}$ must be differentiable, and hence continuous, in that nbd. For $n \geq 1$, $C^{(n)}(I)$ represents the class of all n times continuously differentiable functions on the interval I. In this chapter, we will prove Taylor's theorem (generalized mean-value theorem) and illustrate some of its applications.

Theorem 16.1 (Taylor's Theorem). Let f and its first $n \geq 1$ derivatives be continuous on $[a, b]$, and suppose that $f^{(n+1)}$ exists on (a, b). Then, there exists a point $x_0 \in (a, b)$ such that

$$f(b) = \sum_{k=0}^{n} (b-a)^k \frac{f^{(k)}(a)}{k!} + R_{n+1}^p, \qquad (16.1)$$

where R_{n+1}^p is called the *remainder term* and appears as

$$R_{n+1}^p = (b-a)^p (b-x_0)^{n+1-p} \frac{f^{(n+1)}(x_0)}{n!\,p}, \quad 1 \leq p \leq n+1, \quad p \in \mathcal{N}. \quad (16.2)$$

Proof. Consider the function $F : [a, b] \to \mathcal{R}$ defined as

$$F(x) = f(b) - f(x) - (b-x)f'(x) - (b-x)^2 \frac{f''(x)}{2!} - \cdots$$
$$-(b-x)^n \frac{f^{(n)}(x)}{n!} - C\left(\frac{b-x}{b-a}\right)^p,$$

where C is chosen so that $F(a) = 0$, i.e.,

$$C = f(b) - f(a) - (b-a)f'(a) - (b-a)^2 \frac{f''(a)}{2!} - \cdots - (b-a)^n \frac{f^{(n)}(a)}{n!}. \quad (16.3)$$

Now since the function F is continuous on $[a, b]$, differentiable on (a, b), and $F(a) = F(b) = 0$, Rolle's theorem implies that there exists an $x_0 \in (a, b)$ such that $F'(x_0) = 0$, i.e.,

$$F'(x_0) = \left[-(b-x)^n \frac{f^{(n+1)}(x)}{n!} + Cp \frac{(b-x)^{p-1}}{(b-a)^p} \right]_{x=x_0} = 0$$

and hence $C = R_{n+1}^p$. Substituting this in (16.3), the required formula (16.1) follows.

The representation of R_{n+1}^p in (16.2) is called *Schlömilch-Roche's form of the remainder*. The particular cases $p = n+1$ and $p = 1$, i.e.,

$$R_{n+1}^{n+1} = (b-a)^{n+1} \frac{f^{(n+1)}(x_0)}{(n+1)!} \quad \text{and} \quad R_{n+1}^1 = (b-a)(b-x_0)^n \frac{f^{(n+1)}(x_0)}{n!}$$

are respectively known as *Lagrange's* and *Cauchy's form of the remainders*.

Corollary 16.1. Let f and its first $n \geq 1$ derivatives be continuous on $[a, b]$, and suppose that $f^{(n+1)}$ exists on (a, b). Then, for all distinct points $\alpha, x \in [a, b]$ there exists a point $x_0 = x_0(x) \in (\alpha, x)$ such that

$$f(x) = \sum_{k=0}^{n} (x-\alpha)^k \frac{f^{(k)}(\alpha)}{k!} + R_{n+1}^p, \tag{16.4}$$

where

$$R_{n+1}^p = (x-\alpha)^p (x-x_0)^{n+1-p} \frac{f^{(n+1)}(x_0)}{n!\, p}, \quad 1 \leq p \leq n+1, \quad p \in \mathcal{N}. \tag{16.5}$$

In (16.5) we can take $x_0 = \alpha + \theta(x-\alpha)$, where $\theta \in (0, 1)$, so that

$$R_{n+1}^p = (x-\alpha)^{n+1}(1-\theta)^{n+1-p} \frac{f^{(n+1)}(\alpha + \theta(x-\alpha))}{n!\, p}, \quad 1 \leq p \leq n+1, \ p \in \mathcal{N}. \tag{16.6}$$

Corollary 16.1 with $\alpha = 0$ is known as *Maclaurin's theorem*.

The importance of Taylor's theorem lies in the fact that it relates the values of a function and its first n derivatives at one point α to its approximate value at another point x, i.e.,

$$f(x) \simeq \sum_{k=0}^{n} (x-\alpha)^k \frac{f^{(k)}(\alpha)}{k!}. \tag{16.7}$$

However, in practice, any form of the remainder term in Taylor's formula cannot be calculated exactly, because the specific value of x_0 is often not known. But, for practical purposes all we need is an upper bound on the absolute value of whatever form of the remainder term used. The polynomial in the right side of (16.7), say, $P_n(x)$ is called the *Taylor's polynomial of degree n for the function f at the point a.*

Example 16.1. Consider the function $f(x) = e^x$. Since $f^{(k)}(x) = e^x$, $k = 1, 2, \cdots$, Maclaurin's theorem with (16.6) gives

$$e^x = 1 + \frac{x}{1!} + \frac{x^2}{2!} + \cdots + \frac{x^n}{n!} + R_{n+1}^p, \qquad (16.8)$$

where

$$R_{n+1}^p = x^{n+1}(1-\theta)^{n+1-p}\frac{e^{(\theta x)}}{n!\,p}.$$

Clearly, it follows that $|R_{n+1}^p| \leq |x|^{n+1}e^{|x|}/(n+1)!$. Thus, in particular for all $x \in [-1, 1]$, we have $|R_6^p| < 0.0038$. Hence, for all $x \in [-1, 1]$ the polynomial

$$1 + x + \frac{x^2}{2} + \frac{x^3}{6} + \frac{x^4}{24} + \frac{x^5}{120}$$

differs from e^x by less than 0.0038.

Now we shall show that the number e *is irrational.* For this, (16.8) with $x = 1$, $p = n + 1$ is the same as

$$e = 1 + 1 + \frac{1}{2!} + \cdots + \frac{1}{n!} + \frac{e^\theta}{(n+1)!}, \qquad 0 < \theta < 1.$$

Let $e = a/b$, a, $b \in \mathcal{N}$ be in its lowest terms, and $n > \max\{b, 3\}$, then

$$n!\frac{a}{b} - n! - n! - \frac{n!}{2!} - \cdots - \frac{n!}{n!} = \frac{e^\theta}{n+1}. \qquad (16.9)$$

Clearly, the left side of (16.9) is an integer; however, in view of Problem 8.3(ii) the right side of (16.9) satisfies $0 < e^\theta/(n+1) < 3/4$. Hence, e cannot be a rational number.

Example 16.2. Consider the function $f(x) = \ln(1 + x)$, $x \in [0, 1]$. Since $f^{(k)}(x) = (-1)^{k-1}(k-1)!/(1+x)^k$, $k = 1, 2, \cdots$, Maclaurin's theorem with (16.6) gives

$$\ln(1+x) = x - \frac{x^2}{2} + \frac{x^3}{3} - \cdots + (-1)^{n-1}\frac{x^n}{n} + R_{n+1}^p,$$

where

$$R_{n+1}^p = x^{n+1}(1-\theta)^{n+1-p}\frac{(-1)^n}{(1+\theta x)^{n+1}\,p}.$$

Clearly, it follows that $|R_{n+1}^p| \leq 1/(n+1)$. Now since $1/(n+1) < 0.01$ if $n > 99$, we find that the polynomial

$$x - \frac{x^2}{2} + \frac{x^3}{3} - \cdots - \frac{x^{100}}{100}$$

is required to approximate $\ln(1 + x)$ to differ by less than 0.01.

According to Problem 15.7, if a function $f : I \to \mathcal{R}$ has an extremum at $c \in I^i$ and $f'(c)$ exists, then it is necessary that $f'(c) = 0$. In the following result we provide sufficient conditions for the existence of an extremum.

Theorem 16.2. Let f be such that $f^{(n+1)}(c)$ exists and is different from zero, and $f'(c) = f''(c) = \cdots = f^{(n)}(c) = 0$. Then, if n is odd the function attains its relative (local) maximum or minimum at c according as $f^{(n+1)}(c) < 0$ or $f^{(n+1)}(c) > 0$. If n is even, then f has no extremum at c.

Proof. Since $f^{(n+1)}(c)$ exists, there exists an nbd $[c - \delta_1, c + \delta_1]$, $\delta_1 > 0$ in which $f^{(n)}$ exists and f is $(n - 1)$ times continuously differentiable. Thus, in view of Corollary 16.1 it follows that

$$f(c + h) - f(c) = h^n \frac{f^{(n)}(c + \theta h)}{n!}, \qquad (16.10)$$

where $\theta \in (0, 1)$ and $|h| < \delta_1$. Now, if $f^{(n+1)}(c) > 0$ there exists a $\delta_2 > 0$ such that

$$0 < |h| < \delta_2 \;\Rightarrow\; \left| \frac{f^{(n)}(c+h) - f^{(n)}(c)}{h} - f^{(n+1)}(c) \right| < f^{(n+1)}(c)$$

$$\Rightarrow \begin{cases} 0 = f^{(n)}(c) < f^{(n)}(c + h), & 0 < h < \delta_2 \\ 0 = f^{(n)}(c) > f^{(n)}(c + h), & -\delta_2 < h < 0. \end{cases}$$
$$(16.11)$$

Let $\delta = \min\{\delta_1, \delta_2\}$. If n is odd, then (16.10) and (16.11) imply that $f(c + h) - f(c) > 0$, $h \in (-\delta, 0) \cup (0, \delta)$, i.e., f has a relative minimum at c. If n is even, then (16.10) and (16.11) imply that $f(c + h) - f(c) > 0$, $h \in (0, \delta)$ and $f(c + h) - f(c) < 0$, $h \in (-\delta, 0)$, i.e., f has no extremum at c. The arguments for the case $f^{(n+1)}(c) < 0$ are similar.

Corollary 16.2. Let f be such that $f''(c)$ exists and is different from zero, and $f'(c) = 0$. Then, f attains its relative maximum or minimum at c according as $f''(c) < 0$ or $f''(c) > 0$.

Example 16.3. We shall find the dimensions of a cylinder so that for a given volume V its total surface S is a minimum. Let r be the radius of the base of the cylinder and h the altitude. Clearly, we have

$$S = 2\pi r^2 + 2\pi r h \quad \text{and} \quad V = \pi r^2 h$$

and hence

$$S = 2\pi r^2 + 2\pi r \frac{V}{\pi r^2} = 2 \left(\pi r^2 + \frac{V}{r} \right), \qquad 0 < r < \infty.$$

To find the extremum values of S we need to solve the equation

$$\frac{dS}{dr} = 2 \left(2\pi r - \frac{V}{r^2} \right) = 0,$$

which gives only one value of $r = (V/2\pi)^{1/3} = r_0$, say. Now, since

$$\left.\frac{d^2 S}{dr^2}\right|_{r=r_0} = \left.2\left(2\pi + \frac{2V}{r^3}\right)\right|_{r=r_0} > 0$$

from Corollary 16.2, the function S has a minimum at $r = r_0$. Further, for $r = r_0$ we find

$$h|_{r=r_0} = \frac{V}{\pi r_0^2} = 2r_0.$$

We also note that $\lim_{r\to 0} S = \infty$ and $\lim_{r\to\infty} S = \infty$. Therefore, for the total surface S of a cylinder to be a minimum for a given volume V, the altitude of the cylinder must be equal to its diameter.

Example 16.4 (Laws of Reflection and Refraction of Light).

Fermat's principle in optics states that light travels from one point to another along a path that minimizes the travel time. An immediate consequence of this is that in a homogeneous medium light travels in a straight line, since a straight line gives the shortest distance between two points.

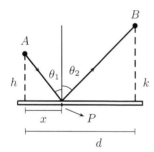

Figure 16.1

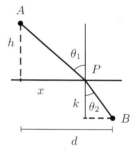

Figure 16.2

Consider a mirror lying horizontally as shown in Figure 16.1. Light travels from a source at point A to point B after reflecting from the mirror at P. We shall find the point of reflection P which requires the light to travel the shortest possible total distance.

From Figure 16.1, it is clear that the total distance D for the light to travel from A to B is

$$D = \sqrt{h^2 + x^2} + \sqrt{k^2 + (d - x)^2}.$$

Thus, we have

$$\frac{dD}{dx} = \frac{x}{\sqrt{h^2 + x^2}} - \frac{d - x}{\sqrt{k^2 + (d - x)^2}},$$

$$\frac{d^2 D}{dx^2} = \frac{h^2}{(h^2 + x^2)^{3/2}} + \frac{k^2}{(k^2 + (d - x)^2)^{3/2}} > 0.$$

Hence, D is minimum when $\sin\theta_1 = \sin\theta_2$, i.e., $\theta_1 = \theta_2$.

The ray of light AP which hits the mirror is called the incident ray, the ray PB is the reflected ray, θ_1 is the angle of incidence, and θ_2 the angle of reflection. Thus, we have shown that the angle of incidence equals the angle of reflection. This well-known law in physics is known as the *reflection law*.

Now consider the problem of refraction of light from a source A in vacuum to point B in medium of refractive index μ. If light travels with velocity v is vacuum, then it travels with velocity v/μ in the second medium. From Figure 16.2 it is clear that the total time T of light to travel from A to B is

$$T = \frac{\sqrt{h^2 + x^2}}{v} + \frac{\sqrt{k^2 + (d-x)^2}}{v/\mu}.$$

Thus, we have

$$v\frac{dT}{dx} = \frac{x}{\sqrt{h^2 + x^2}} - \mu\frac{d-x}{\sqrt{k^2 + (d-x)^2}},$$

$$v\frac{d^2T}{dx^2} = \frac{h^2}{(h^2 + x^2)^{3/2}} + \mu\frac{k^2}{(k^2 + (d-x)^2)^{3/2}}.$$

Hence, T is minimum when $\sin\theta_1 = \mu\sin\theta_2$. This is called *Snell's law* after W. Snell (1591–1626). However, it is now known that this law was first discovered by Ibn Sahl (940–1000) in 984.

Problems

16.1. Let
$$f(x) = \begin{cases} x^5 + ax + b, & x < 1 \\ cx^2 + 2, & x \geq 1. \end{cases}$$
For what values of a, b, c, f is twice differentiable.

16.2. If f is defined in a neighborhood of x_0, f' is continuous there, and $f''(x_0)$ exists, show that

(i). $\lim_{h\to 0} \frac{f(x_0+2h)-2f(x_0+h)+f(x_0)}{h^2} = f''(x_0)$.

(ii). $\lim_{h\to 0} \frac{f(x_0+h)-2f(x_0)+f(x_0-h)}{h^2} = f''(x_0)$.

16.3. Suppose that f and g are n times differentiable at x_0. Use induction to prove the *Leibniz's rule*

$$(fg)^{(n)}(x_0) = \sum_{k=0}^{n} \frac{n!}{k!(n-k)!} f^{(k)}(x_0)g^{(n-k)}(x_0).$$

16.4. Use Rolle's theorem to show that the equation $h(x) = \sin x + x^4 - 6x^3 - 24x^2 + x + 1 = 0$ has at most two roots in $[0, \pi]$.

16.5. Use Maclaurin's theorem to show that

(i). $\sin x = \sum_{k=1}^{n}(-1)^{k-1}\frac{x^{2k-1}}{(2k-1)!} + \frac{x^{2n+1}}{(2n+1)!}\sin\left(\theta x + (2n+1)\frac{\pi}{2}\right)$.

(ii). $\cos x = \sum_{k=1}^{n}(-1)^{k-1}\frac{x^{2(k-1)}}{(2(k-1))!} + \frac{x^{2n}}{(2n)!}\cos\left(\theta x + 2n\frac{\pi}{2}\right)$.

(iii). $(1+x)^m = \sum_{k=1}^{n}\frac{m(m-1)\cdots(m-k+2)}{(k-1)!}x^{k-1}$
$+ \frac{m(m-1)\cdots(m-n+1)}{n!}x^n(1+\theta x)^{m-n}$.

16.6. Show that

(i). $x - \frac{1}{2}x^2 < \ln(1+x) < x - \frac{x^2}{2(1+x)^2}$, $x > 0$.

(ii). $1 + \frac{x}{2} - \frac{x^2}{8} \leq \sqrt{1+x} \leq 1 + \frac{x}{2}$, $x > 0$.

16.7. Assume that the function f is defined by $f(x) = x^r(1-x)^s$, $x \in [0,1]$ where r and s are positive integers greater than 1. Prove the following.

(i). f has a local maximum at $x = r/(r+s)$.

(ii). f has a local minimum at $x = 0$ if r is even.

(iii). f has a local minimum at $x = 1$ if s is even.

16.8. If $\phi(x) = f(x) + f(1-x)$ and $f''(x) < 0$ in $[0,1]$, show that $\phi(x)$ increases in $[0, 1/2]$, decreases in $[1/2, 1]$, and has a maximum at $x = 1/2$.

16.9. (i). Show that $x^x(1-x)^{1-x} \geq \alpha^x(1-\alpha)^{1-x}$ for all $x, \alpha \in (0,1)$, where equality holds only when $x = \alpha$.

(ii). Let $a, b > 0$ and $p, q > 1$ with $1/p + 1/q = 1$. Show that

$$\frac{a^p}{p} + \frac{b^q}{q} \geq ab,$$

where equality holds only when $a^p = b^q$.

(iii). Let $a_1, \cdots, a_n; b_1, \cdots, b_n$ be nonnegative real numbers, and $p, q > 1$ with $1/p + 1/q = 1$. Establish *Hölder's inequality*

$$\left(\sum_{k=1}^{n}a_k^p\right)^{1/p}\left(\sum_{k=1}^{n}b_k^q\right)^{1/q} \geq \sum_{k=1}^{n}a_kb_k,$$

where equality holds only when $a_k^p = \lambda b_k^q$, or when $a_k = 0$ or $b_k = 0$ for $k = 1, \cdots, n$; here, $\lambda > 0$. (For $p = q = 2$ Hölder's inequality is known as *Cauchy's inequality*.)

(iv). Let $a_1, \cdots, a_n; b_1, \cdots, b_n$ be nonnegative real numbers, and $r > 1$. Establish *Minkowski's inequality*

$$\left(\sum_{k=1}^{n}a_k^r\right)^{1/r} + \left(\sum_{k=1}^{n}b_k^r\right)^{1/r} \geq \left[\sum_{k=1}^{n}(a_k + b_k)^r\right]^{1/r},$$

An Introduction to Real Analysis

where equality holds only when $a_k = \lambda b_k$, or when $a_k = 0$ or $b_k = 0$ for $k = 1, \cdots, n$; here, $\lambda > 0$.

16.10. Let n be a positive integer. A function f is said to have a *zero of multiplicity* n at x_0 if f is n times differentiable at x_0, $f^{(k)}(x_0) = 0$, $k = 0, 1, \cdots, n-1$, and $f^{(n)}(x_0) \neq 0$. Show that f has a zero of multiplicity n iff $f(x) = (x - x_0)^n g(x)$, where the function g is continuous at x_0 and n times differentiable in a deleted nbd of x_0, $g(x_0) \neq 0$, and $\lim_{x \to x_0} (x - x_0)^k g^{(k)}(x) = 0$, $k = 1, 2, \cdots, n-1$.

16.11. (i). For the function $f(x) = x^3 \sin x$, $x \in [-\pi/2, \pi/2]$ show that $x = 0$ is a relative minimum.

(ii). Let $f : \mathcal{R} \to \mathcal{R}$ be a function such that $f'(x) = x(x-1)^2(x+2)$. Show that for the function f, $x = 0$ is a relative minimum, $x = -2$ a relative maximum, and $x = 1$ neither maximum nor minimum.

16.12. (i). For the function $f(x) = \ln(x^2 + 1)$, $x \in [-1, 3]$ show that $x = 0$ is a relative minimum.

(ii). For the function $f(x) = (x^2 - 1)/(x^2 + 1)$, $x \in \mathcal{R}$ show that $x = 0$ is a relative minimum.

(iii). For the function $f(x) = 4x/(x^2 + 1)$, $x \in \mathcal{R}$ show that $x = -1$ is a relative minimum and $x = 1$ is a relative maximum.

(iv). For the function $f(x) = (x^4 + 1)/x^2$, $x \in [-2, 0) \cup (0, 2]$ show that $x = -1$ and $x = 1$ are local minimum.

16.13. Find the point on the curve $C : x + y^2 - 2 = 0$ such that it is closest to the point $P(2, -3)$.

16.14. (i). Let f be twice differentiable, and suppose $|f(x)| < A, |f''(x)| < B$ for all $x > a$, where A and B are positive constants. Show that for $x > a$, $|f'(x)| < 2\sqrt{AB}$.

(ii). Let f be twice differentiable on $[0, 4]$, and $|f(x)| \leq 1$, $|f''(x)| \leq 1$ for all $x \in [0, 4]$. Show that $|f'(x)| \leq 2$ for all $x \in [0, 4]$.

Answers or Hints

16.1. $a = 15$, $b = -4$, $c = 10$.

16.2. (i). $\lim_{h \to 0} \frac{f(x_0 + 2h) - 2f(x_0 + h) + f(x_0)}{h^2} = \lim_{h \to 0} \frac{2f'(x_0 + 2h) - 2f'(x_0 + h)}{2h}$
$= \lim_{h \to 0} \left\{ \frac{2(f'(x_0 + 2h) - f'(x_0))}{2h} - \frac{2(f'(x_0 + h) - f'(x_0))}{2h} \right\} = 2f''(x_0) - f''(x_0) = f''(x_0)$.
(ii). Similar to (i).

16.3. The proof is by induction.

16.4. Suppose h has three zeros α, β, γ in $[0, \pi]$ such that $0 \leq \alpha < \beta < \gamma \leq \pi$. Then, Rolle's theorem implies that there exists $\mu \in (\alpha, \beta)$, $\lambda \in (\beta, \gamma)$ such that $h'(\mu) = h'(\lambda) = 0$. Again, since h' is continuous and differentiable on $[\mu, \lambda]$, Rolle's theorem implies that there exists a $x_0 \in (\mu, \lambda)$ such that $h''(x_0) = 0$. However, we have $h''(x) = -\sin x + 12(x+1)(x-4) < 0$, $x \in [0, \pi]$.

16.5. (i). The kth derivative of $\sin x$ is $\sin(x + k\pi/2)$.
(ii). The kth derivative of $\cos x$ is $\cos(x + k\pi/2)$.
(iii). The kth derivative of $(1+x)^m$ is $m(m-1)\cdots(m-k+1)(1+x)^{m-k}$.

16.6. (i). For the left inequality use Example 16.2 (valid for all $x > 0$). For the right inequality show that $f(x) = x - \frac{x^2}{2(1+x)^2} - \ln(1+x)$ is increasing.
(ii). Use Maclaurin's theorem as in Example 16.2.

16.7. $f'(x) = x^{r-1}(1-x)^{s-1}[r - (r+s)x]$, $f'(x) = 0$ at $x = 0, 1, r/(r+s)$. We have $f'(x) > 0$ if r is odd and $x < 0$; $f'(x) < 0$ if r is even and $x < 0$; $f'(x) > 0$ if $0 < x < r/(r+s)$; $f'(x) < 0$ if $r/(r+s) < x < 1$; $f'(x) > 0$ if s is even and $x > 1$; $f'(x) < 0$ if s is odd and $x > 1$. Now apply Corollary 15.3.

16.8. As x varies from 0 to 1, $1 - x$ varies from 1 to 0, thus $f''(1 - x)$ is also negative in $[0, 1]$. Hence, $\phi''(x) = f''(x) + f''(1 - x)$ is negative and $\phi'(x)$ decreases in $(0, 1)$. Now $\phi'(0) = f'(0) - f'(1)$, $\phi'(1/2) = 0$, and $\phi'(1) = f'(1) - f'(0)$, so that $\phi'(x)$ is positive in $(0, 1/2)$ and negative in $(1/2, 1)$, $f'(0)$ being greater than $f'(1)$. It follows that $\phi(x)$ is increasing in $(0, 1/2)$ and decreasing in $(1/2, 1)$, and there is a maximum at $x = 1/2$.

16.9. (i). Let $f(\alpha) = \alpha^x(1-\alpha)^{1-x}$, to obtain $f'(\alpha) = \alpha^x(1-\alpha)^{1-x}[(x-\alpha)/\alpha(1-\alpha)]$. Thus, $f(\alpha)$ is increasing or decreasing according as $0 < \alpha < x$, or $0 < x < \alpha < 1$. But, this implies that $f(x) \geq f(\alpha)$ for all $0 < x, \alpha < 1$, and equality holds only when $x = \alpha$.
(ii). In part (i) let $x = 1/p$ and $\alpha = qa^p/(qa^p + pb^q)$.
(iii). In part (ii) let $a = a_k/(a_1^p + \cdots + a_n^p)^{1/p}, b = b_k/(b_1^q + \cdots + b_n^q)^{1/q}$, $k = 1, \cdots, n$ and sum both the sides.
(iv). Apply part (iii) to sets $a_k, (a_k + b_k)^{p-1}$ and $b_k, (a_k + b_k)^{p-1}$, $p > 1$, to get
$$(\sum_{k=1}^n a_k^p)^{1/p} (\sum_{k=1}^n (a_k + b_k)^p)^{1/q} \geq \sum_{k=1}^n a_k(a_k + b_k)^{p-1},$$
$$(\sum_{k=1}^n b_k^p)^{1/p} (\sum_{k=1}^n (a_k + b_k)^p)^{1/q} \geq \sum_{k=1}^n b_k(a_k + b_k)^{p-1}$$
now add the above inequalities.

16.10. The proof is by induction. For $n = 1$, let f have a simple zero at x_0. Then, Problem 15.1 implies that $f(x) = (x - x_0)g(x)$, where $g(x) = f'(x_0) + E(x)$, where $\lim_{x \to x_0} E(x) = E(x_0) = 0$, and hence g is continuous at x_0 and $\lim_{x \to x_0} g(x) = f'(x_0) \neq 0$. We also note that $g(x) = f(x)/(x - x_0)$ implies that $g(x)$ is differentiable in a deleted nbd of x_0. Conversely, suppose that $f(x) = (x - x_0)g(x)$. Then, $f(x_0) = 0$ and $f'(x_0) = \lim_{x \to x_0} f(x)/(x - x_0) = \lim_{x \to x_0} g(x) = g(x_0) \neq 0$. For the general case we need the following extension of Problem 15.1: If $f^{(n)}(x_0)$ exists, then $f(x) = \sum_{k=0}^{n} f^{(k)}(x_0)(x - x_0)^k/k! + (x - x_0)^n E(x)$, where $\lim_{x \to x_0} E(x) = E(x_0) = 0$.

16.11. Use Theorem 16.2.

16.12. Use Corollary 16.2.

16.13. A point on the curve C has the form $(-y^2 + 2, y)$. Thus, the square of the distance d between this point and P is $d^2 = y^4 + y^2 + 6y + 9$. The required point is $(1, -1)$.

16.14. (i).   From Taylor's formula, for $a < x < x + h$ there exists a $\xi \in (x, x + h)$ such that $f(x + h) = f(x) + hf'(x) + \frac{h^2}{2} f''(\xi)$. Thus, $|hf'(x)| = |f(x + h) - f(x) - \frac{h^2}{2} f''(\xi)| \leq |f(x + h)| + |f(x)| + \frac{h^2}{2}|f''(\xi)| < 2A + \frac{h^2}{2}B$. Hence,

$$|f'(x)| < \frac{2A}{h} + \frac{h}{2}B = \left(\sqrt{\frac{2A}{h}} - \sqrt{\frac{hB}{2}} \right)^2 + 2\sqrt{AB}. \qquad (16.12)$$

Note that $\sqrt{\frac{2A}{h}} - \sqrt{\frac{hB}{2}}$ is a continuous function of h for $h > 0$, and it is positive for small h, whereas negative for large h, and so there exists a $h_0 > 0$ where it is zero. Since the inequality (16.12) is valid for all $h > 0$, we have $|f'(x)| < 2\sqrt{AB}$.

(ii).   As in (i), as long as x and $x+h$ are in $[0, 4]$ we have $|f'(x)| \leq \frac{2}{|h|} + \frac{|h|}{2}$. Now we note that for any given $x \in [0, 4]$ there exists a h such that $|h| = 2$ and $x + h \in [0, 4]$, hence $|f'(x)| \leq 2$.

Chapter 17

Convex Functions

In this chapter, we will define convexity of a function at a point and in an interval, and prove some fundamental properties of convex functions. The collection of convex functions is important for a variety of applications, especially in Fourier analysis, numerical analysis, probability theory, and optimization theory.

A curve $y = f(x)$, or just the function f, is said to be *convex upward* (*downward*) at a point x_0 if there is an nbd of x_0 such that the tangent to the curve at the point x_0 lies above (below) the points of the curve for all x in that nbd. Recall that if f is differentiable, then the tangent line at the point x_0 is $y = f(x_0) + f'(x_0)(x - x_0)$. Thus, the curve $y = f(x)$ is convex upward (downward) if $f(x) < (>) f(x_0) + f'(x_0)(x - x_0)$ in some deleted nbd of x_0. A point x_0 is called a *point of inflection* of the curve $y = f(x)$ if there exists a small $\delta > 0$ such that the curve lies on one side of the tangent line at the point x_0 for all $x \in (x_0 - \delta, x_0)$ and on the other side for all $x \in (x_0, x_0 + \delta)$. The following result provides sufficient conditions for convexity of a curve $y = f(x)$ at a point x_0.

Theorem 17.1. Let f be $(n + 1)$-times continuously differentiable at x_0, and $f^{(k)}(x_0) = 0$, $k = 2, 3, \cdots, n$ and $f^{(n+1)}(x_0) \neq 0$. Then, if n is an odd integer the curve $y = f(x)$ is convex upward or downward at x_0 depending on whether $f^{(n+1)}(x_0) < 0$ or $f^{(n+1)}(x_0) > 0$; if n is an even integer then x_0 is a point of inflection of the curve.

Proof. By Taylor's formula with residue (16.6) and $p = n + 1$, we find

$$f(x) = f(x_0) + (x - x_0)f'(x_0) + \frac{(x - x_0)^{n+1}}{(n + 1)!} f^{(n+1)}(x_0 + \theta(x - x_0)), \quad 0 < \theta < 1.$$

Now since $f^{(n+1)}$ is continuous at x_0, if $f^{(n+1)}(x_0) \neq 0$ then $f^{(n+1)}(x_0 + \theta(x - x_0)) \neq 0$, for all x in a sufficiently small nbd $I_\delta(x_0)$ of the point x_0. Thus, $f^{(n+1)}(x_0 + \theta(x - x_0))$ retains the sign of $f^{(n+1)}(x_0)$ throughout the nbd $I_\delta(x_0)$. Therefore, if $f^{(n+1)}(x_0) > 0$ and n is odd, then for all $x \in I_\delta(x_0)$ it follows that

$$f(x) - f(x_0) - f'(x_0)(x - x_0) = \frac{(x - x_0)^{n+1}}{(n + 1)!} f^{(n+1)}(x_0 + \theta(x - x_0)) > 0,$$

i.e., f is convex downward at x_0. Similarly, if $f^{(n+1)}(x_0) < 0$ and n is odd, then f is convex upward at x_0. Also, if n is even, then since the factor $(x - x_0)^{n+1}$ changes sign as x passes through x_0, it follows that x_0 is a point of inflection of f.

Corollary 17.1. (1). If a function f has the continuous second derivative at x_0 and $f''(x_0) < 0 \ (> 0)$, then f is convex upward (downward) at x_0.

(2). If a function f has the continuous third derivative at x_0 and $f''(x_0) = 0$, $f'''(x_0) \neq 0$, then f has a point of inflection at x_0.

Example 17.1. The curve $f(x) = x^2$ is convex downward at each $x \in \mathcal{R}$. The function $f(x) = x^3$ is convex upward at each $x \in (-\infty, 0)$, convex downward at each $x \in (0, \infty)$, and it has a point of inflection at $x = 0$. Obviously, the above definition does not apply to the function $f(x) = |x|$ at the point $x = 0$.

A curve $y = f(x)$, or just the function f, is said to be *convex upward* (*downward*) on an interval I if any arc of this curve with endpoints having abscissas x_0, $x_1 \in I$ with $x_0 < x_1$ lies not lower (not higher) than the chord subtending it, i.e.,

$$f((1-t)x_0 + tx_1) \geq (\leq) \ (1-t)f(x_0) + tf(x_1), \quad 0 \leq t \leq 1.$$

Clearly, if f is convex upward on I, then $-f$ is convex downward on I. Thus, in what follows we shall discuss only convex downward functions, and the word downward will be dropped.

Since $|(1-t)x_0 + tx_1| \leq (1-t)|x_0| + t|x_1|$ for all $t \in [0,1]$, the function $|x|$ is convex in $[-a, a]$, $a > 0$. Thus, convex functions need not be differentiable; however, the following result shows that they are always continuous.

Theorem 17.2. If $f : (a,b) \to \mathcal{R}$ is convex, then f is continuous on (a, b).

Proof. Let $x_0 \in (a, b)$, and let $\delta > 0$ be such that $[x_0 - \delta, x_0 + \delta] \subset (a, b)$. If $x \in [x_0, x_0 + \delta]$, then $x = (1-t)x_0 + t(x_0 + \delta)$, where $t = (x - x_0)/\delta$, $0 \leq t \leq 1$. Now since f is convex, we have $f(x) \leq (1 - t)f(x_0) + tf(x_0 + \delta)$, and hence

$$f(x) - f(x_0) \ \leq \ t(f(x_0 + \delta) - f(x_0)) \ \leq \ \frac{(x - x_0)}{\delta}(M - f(x_0)), \quad (17.1)$$

where $M = \max\{f(x_0 + \delta), f(x_0 - \delta)\}$.

Next, since $x_0 - \delta < x_0 \leq x$, we can write $x_0 = (1 - s)(x_0 - \delta) + sx$, where $s = \delta/(x - x_0 + \delta)$, $1/2 \leq s \leq 1$. Thus, we have $f(x_0) \leq (1-s)f(x_0 - \delta) + sf(x)$, which gives

$$f(x) - f(x_0) \ \geq \ \frac{(1 - s)}{s}[f(x_0) - f(x_0 - \delta)].$$

However, since $(1-s)/s = (x-x_0)/\delta$ and $M \geq f(x_0 - \delta)$, the preceding inequality gives

$$f(x) - f(x_0) \geq \frac{(x-x_0)}{\delta}(f(x_0) - M). \tag{17.2}$$

Finally, since $M \geq f(x_0)$ from (17.1) and (17.2) it follows that

$$|f(x) - f(x_0)| \leq \frac{|x-x_0|}{\delta}(M - f(x_0)). \tag{17.3}$$

If $x \in [x_0 - \delta, x_0]$, a similar argument will give the same estimate. Thus, (17.3) holds for all $x \in [x_0, x_0 + \delta]$. Hence, f is continuous at x_0.

Theorem 17.2 does not hold on the closed interval $[a, b]$. Indeed, the function $f(x) = x^2$, $x \in (0,1)$, $f(0) = 1$, $f(1) = 1$ is convex on $[0,1]$, but not continuous.

We note that the inequality $f((1-t)x_0 + tx_1) \leq (1-t)f(x_0) + tf(x_1)$ with $x = (1-t)x_0 + tx_1$, i.e., $t = (x-x_0)/(x_1 - x_0)$ is the same as

$$f(x) \leq \frac{x_1 - x}{x_1 - x_0}f(x_0) + \frac{x - x_0}{x_1 - x_0}f(x_1), \quad x \in [x_0, x_1] \tag{17.4}$$

which can be written as

$$\frac{f(x) - f(x_0)}{x - x_0} \leq \frac{f(x_1) - f(x)}{x_1 - x}, \quad x \in (x_0, x_1). \tag{17.5}$$

Thus, if f is convex, then the slope of the chord always increases, i.e., $x_0 < x < x_1$ implies (17.5). This simple observation provides the following characterization of differentiable convex functions.

Theorem 17.3. If f is differentiable on (a,b), then f is convex on (a,b) iff f' is increasing on (a,b).

Proof. Let x_0, $x_1 \in (a,b)$ with $x_0 < x_1$. Choose $h > 0$ so small that $a < x_0 < x_0 + h < x_1 < x_1 + h < b$, then we have

$$\frac{f(x_0 + h) - f(x_0)}{h} \leq \frac{f(x_1 + h) - f(x_1)}{h},$$

which implies $f'(x_0^+) \leq f'(x_1^+)$. Similarly, we have $f'(x_0^-) \leq f'(x_1^-)$.

Conversely, suppose that f' is increasing. Let $a < x_0 < x < x_1 < b$. Then, by the mean-value theorem there exists $c \in (x_0, x)$ and $d \in (x, x_1)$ such that

$$\frac{f(x) - f(x_0)}{x - x_0} = f'(c) \quad \text{and} \quad \frac{f(x_1) - f(x)}{x_1 - x} = f'(d).$$

Since $c < d$ we have $f'(c) \leq f'(d)$, which implies that the inequality (17.5) holds, and hence f is convex on (a,b).

Corollary 17.2. If f is twice differentiable on (a,b), then f is convex on (a,b) iff $f''(x) \geq 0$, $x \in (a,b)$.

We shall now state and prove a result which improves Theorem 17.2.

Theorem 17.4. If $f : (a,b) \to \mathcal{R}$ is convex, then f is Lipschitz continuous in any closed interval $[c,d] \subset (a,b)$. Thus, f is absolutely continuous in $[c,d]$ and continuous in (a,b).

Proof. Let $\epsilon > 0$ be such that $[c - \epsilon, d + \epsilon] \subset (a,b)$. Then, in view of Problem 17.7 there exists m and M so that $m \leq f(x) \leq M$, $x \in [c - \epsilon, d + \epsilon]$. Now let x and y be any two distinct points in $[c,d]$, and let

$$z = y + \epsilon \frac{y - x}{|y - x|} \quad \text{and} \quad \lambda = \frac{|y - x|}{\epsilon + |y - x|}.$$

Clearly, it follows that $z \in [c - \epsilon, d + \epsilon]$ and $y = \lambda z + (1 - \lambda)x$. Thus, the convexity implies that

$$f(y) \leq \lambda f(z) + (1 - \lambda)f(x) = \lambda[f(z) - f(x)] + f(x),$$

which is the same as

$$f(y) - f(x) \leq \lambda(M - m) < \frac{|y - x|}{\epsilon}(M - m) = L|y - x|,$$

where $L = (M - m)/\epsilon$.

The following result complements Theorem 17.3.

Theorem 17.5. If $f : (a,b) \to \mathcal{R}$ is convex, then $f'(x^-)$ and $f'(x^+)$ exist and are increasing in (a,b).

Proof. Let $w < x < y < z$ be four points in (a,b). Then, from (17.5) it follows that

$$\frac{f(x) - f(y)}{x - y} \leq \frac{f(z) - f(y)}{z - y}. \tag{17.6}$$

As we have noted earlier in (17.6) the left side of the inequality increases as $x \to y$ and the left side decreases as $z \to y$. But, this implies that $f'(y^-)$ and $f'(y^+)$ exist for all $y \in (a,b)$, and satisfy

$$f'(y^-) \leq f'(y^+). \tag{17.7}$$

Similarly, we find

$$f'(w^+) \leq \frac{f(x) - f(w)}{x - w} \leq \frac{f(y) - f(x)}{y - x} \leq f'(y^-). \tag{17.8}$$

A combination of (17.7) and (17.8) immediately gives $f'(w^-) \leq f'(w^+) \leq f'(y^-) \leq f'(y^+)$.

Corollary 17.3. If $f : (a, b) \to \mathcal{R}$ is convex, then f is differentiable on (a, b), except perhaps at countably many points of (a, b).

Finally, in this chapter we state the following important theorem which provides a characterization of convex functions.

Theorem 17.6. A function $f : (a, b) \to \mathcal{R}$ is convex if there exists an increasing function $g : (a, b) \to \mathcal{R}$ and a point $c \in (a, b)$ such that for all $x \in (a, b)$,

$$f(x) - f(c) = \int_c^x g(t)dt;$$

here, the integral is in the Riemann sense which is the subject matter of Chapters 19 and 20.

Example 17.2. Consider the functions $f(x) = x^2$, $g(x) = 2x$, $x \in (1, 3)$ and the point $c = 2 \in (1, 3)$. Clearly, the function g is increasing in $(1, 3)$, and $f(x) - f(2) = x^2 - 4 = \int_2^x 2t dt$. Thus, in view of Theorem 17.6, the function $f(x) = x^2$ is convex in $(1, 3)$.

Problems

17.1. Show that the only function which is both convex upward and convex downward in any interval $[a, b]$ is the linear function $f(x) = \alpha x + \beta$.

17.2. Let f and g be convex functions on I. Show that for $a, b \geq 0$, the function $af(x) + bg(x)$ is convex.

17.3. Suppose that f and g are convex on (a, b). Show that the function $h(x) = \max\{f(x), g(x)\}$ is also convex on (a, b).

17.4. Let f be differentiable on (a, b). Show that f is convex iff for all $x, y \in (a, b)$, with $y > x$, the following holds $f(y) - f(x) \geq (y - x)f'(y)$.

17.5. Show that the following functions are convex.

(i). e^x, $x \in \mathcal{R}$.

(ii). x^α, $\alpha < 0$, $\alpha > 1$, $x \in (0, \infty)$.

(iii). $-x^\alpha$, $0 < \alpha \leq 1$, $x \in (0, \infty)$.

(iv). $-\ln x$, $x \in (0, \infty)$.

(v). $x \ln x$, $x \in (0, \infty)$.

17.6. A function f is said to have a *proper maximum* (*proper minimum*) at x_0 if there exists a $\delta > 0$ such that $f(x) < f(x_0)$ ($f(x) > f(x_0)$) for all $0 < |x - x_0| < \delta$. Show that

(i). If f is convex on (a, b), then f has no proper maximum on (a, b).

(ii). If f is convex on $[0, \infty)$ and has a proper minimum, then $f(x) \to \infty$ as $x \to \infty$.

17.7. Show that a convex function on a closed interval $[a, b]$ is bounded.

17.8. Let f be a convex function in (a, b) and $x_1, \cdots, x_n \in (a, b)$, and $\lambda_1, \cdots, \lambda_n$ be positive numbers. Prove *Jensen's inequality*

$$f\left(\frac{\lambda_1 x_1 + \cdots + \lambda_n x_n}{\lambda_1 + \cdots + \lambda_n}\right) \leq \frac{\lambda_1 f(x_1) + \cdots \lambda_n f(x_n)}{\lambda_1 + \cdots + \lambda_n}.$$

Further, if f is convex upward in (a, b), then the inequality above is reversed.

17.9. Prove the *arithmetic-geometric mean inequality*2pt

$$(x_1 \cdots x_n)^{1/n} \leq \frac{x_1 + \cdots + x_n}{n}, 2pt$$

where $x_1, \cdots, x_n$ are positive numbers.

17.10. Prove the following inequality2pt

$$(x_1 + \cdots + x_n)^\alpha \leq n^{\alpha-1}(x_1^\alpha + \cdots + x_n^\alpha), \quad \alpha \geq 12pt$$

where $x_1, \cdots, x_n$ are positive numbers.

Answers or Hints

17.1. From the definition, the function f is convex upward and downward on an interval $[a, b]$, if and only if $f((1-t)a+tb) = (1-t)f(a)+tf(b)$, $t \in [0, 1]$. Note that any point $x \in [a, b]$ can be uniquely written as $x = (1 - t)a + tb$, where $t = (x-a)/(b-a)$, i.e., $x = \left(1 - \frac{x-a}{b-a}\right)a + \left(\frac{x-a}{b-a}\right)b$. Thus, it follows that $f(x) = \left(1 - \frac{x-a}{b-a}\right)f(a) + \left(\frac{x-a}{b-a}\right)f(b)$, which is the same as $f(x) = \frac{f(b)-f(a)}{b-a}x + \frac{bf(a)-af(b)}{b-a}$.

17.2. Since $f((1-t)x_1+tx_2) \geq (1-t)f(x_1)+tf(x_2)$ and $g((1-t)x_1+tx_2) \geq (1-t)g(x_1)+tg(x_2), t \in [0, 1]$, we have $(af+bg)((1-t)x_1+tx_2) = af((1-t)x_1+tx_2)+bg((1-t)x_1+tx_2) \geq a[(1-t)f(x_1)+tf(x_2)]+b[(1-t)g(x_1)+tg(x_2)] = (1 - t)(af + bg)(x_1) + t(af + bg)(x_2)$.

17.3. We have $h((1 - t)x_1 + tx_2) = \max(f((1 - t)x_1 + tx_2), g((1 - t)x_1 + tx_2)) \leq \max((1 - t)f(x_1) + tf(x_2), (1 - t)g(x_1) + tg(x_2)) \leq \max((1 - t)\max\{f(x_1), g(x_1)\} + t\max\{f(x_2), g(x_2)\} \leq (1 - t)\max\{f(x_1), g(x_1)\} + t\max\{f(x_1), g(x_1)\} = (1 - t)h(x_1) + th(x_2)$.

17.4. Follows from (17.6).

17.5. (i). $f''(x) = e^x > 0$, thus from Corollary 17.2, f is convex.

(ii). $f''(x) = \alpha(\alpha-1)x^{\alpha-2} > 0$, thus from Corollary 17.2, f is convex.
(iii). $f''(x) = -\alpha(\alpha-1)x^{\alpha-2} > 0$, thus from Corollary 17.2, f is convex.
(iv). $f''(x) = \frac{1}{x^2} > 0$, thus from Corollary 17.2, f is convex.
(v). $f''(x) = \frac{1}{x} > 0$, thus from Corollary 17.2, f is convex.

17.6. (i). Suppose that x_0 is a proper maximum. Let δ be as in the statement of the problem, and let $\delta' < \delta$ and $[x_0 - \delta', x_0 + \delta'] \subset (x_0 - \delta, x_0 + \delta)$. Since, $x_0 = \frac{1}{2}(x_0 - \delta') + \frac{1}{2}(x_0 + \delta')$, and f is convex, we have $f(x_0) \le \frac{1}{2}f(x_0 - \delta') + \frac{1}{2}f(x_0 + \delta') < \frac{1}{2}f(x_0) + \frac{1}{2}f(x_0) = f(x_0)$, a contradiction.
(ii). Let $I = [0, \infty) \subset \mathcal{R}$ be an interval and f has a proper minimum at $x_0 \in I$, i.e., $f(x) > f(x_0)$ in an nbd of x_0. We will show that $f(x) > f(x_0)$ for all $x \in I$. Let $w \in I, w > x_0$. Since f is convex, for $y \in (x_0, w)$ close to x_0 and in the nbd of x_0 described above, from (17.5) it follows that $\frac{f(y)-f(x_0)}{y-x_0} \le \frac{f(w)-f(y)}{w-y}$, and hence $f(w) > f(y)$ and so $f(w) > f(x_0)$. The same inequality follows for $w < x_0$. For $[x_0, \infty)$, let $x_0 < x_1 < x$. Then, from (17.5), we have $\frac{f(x_1)-f(x_0)}{x_1-x_0} \le \frac{f(x)-f(x_1)}{x-x_1}$, which implies that $f(x) \ge f(x_1) + (x-x_1)\frac{f(x_1)-f(x_0)}{x_1-x_0}$. From this it is clear that $f(x) \to \infty$ as $x \to \infty$.

17.7. In (17.4) let $x_0 = a$ and $x_1 = b$, to get $f(x) \le \max\{f(a), f(b)\} = M$. Next, let $v \in [a, b]$. Then, there exists $t \in [-(b-a)/2, (b-a)/2]$ with $v = (a+b)/2 + t$; note that $(a+b)/2 - t \in [a, b]$. Also, $(a+b)/2 = \frac{1}{2}v + \frac{1}{2}\left(\frac{a+b}{2} - t\right)$. From the above we have $f\left(\frac{a+b}{2} - t\right) \le M$, and so
$$f\left(\frac{a+b}{2}\right) \le \frac{1}{2}f(v) + \frac{1}{2}f\left(\frac{a+b}{2} - t\right) \le \frac{1}{2}f(v) + \frac{1}{2}M.$$
As a result $f(v) \ge 2f\left(\frac{a+b}{2}\right) - M$, i.e., f is bounded below by $2f\left(\frac{a+b}{2}\right) - M = m$. Thus, $m \le f(x) \le M$, $x \in [a, b]$.

17.8. The proof is by induction. In (17.4), let $x_0 = x_1, x_1 = x_2$ and $x = (\lambda_1 x_1 + \lambda_2 x_2)/(\lambda_1 + \lambda_2)$, to obtain
$$f\left(\frac{\lambda_1 x_1 + \lambda_2 x_2}{\lambda_1 + \lambda_2}\right) \le \frac{\lambda_1 f(x_1) + \lambda_2 f(x_2)}{\lambda_1 + \lambda_2}.$$
If the inequality is true for n, then we have
$$f\left(\frac{\sum_{i=1}^n \lambda_i x_i}{\sum_{i=1}^{n+1} \lambda_i} + \frac{\lambda_{n+1} x_{n+1}}{\sum_{i=1}^{n+1} \lambda_i}\right) = f\left(\frac{(\sum_{i=1}^n \lambda_i)\frac{\sum_{i=1}^n \lambda_i x_i}{\sum_{i=1}^n \lambda_i}}{\sum_{i=1}^{n+1} \lambda_i} + \frac{\lambda_{n+1} x_{n+1}}{\sum_{i=1}^{n+1} \lambda_i}\right)$$
$$\le \frac{\sum_{i=1}^n \lambda_i}{\sum_{i=1}^{n+1} \lambda_i} f\left(\frac{\sum_{i=1}^n \lambda_i x_i}{\sum_{i=1}^n \lambda_i}\right) + \frac{\lambda_{n+1}}{\sum_{i=1}^{n+1} \lambda_i} f(x_{n+1})$$
$$\le \frac{\sum_{i=1}^n \lambda_i}{\sum_{i=1}^{n+1} \lambda_i} \frac{\sum_{i=1}^n \lambda_i f(x_i)}{\sum_{i=1}^n \lambda_i} + \frac{\lambda_{n+1}}{\sum_{i=1}^{n+1} \lambda_i} f(x_{n+1}) = \frac{\sum_{i=1}^{n+1} \lambda_i f(x_i)}{\sum_{i=1}^{n+1} \lambda_i}.$$

17.9. In Problem 17.8 take the convex function $f(x) = -\ln x$, $x > 0$ and $\lambda_1 = \cdots = \lambda_n = 1$ (for an alternative answer see Problem 1.10).

17.10. In Problem 17.8 take the convex function $f(x) = x^\alpha, x > 0, \alpha \ge 1$ and $\lambda_1 = \cdots = \lambda_n = 1$.

Chapter 18

Indeterminate Forms

From the results of the previous chapters, we do not know how to find $\lim_{x \to x_0} f(x)/g(x)$ when $\lim_{x \to x_0} f(x) = \lim_{x \to x_0} g(x) = 0$. In this chapter, we will discuss L'Hôpital's rule (originally due to John Bernoulli) which enables us to determine limits of functions that are not only in an *indeterminate form* $0/0$, but also ∞/∞, and perhaps even $\infty - \infty$, $0 \cdot \infty$, 0^0, ∞^0, and 1^∞. Notice that two forms of L'Hôpital's rule for sequences have already been discussed in Theorem 8.4 and Problem 8.15.

Theorem 18.1 (L'Hôpital's Rule, $0/0$ Case). Assume that f' and g' exist on a deleted nbd $I_\delta^*(x_0)$ of x_0, where $g'(x) \neq 0$, and $\lim_{x \to x_0} f(x) = \lim_{x \to x_0} g(x) = 0$ and $\lim_{x \to x_0}[f'(x)/g'(x)]$ exists (finite or infinite). Then, $\lim_{x \to x_0}[f(x)/g(x)]$ exists and is the same as $\lim_{x \to x_0}[f'(x)/g'(x)]$.

Proof. If necessary, we define $f(x_0) = g(x_0) = 0$, so that the functions f and g are continuous on $I_\delta(x_0)$. If $x_0 < x < x_0 + \delta$, then f and g are continuous on $[x_0, x]$, differentiable on (x_0, x) and $g'(t) \neq 0$, $t \in (x_0, x)$. Also $g(x) \neq 0$, in fact, if $g(x) = 0$ then since $g(x_0) = 0$ by Rolle's theorem there exists a point $c \in (x_0, x)$ such that $g'(c) = 0$. But this contradicts the fact that $g'(t) \neq 0$, $t \in (x_0, x)$. Thus, by Cauchy's mean-value theorem there exists a point t_x such that

$$\frac{f(x)}{g(x)} = \frac{f(x) - f(x_0)}{g(x) - g(x_0)} = \frac{f'(t_x)}{g'(t_x)}.$$

Now as $x \to x_0^+$, $t_x \to x_0^+$, and since $\lim_{x \to x_0}[f'(x)/g'(x)]$ exists, it follows that

$$\lim_{x \to x_0^+} \frac{f(x)}{g(x)} = \lim_{x \to x_0^+} \frac{f'(t_x)}{g'(t_x)} = \lim_{x \to x_0} \frac{f'(x)}{g'(x)}.$$

The same argument shows that

$$\lim_{x \to x_0^-} \frac{f(x)}{g(x)} = \lim_{x \to x_0} \frac{f'(x)}{g'(x)}.$$

Thus, we have

$$\lim_{x \to x_0} \frac{f(x)}{g(x)} = \lim_{x \to x_0} \frac{f'(x)}{g'(x)}.$$

Corollary 18.1. Assume that f' and g' exist on an nbd $(x_0, x_0 + \delta)$, where $g'(x) \neq 0$, and $\lim_{x \to x_0^+} f(x) = \lim_{x \to x_0^+} g(x) = 0$, and $\lim_{x \to x_0^+} [f'(x)/g'(x)]$ exists (finite or infinite). Then, $\lim_{x \to x_0^+} [f(x)/g(x)]$ exists and is the same as $\lim_{x \to x_0^+} [f'(x)/g'(x)]$.

Corollary 18.2. Assume that f' and g' exist on an nbd $(x_0 - \delta, x_0)$, where $g'(x) \neq 0$, and $\lim_{x \to x_0^-} f(x) = \lim_{x \to x_0^-} g(x) = 0$, and $\lim_{x \to x_0^-} [f'(x)/g'(x)]$ exists (finite or infinite). Then, $\lim_{x \to x_0^-} [f(x)/g(x)]$ exists and is the same as $\lim_{x \to x_0^-} [f'(x)/g'(x)]$.

Corollary 18.3. Assume that $f^{(n)}$ and $g^{(n)}$ exist on a deleted nbd $I_\delta^*(x_0)$ of x_0, where $g^{(n)}(x) \neq 0$, and $\lim_{x \to x_0} f^{(k)}(x) = \lim_{x \to x_0} g^{(k)}(x) = 0$, $k = 0, 1, \cdots, n-1$ and $\lim_{x \to x_0} [f^{(n)}(x)/g^{(n)}(x)]$ exists (finite or infinite). Then, $\lim_{x \to x_0} [f(x)/g(x)]$ exists and is the same as $\lim_{x \to x_0} [f^{(n)}(x)/g^{(n)}(x)]$.

Example 18.1. From Corollary 18.3 it follows that

$$
\begin{aligned}
\lim_{x \to 0} \frac{\sin x - x \cos x}{x^2 \sin x} &= \lim_{x \to 0} \frac{x \sin x}{2x \sin x + x^2 \cos x} \\
&= \lim_{x \to 0} \frac{\sin x}{2 \sin x + x \cos x} = \lim_{x \to 0} \frac{\cos x}{3 \cos x - x \sin x} = \frac{1}{3}.
\end{aligned}
$$

Theorem 18.2. Assume that f' and g' exist on (M, ∞) for some $M \in R$, where $g'(x) \neq 0$, and $\lim_{x \to \infty} f(x) = \lim_{x \to \infty} g(x) = 0$ and $\lim_{x \to \infty} [f'(x)/g'(x)]$ exists (finite or infinite). Then, $\lim_{x \to \infty} [f(x)/g(x)]$ exists and is the same as $\lim_{x \to \infty} [f'(x)/g'(x)]$.

Proof. Let $t = 1/x$ so that $x \to \infty$ iff $t \to 0^+$. Thus, $\lim_{x \to \infty} f(x) = \lim_{x \to \infty} g(x) = 0$ are the same as $\lim_{t \to 0^+} f(1/t) = \lim_{t \to 0^+} g(1/t) = 0$. Now from Corollary 18.1 it follows that

$$
\lim_{t \to 0^+} \frac{f(1/t)}{g(1/t)} = \lim_{t \to 0^+} \frac{-f'(1/t)/t^2}{-g'(1/t)/t^2} = \lim_{t \to 0^+} \frac{f'(1/t)}{g'(1/t)}
$$

provided the limit on the right exists, i.e.,

$$
\lim_{x \to \infty} \frac{f(x)}{g(x)} = \lim_{x \to \infty} \frac{f'(x)}{g'(x)}
$$

provided the limit on the right exists.

Replacing x by $-x$ a result similar to Theorem 18.2 holds when $x \to -\infty$.

Example 18.2. Two applications of Theorem 18.2 gives

$$
\lim_{x \to \infty} x^2 \ln^2 \left(1 + \frac{1}{x}\right) = \lim_{x \to \infty} \frac{\ln^2(1 + 1/x)}{1/x^2} = \lim_{x \to \infty} \frac{(-2/x^2) \ln(1 + 1/x)}{-2/x^3}
$$

$$
= \lim_{x \to \infty} \frac{\ln(1 + 1/x)}{1/x} = \lim_{x \to \infty} \frac{(-1/x^2)[1/(1 + 1/x)]}{-1/x^2} = \lim_{x \to \infty} \frac{x}{x+1} = 1.
$$

Theorem 18.3 (L'Hôpital's Rule, ∞/∞ Case). Assume that f' and g' exist on $(x_0, x_0 + \delta)$ where $g'(x) \neq 0$, and $\lim_{x \to x_0^+} f(x) = \lim_{x \to x_0^+} g(x) = \infty$ and $\lim_{x \to x_0^+} [f'(x)/g'(x)]$ exists (finite or infinite). Then, $\lim_{x \to x_0^+} [f(x)/g(x)]$ exists and is the same as $\lim_{x \to x_0^+} [f'(x)/g'(x)]$.

Proof. Suppose that $\lim_{x \to x_0^+} [f'(x)/g'(x)] = L$ is finite. Let $h(x) = f(x) - Lg(x)$, $x \in (x_0, x_0 + \delta)$. Then, $\lim_{x \to x_0^+} [h'(x)/g'(x)] = 0$. Therefore, since $g(x) \to \infty$ as $x \to x_0^+$, for $\epsilon > 0$ there exists a $\delta_1 > 0$ such that

$$0 < x - x_0 < \delta_1 < \delta \implies g(x) > 0 \quad \text{and} \quad \left| \frac{h'(x)}{g'(x)} \right| < \frac{\epsilon}{2}. \tag{18.1}$$

If $x_0 < x < x_0 + \delta_1/2$, then g and h satisfy the conditions of Cauchy's mean-value theorem on $[x, x_0 + \delta_1/2]$, and we have

$$\left| \frac{h(x_0 + \delta_1/2) - h(x)}{g(x_0 + \delta_1/2) - g(x)} \right| = \left| \frac{h'(t_x)}{g'(t_x)} \right| < \frac{\epsilon}{2}, \quad t_x \in (x, x_0 + \delta_1/2). \tag{18.2}$$

Since $g(x) \to \infty$ as $x \to x_0^+$, there exists $\delta_2 < \delta_1/2$ such that

$$0 < x - x_0 < \delta_2 \implies g(x) > g(x_0 + \delta_1/2) > 0. \tag{18.3}$$

Combining (18.1) to (18.3), we get

$$0 < x - x_0 < \delta_2 \implies \frac{|h(x) - h(x_0 + \delta_1/2)|}{g(x)} < \frac{\epsilon}{2}.$$

Again, since $g(x) \to \infty$ as $x \to x_0^+$, there exists $\delta_3 \in (0, \delta_2)$ such that

$$0 < x - x_0 < \delta_3 \implies g(x) > 2|h(x_0 + \delta_1/2)|/\epsilon.$$

Therefore, for $0 < x - x_0 < \delta_3$, we have

$$\left| \frac{h(x)}{g(x)} \right| \leq \frac{|h(x) - h(x_0 + \delta_1/2)|}{g(x)} + \frac{|h(x_0 + \delta_1/2)|}{g(x)} < \frac{\epsilon}{2} + \frac{\epsilon}{2} = \epsilon.$$

This shows that $\lim_{x \to x_0^+} [h(x)/g(x)] = 0$, i.e., $\lim_{x \to x_0^+} [f(x)/g(x)] = L$.

If $L = +\infty$, then $\lim_{x \to x_0^+} [g'(x)/f'(x)] = 0$. Thus, $\lim_{x \to x_0^+} [g(x)/f(x)] = 0$. Now since $\lim_{x \to x_0^+} f(x) = \lim_{x \to x_0^+} g(x) = \infty$ there exists a δ such that $f(x)$ and $g(x)$ are positive on $(x_0, x_0 + \delta)$. Hence, we must have $\lim_{x \to x_0^+} [f(x)/g(x)] = +\infty$. The case $L = -\infty$ cannot arise.

In Theorem 18.3, by replacing f and g by $-f$ and $-g$, respectively, we get the following corollary.

Corollary 18.4. Assume that f' and g' exist on $(x_0, x_0 + \delta)$ where $g'(x) \neq 0$, and $\lim_{x \to x_0^+} f(x) = \lim_{x \to x_0^+} g(x) = -\infty$, and $\lim_{x \to x_0^+} [f'(x)/$

$g'(x)]$ exists (finite or infinite). Then, $\lim_{x \to x_0^+}[f(x)/g(x)]$ exists and is the same as $\lim_{x \to x_0^+}[f'(x)/g'(x)]$.

It is clear that in Theorem 18.3 as well as in Corollary 18.4 with appropriate changes we can replace $x \to x_0^+$ by $x \to x_0^-$, $x \to x_0$, or $x \to \infty$. We also remark that the other indeterminate forms $\infty - \infty$, $0 \cdot \infty$, 0^0, ∞^0 and 1^∞, often can be reduced to the forms $0/0$ and ∞/∞. For example, if $f \to \infty$ and $g \to \infty$ we can write $f - g = (1/g - 1/f)/(1/fg)$, which leads to a form $0/0$.

Example 18.3. To find $\lim_{x \to \infty}(\sqrt{x+1} - \sqrt{x})$, which assumes the form $\infty - \infty$, we note that

$$
\begin{aligned}
\lim_{x \to \infty}(\sqrt{x+1} - \sqrt{x}) &= \lim_{x \to \infty}(\sqrt{x+1} - \sqrt{x})\frac{(\sqrt{x+1}+\sqrt{x})}{(\sqrt{x+1}+\sqrt{x})} \\
&= \lim_{x \to \infty}\frac{x+1-x}{\sqrt{x+1}+\sqrt{x}} = \lim_{x \to \infty}\frac{1}{\sqrt{x+1}+\sqrt{x}} = 0.
\end{aligned}
$$

Example 18.4. To find $\lim_{x \to \pi} \sin x \ln(|\tan x|)$, which is of the form $0 \cdot \infty$, we note that

$$
\begin{aligned}
\lim_{x \to \pi} \sin x \ln(|\tan x|) &= \lim_{x \to \pi}\frac{\ln(|\tan x|)}{\csc x} = \lim_{x \to \pi}\frac{\sec^2 x/\tan x}{-\csc^2 x \cot x} \\
&= -\lim_{x \to \pi}\tan^2 x = 0.
\end{aligned}
$$

Example 18.5. To find $\lim_{x \to 0}|\sin x|^x$, which is of the form 0^0, we note that $\lim_{x \to 0}|\sin x|^x = \lim_{x \to 0}\exp(x \ln(|\sin x|))$. Now since

$$
\begin{aligned}
\lim_{x \to 0} x \ln(|\sin x|) &= \lim_{x \to 0}\frac{\ln(|\sin x|)}{1/x} = \lim_{x \to 0}\frac{\cos x/\sin x}{-1/x^2} \\
&= -(\lim_{x \to 0}\cos x)\left(\lim_{x \to 0}\frac{x^2}{\sin x}\right) = -\lim_{x \to 0}\frac{2x}{\cos x} = 0,
\end{aligned}
$$

it follows that $\lim_{x \to 0}|\sin x|^x = 1$.

Example 18.6. To find $\lim_{x \to 0^+}(1/x)^{\sin x}$, which assumes the form ∞^0, we set $f(x) = (1/x)^{\sin x}$ which is defined for all $x > 0$. Then, $\ln f(x) = (\sin x)\ln(1/x) = -(\sin x)(\ln x) = -\ln x/(1/\sin x)$, which is of the form ∞/∞. Now since

$$
\begin{aligned}
\lim_{x \to 0^+}\frac{-\ln x}{1/\sin x} &= \lim_{x \to 0^+}\frac{-1/x}{-\cos x/\sin^2 x} = \lim_{x \to 0^+}\frac{\sin^2 x}{x \cos x} \\
&= \lim_{x \to 0^+}\frac{2\sin x \cos x}{-x \sin x + \cos x} = 0,
\end{aligned}
$$

it follows that $\lim_{x \to 0^+}(1/x)^{\sin x} = \lim_{x \to 0^+} f(x) = \lim_{x \to 0^+} e^{\ln(f(x))} = e^0 = 1$.

Example 18.7. To find $\lim_{x \to \pi/2}(\sin x)^{\tan x}$, which is of the form 1^∞, we set $f(x) = (\sin x)^{\tan x}$. Clearly, $f(x) = \exp(\tan x \ln(\sin x))$, and since

$$
\begin{aligned}
\lim_{x \to \pi/2}\tan x \ln(\sin x) &= \lim_{x \to \pi/2}\frac{\ln(\sin x)}{\cot x} = \lim_{x \to \pi/2}\frac{\cos x/\sin x}{-\csc^2 x} \\
&= -\lim_{x \to \pi/2}\sin x \cos x = 0,
\end{aligned}
$$

it follows that $\lim_{x \to \pi/2}(\sin x)^{\tan x} = 1$.

Problems

18.1. Find the following limits

(i). $\lim_{x \to 2} \frac{x^2-4}{x^2+x-6}$.

(ii). $\lim_{x \to \pi/2} \sec 3x \cos 5x$.

(iii). $\lim_{x \to 0} \left(\cot x - \frac{1}{x} \right)$.

(iv). $\lim_{x \to 1} \frac{x^c - cx + c - 1}{(x-1)^2}$ for any $c \in \mathcal{Q}$.

(v). $\lim_{x \to 0} \frac{x^2 \sin \frac{1}{x}}{\tan x}$.

(vi). $\lim_{x \to 0} \frac{(\sin 4x)(\sin 3x)}{x \sin 2x}$.

(vii). $\lim_{x \to 0} \frac{a^{2x} - b^{2x}}{x}$, $\quad a > 0, \quad b > 0$.

(viii). $\lim_{x \to \infty} \frac{x + \sin x}{x}$.

(ix). $\lim_{x \to \infty} x \ln \left(1 + \frac{a}{x} \right)$, $\ a \in \mathcal{R}$, and hence deduce $\lim_{x \to \infty} \left(1 + \frac{a}{x} \right)^x$.

(x). $\lim_{x \to \infty} \frac{x^n}{e^x}$, $\ n \geq 1$ is an integer, and hence deduce $\lim_{x \to \infty} \frac{x^\alpha}{e^x}$, where α is a positive real number.

18.2. Let g be a function on $[-1, 1]$ such that $g(0) = 0 = g'(0)$ and $g''(0) = 17$. Define the function f on $[-1, 1]$ by

$$f(x) = \begin{cases} \frac{g(x)}{x} & \text{if } x \neq 0 \\ 0 & \text{if } x = 0. \end{cases}$$

Find $f'(0)$.

18.3. Find a pair of functions f and g for which $f(0) = g(0) = 0$ and $\lim_{x \to 0} f(x)/g(x)$ exists, but $\lim_{x \to 0} f'(x)/g'(x)$ does not exist.

18.4. Show that if ϕ is continuous at x_0, differentiable on a deleted nbd $I_\delta^*(x_0)$ of x_0, and $\lim_{x \to x_0} \phi'(x)$ exists, then $\phi'(x_0)$ exists and $\phi'(x_0) = \lim_{x \to x_0} \phi'(x)$, i.e., ϕ' is continuous at x_0.

18.5. Let f be n-times differentiable at x_0 and let $P_n(x)$ be Taylor's polynomial of degree n for the function f at the point x_0. Show that

$$\lim_{x \to x_0} \frac{f(x) - P_n(x)}{(x - x_0)^n} = 0.$$

18.6 (Cauchy's Rule). Let the functions f, g be defined on an interval I and $x_0 \in I$. If f and g are differentiable at x_0, $f(x_0) = g(x_0) = 0$ and

$g'(x_0) \neq 0$, then there exists an nbd N of x_0 such that $g(x) \neq 0$ for all $x \in N\backslash\{x_0\}$ and $\lim_{x \to x_0} f(x)/g(x) = f'(x_0)/g'(x_0)$.

18.7. For the functions $f(x) = \begin{cases} x^2, & \text{if } x \text{ is rational} \\ 0, & \text{if } x \text{ is irrational} \end{cases}$, $g(x) = \sin x$ use Problem 18.6 to find $\lim_{x \to 0} f(x)/g(x)$.

18.8. For the functions $f(x) = \begin{cases} x^2 \sin \frac{1}{x}, & x \neq 0 \\ 0, & x = 0 \end{cases}$, $g(x) = x$ use Problem 18.6 to find $\lim_{x \to 0} f(x)/g(x)$.

Answers or Hints

18.1. (i). $4/5$.
(ii). $-5/3$.
(iii). 0.
(iv). $c(c-1)/2$.
(v). 0.
(vi). 6.
(vii). $\ln(a/b)^2$.
(viii). 1.
(ix). a; e^a.
(x). 0; let $n \geq \alpha$, then for $x \geq 1$, $0 < \frac{x^\alpha}{e^x} \leq \frac{x^n}{e^x}$, and hence $\lim_{x \to \infty} \frac{x^\alpha}{e^x} = 0$.

18.2. $f'(0) = \lim_{h \to 0}(f(h) - f(0))/h = \lim_{h \to 0} f(h)/h = \lim_{h \to 0} g(h)/h^2 = \lim_{h \to 0} g'(h)/2h = \lim_{h \to 0}(g'(h) - g'(0))/2h = (1/2)g''(0) = 17/2$.

18.3. Let $f(x) = \begin{cases} x^2 \sin \frac{1}{x}, & x \neq 0 \\ 0, & x = 0 \end{cases}$ and $g(x) = x$. Then, $f'(x) = \begin{cases} 2x \sin \frac{1}{x} - \cos \frac{1}{x}, & x \neq 0 \\ 0, & x = 0 \end{cases}$ and $g'(x) = 1$. Now, $\lim_{x \to 0} f(x)/g(x) = \lim_{x \to 0} x \sin \frac{1}{x} = 0$, but $\lim_{x \to 0} f'(x)/g'(x) = \lim_{x \to 0} \left[2x \sin \frac{1}{x} - \cos \frac{1}{x}\right]$ does not exist.

18.4. Let $f(x) = \phi(x) - \phi(x_0)$ and $g(x) = x - x_0$. Then, from Theorem 18.1, we have $\lim_{x \to x_0}(\phi(x) - \phi(x_0))/(x - x_0) = \lim_{x \to x_0} \phi'(x)$, i.e., $\phi'(x_0) = \lim_{x \to x_0} \phi'(x)$.

18.5. Use Corollary 18.3, $(n-1)$-times, and then use the definition of $f^{(n)}(x_0)$.

18.6. Since $\lim_{x \to x_0}(g(x) - g(x_0))/(x - x_0) = g'(x_0) \neq 0$, there is an nbd N of x_0 such that $(g(x) - g(x_0))/(x - x_0) \neq 0$ for all $x \in N\backslash\{x_0\}$. Now, we have $\lim_{x \to x_0} \frac{f(x)}{g(x)} = \frac{\lim_{x \to x_0}(f(x) - f(x_0))/(x - x_0)}{\lim_{x \to x_0}(g(x) - g(x_0))/(x - x_0)} = \frac{f'(x_0)}{g'(x_0)}$.

18.7. Note that f is differentiable only at $x = 0$ and $f'(0) = 0$, whereas g is differentiable on $\mathcal{R}$ and $g'(x) = \cos x$, $g'(0) = 1$. Clearly, L'Hôpital's rule (Theorem 18.1) is not applicable; however, Cauchy's rule gives $\lim_{x \to 0} \frac{f(x)}{g(x)}$ $= \frac{f'(0)}{g'(0)} = \frac{0}{1} = 0$.

18.8. In view of the solution to Problem 18.3 note that for these functions L'Hôpital's rule (Theorem 18.1) is not applicable; however, Cauchy's rule gives $\lim_{x \to 0} \frac{f(x)}{g(x)} = \frac{f'(0)}{g'(0)} = \frac{0}{1} = 0$.

Chapter 19

Riemann Integration

In this and the next chapters, we will provide a rigorous treatment of the Riemann integral which is usually encountered in beginning calculus courses where the emphasis is placed on computational and operational aspects of the integral. The main purpose of this chapter is to provide necessary and/or sufficient conditions for a function to be Riemann integrable. We begin with some definitions.

Recall that a collection of points $P = \{x_0, x_1, \cdots, x_n\}$ satisfying $a = x_0 < x_1 < \cdots < x_n = b$ is called a *partition* of the interval $[a, b]$. The partition points of P give rise to the partition subintervals $[x_0, x_1], [x_1, x_2], \cdots, [x_{n-1}, x_n]$. The norm of P, denoted as $\|P\|$, is defined by $\|P\| = \max_{1 \leq i \leq n} \{x_i - x_{i-1}\}$, i.e., $\|P\|$ is the length of the largest subinterval determined by P. A partition P' is called a refinement of P if $P' \subset P$. The smallest partition of an interval $[a, b]$ is $P = \{a, b\}$. A refinement of this partition is $P' = \{a, (a + b)/2, b\}$.

Let $f : [a, b] \to \mathcal{R}$, $c_i \in [x_{i-1}, x_i]$, $i = 1, 2, \cdots, n$ be arbitrary points, and let $\Delta x_i = x_i - x_{i-1}$. The sum

$$S(f, P) = \sum_{i=1}^{n} f(c_i) \Delta x_i$$

is called a *Riemann sum* corresponding to the interval $[a, b]$. Clearly, for a given partition, there are infinitely many Riemann sums because there are infinitely many ways of choosing the points c_i's.

A function f is *Riemann integrable* (integrable, hereafter) on $[a, b]$ if there exists a number L such that given $\epsilon > 0$ there exists a $\delta > 0$ such that any partition P of $[a, b]$ with $\|P\| < \delta$ and any Riemann sum $S(f, P)$ corresponding to f and P (i.e., for any choices $c_i \in [x_{i-1}, x_i], i = 1, 2, \cdots, n$) satisfies $|S(f, P) - L| < \epsilon$. The number L, if it exists, is called the *Riemann integral* of f and we write it as

$$L = \int_a^b f(x) dx.$$

Example 19.1. We shall apply the above definition to show that $\int_0^1 x dx = 1/2$. Let $\epsilon > 0$ be given and let P be a partition of $[0, 1]$ such that $\|P\| < \epsilon$ (in other words, we take $\delta = \epsilon$). Choose arbitrary points $c_i \in [x_{i-1}, x_i], i = 1, 2, \cdots, n$. Then, we have

$$\left| \sum_{i=1}^{n} c_i \Delta x_i - \frac{1}{2} \right| = \left| \sum_{i=1}^{n} \bar{x}_i \Delta x_i - \frac{1}{2} + \sum_{i=1}^{n} (c_i - \bar{x}_i) \Delta x_i \right|, \qquad (19.1)$$

where $\bar{x}_i = (x_i + x_{i-1})/2$. Now, the first two terms on the right of (19.1) give

$$\begin{aligned} \sum_{i=1}^{n} \bar{x}_i \Delta x_i - \frac{1}{2} &= \sum_{i=1}^{n} \frac{x_i + x_{i-1}}{2} (x_i - x_{i-1}) - \frac{1}{2} \\ &= \sum_{i=1}^{n} \frac{x_i^2 - x_{i-1}^2}{2} - \frac{1}{2} = \frac{1}{2} (x_n^2 - x_0^2) - \frac{1}{2} = 0 \end{aligned}$$

since $x_n = 1, x_0 = 0$. Therefore, it follows that

$$\left| \sum_{i=1}^{n} c_i \Delta x_i - \frac{1}{2} \right| = \left| \sum_{i=1}^{n} (c_i - \bar{x}_i) \Delta x_i \right| \leq \sum_{i=1}^{n} |c_i - \bar{x}_i| \Delta x_i \leq \epsilon \sum_{i=1}^{n} \Delta x_i = \epsilon.$$

The following result asserts that the integral of a function if it exists then it is unique.

Theorem 19.1. If f is integrable on $[a, b]$ then its integral L is unique.

Proof. Assume that L and M are integrals of f on $[a, b]$. Let $\epsilon > 0$. Since L is an integral of f, there exists a $\delta_L > 0$ such that for any partition P of $[a, b]$ with $\|P\| < \delta_L$ and any Reimann sum $S(f, P)$, we have

$$|S(f, P) - L| < \epsilon. \qquad (19.2)$$

Similarly, there exists a $\delta_M > 0$ such that for any partition P of $[a, b]$ with $\|P\| < \delta_M$ and any Reimann sum $S(f, P)$, we have

$$|S(f, P) - M| < \epsilon. \qquad (19.3)$$

Now choose a partition P_1 of $[a, b]$ such that $\|P_1\| < \min\{\delta_L, \delta_M\}$ and a Reimann sum $S(f, P_1)$. Then, (19.2) and (19.3) are simultaneously satisfied. Therefore, it follows that

$$|L - M| \leq |L - S(f, P_1)| + |S(f, P_1) - M| \leq 2\epsilon.$$

Since ϵ is arbitrary, $L = M$ and thus the integral is unique.

For $f : [a, b] \to \mathcal{R}$ to be integrable it is necessary that f must be bounded.

Theorem 19.2. If f is unbounded on $[a, b]$ then f is not integrable on $[a, b]$.

Proof. Unboundedness of f implies that there exists a sequence $\{d_i\}_{i=1}^{\infty} \subset [a, b]$ such that, without loss of generality, $f(d_i) \to \infty$ as $i \to \infty$. The Heine-Borel Theorem 4.4 implies that a subsequence $\{d_{i_k}\}_{k=1}^{\infty}$ converges to some $d \in [a, b]$. To show that f is not integrable on $[a, b]$ we need to show that there exists an $\epsilon_0 > 0$ such that for any real number L and any partition $P = \{a = x_0 < x_1 < \cdots < x_n = b\}$ of $[a, b]$ there is a choice of points $c_i \in [x_{i-1}, x_i]$, $i = 1, 2, \cdots, n$ such that $|\sum_{i=1}^n f(c_i)\Delta x_i - L| \geq \epsilon_0$. For this, we take $\epsilon_0 = 1$. Let $P = \{a = x_0 < x_1 < \cdots < x_n = b\}$ be any partition of $[a, b]$ and L any real number. Then, there exists a k_0 such that $\{d_{i_k}\}_{k=1}^\infty \cap [x_{k_0-1}, x_{k_0}]$ contains infinitely many elements. Put $\{e_i\}_{i=1}^\infty = \{d_{i_k}\}_{k=1}^\infty \cap [x_{k_0-1}, x_{k_0}]$ and observe that $f(e_i) \to \infty$ as $i \to \infty$. Choose any points $c_i \in [x_{i-1}, x_i]$, $i = 1, 2, \cdots, n$, $i \neq i_0$ and choose $c_{k_0} = e_m$, where

$$f(e_m) > \frac{1}{\Delta x_{k_0}} \left(1 + L - \sum_{i \neq k_0} f(c_i)\Delta x_i\right).$$

Then, $\sum_{i=1}^n f(c_i)\Delta x_i - L = \sum_{i \neq k_0}^n f(c_i)\Delta x_i + f(c_{k_0})\Delta x_{k_0} > 1$. Therefore, L cannot be the integral of f. Since L is an arbitrary real number, f is not integrable.

We note that working directly with arbitrary Riemann sums may be cumbersome and usually very technical as we have experienced in Example 19.1. To minimize this technicality, in what follows, we shall introduce *upper and lower Riemann sums (Darboux sums)* and *upper and lower Riemann integrals (Darboux integrals)*.

Suppose f is bounded on $[a, b]$ and $P = \{a = x_0 < x_1 < \cdots < x_n = b\}$ is a partition of $[a, b]$. *The upper Riemann sum* $\overline{S}(f, P)$ corresponding to f and P is defined by

$$\overline{S}(f, P) = \sum_{i=1}^n M_i\Delta x_i,$$

where $M_i = \sup_{x \in [x_{i-1}, x_i]} f(x)$, $i = 1, 2, \cdots, n$. *The lower Riemann sum* $\underline{S}(f, P)$ corresponding to f and P is defined by

$$\underline{S}(f, P) = \sum_{i=1}^n m_i\Delta x_i,$$

where $m_i = \inf_{x \in [x_{i-1}, x_i]} f(x)$, $i = 1, 2, \cdots, n$. The upper Riemann integral $\overline{\int_a^b} f(x)dx$ of f is defined by

$$\overline{\int_a^b} f(x)dx = \inf_P \overline{S}(f, P).$$

The lower Riemann integral $\underline{\int_a^b} f(x)dx$ of f is defined by

$$\underline{\int_a^b} f(x)dx = \sup_P \underline{S}(f, P).$$

Clearly, any Riemann sum $S(f, P)$ satisfies

$$m(b-a) \leq \underline{S}(f, P) \leq S(f, P) \leq \overline{S}(f, P) \leq M(b-a),$$

where $m = \inf_{x \in [a,b]} f(x)$ and $M = \sup_{x \in [a,b]}$. Also, $\overline{S}(-f, P) = -\underline{S}(f, P)$ and $\underline{S}(-f, P) = -\overline{S}(f, P)$.

Example 19.2. Consider the function defined in Problem 11.3(i) on $[a, b]$. Let $P = \{a = x_0 < x_1 < \cdots < x_n = b\}$ be any partition of $[a, b]$. Since any of the subintervals $[x_{i-1}, x_i]$ contains a rational number and an irrational number, we have

$$\underline{S}(f, P) = \sum_{i=1}^{n} 0 \cdot \Delta x_i = 0, \quad \overline{S}(f, P) = \sum_{i=1}^{n} 1 \cdot \Delta x_i = 1,$$

$$\overline{\int_a^b} f(x)dx = 1, \quad \underline{\int_a^b} f(x)dx = 0.$$

Theorem 19.3. Let $f : [a, b] \to \mathcal{R}$ be a bounded function and P_1 and P_2 are partitions of $[a, b]$ such that $P_1 \subseteq P_2$ then

$$\overline{S}(f, P_1) \geq \overline{S}(f, P_2) \quad \text{and} \quad \underline{S}(f, P_1) \leq \underline{S}(f, P_2).$$

Proof. Let $P_1 = \{a = x_0 < x_1 < \cdots < x_{n_1} = b\}$. Since for $i = 1, 2, \cdots, n_1, x_{i-1}, x_i \in P_1 \subseteq P_2$, we may label the points in P_2 so that $x_{i-1} = x_{i1} < x_{i2} < \cdots < x_{in_i} = x_i$. Denote by M_i the sup of f on $[x_{i-1}, x_i]$ and by M_{ij} the sup of f on $[x_{i,j-1}, x_{ij}]$. Observe that $M_i \geq M_{ij}$ for each $j = 1, 2, \cdots, n_i$ and $\Delta x_i = \sum_{j=1}^{n_i} \Delta x_{ij}$ for $i = 1, 2, \cdots, n_1$. Thus, it follows that

$$\begin{aligned}
\overline{S}(f, P_1) &= \sum_{i=1}^{n_1} M_i \Delta x_i = \sum_{i=1}^{n_1} M_i \sum_{j=1}^{n_i} \Delta x_{ij} \\
&= \sum_{i=1}^{n_1} \sum_{j=1}^{n_i} M_i \Delta x_{ij} \geq \sum_{i=1}^{n_1} \sum_{j=1}^{n_i} M_{ij} \Delta x_{ij} = \overline{S}(f, P_2).
\end{aligned}$$

Similarly, we can show that $\underline{S}(f, P_1) \leq \underline{S}(f, P_2)$.

Remark 19.1. If in Theorem 19.3, P_1 and P_2 are arbitrary partitions of $[a, b]$, then we have

$$\underline{S}(f, P_1) \leq \underline{S}(f, P_1 \cup P_2) \leq \overline{S}(f, P_1 \cup P_2) \leq \overline{S}(f, P_2).$$

Corollary 19.1. Let $f : [a, b] \to \mathcal{R}$ be a bounded function. Then, $\underline{\int_a^b} f(x)dx \leq \overline{\int_a^b} f(x)dx$.

Proof. From Remark 19.1, we have $\underline{S}(f, P_1) \leq \overline{S}(f, P_2)$, where P_1 and P_2 are arbitrary partitions of $[a, b]$. This means that $\overline{S}(f, P_2)$ is an upper bound for the set $Q = \{\underline{S}(f, P_1) : P_1 \text{ is a partition of } [a, b]\}$. Thus, $\overline{S}(f, P_2)$

is at least as large as $\sup Q = \underline{\int_a^b} f(x)dx$, i.e., $\underline{\int_a^b} f(x)dx \leq \overline{S}(f, P_2)$ for every partition P_2 of $[a, b]$. But, this simply implies that $\underline{\int_a^b} f(x)dx \leq \inf\{\overline{S}(f, P_2) : P_2 \text{ is a partition of } [a, b]\} = \overline{\int_a^b} f(x)dx$.

In our next result we provide relationships between the Riemann integral of a function f and its lower and upper integrals.

Theorem 19.4. Let $f : [a, b] \to \mathcal{R}$ be a bounded function.

(1). If f is integrable on $[a, b]$, then $\int_a^b f(x)dx = \overline{\int_a^b} f(x)dx = \underline{\int_a^b} f(x)dx$.

(2). If $\overline{\int_a^b} f(x)dx \leq \underline{\int_a^b} f(x)dx$, then f is integrable on $[a, b]$.

Proof. (1). Assume that f is integrable on $[a, b]$. Let $\epsilon > 0$. There exists a partition $P = \{a = x_0 < x_1 < \cdots < x_n = b\}$ of $[a, b]$ such that

$$\left| S(f, P) - \int_a^b f(x)dx \right| < \epsilon \tag{19.4}$$

for any Riemann sum $S(f, P)$ corresponding to f and P. For each $i = 1, 2, \cdots, n$ choose a point c_i in $[x_{i-1}, x_i]$ such that $f(c_i) \geq M_i - \epsilon/(b-a)$ and let $S(f, P)$ be the Riemann sum corresponding to these points. Then, it follows that

$$S(f, P) = \sum_{i=1}^n f(c_i)\Delta x_i \geq \sum_{i=1}^n \left(M_i - \frac{\epsilon}{b-a} \right) \Delta x_i = \overline{S}(f, P) - \epsilon. \tag{19.5}$$

From (19.4) and (19.5), we find

$$\overline{\int_a^b} f(x)dx \leq \overline{S}(f, P) \leq \int_a^b f(x)dx + 2\epsilon. \tag{19.6}$$

Repeating the above steps for the Riemann lower sums, we get a Riemann sum $S(f, P)$ such that

$$S(f, P) \leq \underline{S}(f, P) + \epsilon,$$

which, together with (19.4) again give

$$\int_a^b f(x)dx \leq S(f, P) + \epsilon \leq \underline{S}(f, P) + 2\epsilon \leq \underline{\int_a^b} f(x)dx + 2\epsilon. \tag{19.7}$$

Combining (19.6) and (19.7), we obtain

$$\overline{\int_a^b} f(x)dx \leq \int_a^b f(x)dx + 2\epsilon \leq \underline{\int_a^b} f(x)dx + 4\epsilon.$$

Since ϵ is arbitrary, Corollary 19.1 implies that

$$\overline{\int_a^b} f(x)dx \;\leq\; \int_a^b f(x)dx \;\leq\; \underline{\int_a^b} f(x)dx \;\leq\; \overline{\int_a^b} f(x)dx.$$

But, this means equality must hold throughout.

(2). Assume that $\overline{\int_a^b} f(x)dx \leq \underline{\int_a^b} f(x)dx$. Let $\epsilon > 0$ be arbitrary. Then, there exists a partition P_1 of $[a,b]$ such that $\overline{S}(f,P_1) < \overline{\int_a^b} f(x)dx + \epsilon$ and there exists a partition P_2 such that $\underline{S}(f,P_2) > \underline{\int_a^b} f(x)dx - \epsilon$. Let $P = P_1 \cup P_2$ and $L = \overline{\int_a^b} f(x)dx$. For any Riemann sum corresponding to f, P in view of Theorem 19.3, we have

$$
\begin{aligned}
L - \epsilon \;&=\; \overline{\int_a^b} f(x)dx - \epsilon \;\leq\; \underline{\int_a^b} f(x)dx - \epsilon \;<\; \underline{S}(f,P_2) \;\leq\; \underline{S}(f,P) \\
&\leq\; S(f,P) \;\leq\; \overline{S}(f,P) \;\leq\; \overline{S}(f,P_1) \;<\; \overline{\int_a^b} f(x)dx + \epsilon \;=\; L + \epsilon.
\end{aligned}
$$

Thus, it follows that $|S(f,P) - L| < \epsilon$, i.e., f is integrable on $[a,b]$.

Theorem 19.5. Let $f : [a,b] \to \mathcal{R}$ be a bounded function. Then, f is integrable on $[a,b]$ iff for each $\epsilon > 0$ there exists a partition P of $[a,b]$ such that

$$\overline{S}(f,P) - \underline{S}(f,P) \;<\; 2\epsilon.$$

Proof. Suppose f is integrable, then from Theorem 19.4(1), we have $\overline{\int_a^b} f(x)dx = \underline{\int_a^b} f(x)dx$. Now let P_1, P_2, and P be as in Theorem 19.4(2). Then, it follows that

$$
\begin{aligned}
\overline{S}(f,P) - \underline{S}(f,P) \;&\leq\; \overline{S}(f,P_1) - \underline{S}(f,P_2) \\
&\leq\; \left(\overline{\int_a^b} f(x)dx + \epsilon\right) - \left(\underline{\int_a^b} f(x)dx - \epsilon\right) \;=\; 2\epsilon.
\end{aligned}
$$

Conversely, given $\epsilon > 0$, suppose there exists a partition P of $[a,b]$ such that $\overline{S}(f,P) < \underline{S}(f,P) + 2\epsilon$. Then, we have

$$\overline{\int_a^b} f(x)dx \;\leq\; \overline{S}(f,P) \;<\; \underline{S}(f,P) + 2\epsilon \;\leq\; \underline{\int_a^b} f(x)dx + 2\epsilon.$$

But, since ϵ is arbitrary, we must have $\overline{\int_a^b} f(x)dx \leq \underline{\int_a^b} f(x)dx$. The result now follows from Theorem 19.4(2).

In view of Theorem 19.5, Example 19.2 suggests that not every bounded function is integrable. In Problems 19.5 to 19.7 we provide some easily verifiable conditions which ensure the existence of Riemann integration of functions.

Problems

19.1. Let $f : [a, b] \to \mathcal{R}$ be integrable. Show that

(i). $\lim_{n\to\infty} \sum_{i=1}^{n} hf(a + ih) = \int_a^b f(x)dx$, where $h = (b - a)/n$.

(ii). $\lim_{n\to\infty} \frac{1}{n^3}(1 + 4 + 9 + \cdots + n^2) = \frac{1}{3}$.

(iii). $\lim_{n\to\infty} \sum_{i=1}^{n} \frac{1}{n} \sin^2\left(\frac{\pi i}{3n}\right) = \frac{\pi}{6} - \frac{\sqrt{3}}{8}$.

(iv). $\lim_{n\to\infty} n\left(\frac{1}{1+n^2} + \frac{1}{2^2+n^2} + \cdots + \frac{1}{n^2+n^2}\right) = \frac{\pi}{4}$.

(v). $\lim_{n\to\infty} \frac{3}{n^6}(2^5 + 3^5 + \cdots + n^5) = \frac{1}{2}$.

19.2. Let $f : [a, b] \to \mathcal{R}$ be integrable. Show that

$$\lim_{n\to\infty} \sum_{i=1}^{n} f(ah^i)(ah^i - ah^{i-1}) = \int_a^b f(x)dx,$$

where $a, b > 0$ and $h = (b/a)^{1/n}$. In particular, calculate $\int_2^3 x^2 dx$ by considering a partition P which divides the interval $[2, 3]$ into n parts in geometric progression at the points $2, 2h, 2h^2, 2h^3, \cdots, 2h^{n-1}, 2h^n = 3$.

19.3. For the function $f(x) = [x]$ = the largest integer $\leq x$, evaluate $\int_0^2 f(x)dx$ using upper and lower sums.

19.4. Let $f, g : [a, b] \to \mathcal{R}$ be bounded functions. Show that for any partition P of $[a, b]$,

$$\overline{S}(f + g, P) \leq \overline{S}(f, P) + \overline{S}(g, P),$$
$$\underline{S}(f + g, P) \geq \underline{S}(f, P) + \underline{S}(g, P).$$

19.5. Show that if $f : [a, b] \to \mathcal{R}$ is monotone on $[a, b]$, then f is integrable on $[a, b]$.

19.6. Show that if $f : [a, b] \to \mathcal{R}$ is continuous on $[a, b]$, then f is integrable on $[a, b]$.

19.7. Show that if $f : [a, b] \to \mathcal{R}$ is bounded and has finitely many discontinuities on $[a, b]$, then f is integrable on $[a, b]$. In particular, evaluate $\int_0^1 f(x)dx$ where

$$f(x) = \begin{cases} 1, & \text{if } x = \frac{1}{1000}, \frac{2}{1000}, \cdots, 1 \\ 0, & \text{otherwise.} \end{cases}$$

19.8. Prove that the following function is integrable on $[0, 1]$ even though it has infinitely many discontinuities on $[0, 1]$

$$f(x) = \begin{cases} 1 & \text{if } x = \frac{1}{n}, \quad n = 1, 2, \cdots \\ 0 & \text{otherwise.} \end{cases}$$

What is $\int_0^1 f(x)dx$?

19.9. Give an example of a function $f : [0,1] \to \mathcal{R}$ which is

(i). Bounded on $[0,1]$, but not integrable.

(ii). Integrable on $[0,1]$, but not continuous on $[0,1]$.

(iii). Integrable on $[0,1]$, but not monotone on $[0,1]$.

(iv). Integrable on $[0,1]$, but neither continuous nor monotone on $[0,1]$.

19.10. Determine whether $f : [0,1] \to \mathcal{R}$ is integrable.

(i). $f(x) = \left| x - \frac{1}{2} \right| \cos x$.

(ii). $f(x) = \begin{cases} x^4/(x-1) & \text{for } x \neq 1 \\ 1 & \text{for } x = 1. \end{cases}$

(iii). $f(x) = \begin{cases} 1/n & \text{if } 1/2^n < x \leq 1/2^{n-1}, \quad n = 1, 2, \cdots \\ 0 & \text{for } x = 0. \end{cases}$

(iv). $f(x) = \begin{cases} 1/\sin x & \text{for } x \neq 0 \\ 1 & \text{for } x = 0. \end{cases}$

(v). $f(x) = \begin{cases} 3x & \text{if } 0 \leq x \leq 1/3 \\ 1 + (3x-1)^2 & \text{if } 1/3 < x \leq 1/2 \\ 5x & \text{if } 1/2 < x \leq 1. \end{cases}$

(vi). $f(x) = (70x)^{20} - \sqrt{20x} + \tan x^2$.

Answers or Hints

19.1. (i). $\sum_{i=1}^n hf(a+ih)$ is a Riemann sum, $\|P\| = (b-a)/n \to 0$ as $n \to \infty$.

(ii). $\lim_{n\to\infty} \sum_{i=1}^n \frac{1}{n}\frac{i^2}{n^2} = \int_0^1 x^2 dx = \frac{1}{3}$.

(iii). $\lim_{n\to\infty} \sum_{i=1}^n \frac{1}{n} \sin^2\left(\frac{\pi i}{3n}\right) = \int_0^{\pi/3} \sin^2 x dx = \frac{\pi}{6} - \frac{\sqrt{3}}{8}$.

(iv). $\lim_{n\to\infty} \frac{1}{n}\left(\frac{1}{(1/n)^2+1} + \cdots + \frac{1}{(n/n)^2+1}\right) = \int_0^1 \frac{dx}{1+x^2} = \frac{\pi}{4}$.

(v). $\lim_{n\to\infty} \frac{3}{n^6}(1^5 + 2^5 + 3^5 + \cdots + n^5) = \int_0^1 3x^5 dx = \frac{1}{2}$.

19.2. $\sum_{i=1}^n f(ah^i)(ah^i - ah^{i-1})$ is a Riemann sum, $h \to 1$ and so $\|P\| \to 0$ as $n \to \infty$. For the given integral, we have $\sum_{i=1}^n (2h^i)^2(2h^i - 2h^{i-1}) = 8(h-1)\sum_{i=1}^n h^{3i-1} = 8(h^{3n}-1)h^2(h-1)/(h^3-1) = 8((3/2)^3 - 1)h^2/(h^2+h+1) \to 19/3$ as $h \to 1$.

19.3. $f(x) = \begin{cases} 0, & 0 \leq x < 1 \\ 1, & 1 \leq x < 2 \\ 2, & x = 2. \end{cases}$ Consider the partition $P_n : 0 < \frac{1}{n} < \frac{2}{n} < \cdots < \frac{2n-1}{n} < 2$. Then, $\underline{S}(f, P_n) = \frac{1}{n}(n \times 1) = 1$, $\overline{S}(f, P_n) = \frac{1}{n}(n \times 1 + 2) = $

$1 + \frac{2}{n}$. Thus, $\lim_{n\to\infty} \underline{S}(f, P_n) \leq \underline{\int_0^2} f(x)dx \leq \overline{\int_0^2} f(x)dx \leq \lim_{n\to\infty} \overline{S}(f, P_n)$ implies $\int_0^2 f(x)dx = 1$.

19.4. Use $\sup_{x\in[x_{i-1},x_i]}(f(x)+g(x)) \leq \sup_{x\in[x_{i-1},x_i]} f(x) + \sup_{x\in[x_{i-1},x_i]} g(x)$, and $\inf_{x\in[x_{i-1},x_i]}(f(x)+g(x)) \geq \inf_{x\in[x_{i-1},x_i]} f(x) + \inf_{x\in[x_{i-1},x_i]} g(x)$.

19.5. Assume that f is nondecreasing (the other case can be considered similarly). If f is a constant c, then it is easy to show that $\int_a^b f(x)dx = c(b-a)$. So, we may now assume that $f(b) > f(a)$. Let $\epsilon > 0$ and $P = \{a = x_0 < x_1 < \cdots < x_n = b\}$ be a partition of $[a, b]$ such that $\|P\| \leq \epsilon/(f(b) - f(a))$. Then, we have

$$
\begin{aligned}
\int_a^b f(x)dx &\leq \overline{S}(f, P) = \sum_{i=1}^n f(x_i)\Delta x_i \\
&= \underline{S}(f, P) + \sum_{i=1}^n (f(x_i) - f(x_{i-1}))\Delta x_i \\
&\leq \underline{S}(f, P) + \|P\| \sum_{i=1}^n (f(x_i) - f(x_{i-1})) \\
&= \underline{S}(f, P) + \|P\|(f(b) - f(a)) \leq \underline{S}(f, P) + \epsilon \leq \int_a^b f(x)dx + \epsilon.
\end{aligned}
$$

The conclusion follows from Theorem 19.4(2) since ϵ is arbitrary.

19.6. Since f is continuous on the closed and bounded interval $[a, b]$, f is uniformly continuous there. Let $\epsilon > 0$. There exists a $\delta > 0$ such that if $x, y \in [a, b]$ and $|x - y| < \delta$, then $|f(x) - f(y)| < \epsilon/(b-a)$. Let $P = \{a = x_0 < x_1 < \cdots < x_n = b\}$ be a partition of $[a, b]$ such that $\|P\| \leq \delta$. Since f is continuous there exists t_i and s_i in $[x_{i-1}, x_i]$ such that $f(t_i) = \sup_{x\in[x_{i-1},x_i]} f(x)$ and $f(s_i) = \inf_{x\in[x_{i-1},x_i]} f(x)$ for all $i = 1, \cdots, n$. Thus, it follows that

$$
\begin{aligned}
\overline{S}(f, P) - \underline{S}(f, P) &= \sum_{i=1}^n [f(t_i) - f(s_i)](x_i - x_{i-1}) \\
&< \frac{\epsilon}{(b-a)} \sum_{i=1}^n (x_i - x_{i-1}) = \epsilon.
\end{aligned}
$$

The conclusion follows from Theorem 19.5 since ϵ is arbitrary.

19.7. Suppose there is a single discontinuity which occurs at $c \in [a, b]$, and that $|f(x)| \leq M$ for all $x \in [a, b]$. Let $\epsilon > 0$. Since f is continuous on $\{x : x \in [a, b], |x - c| \geq \epsilon/8M\} = J$, there exists a partition P of J such that $\overline{S}(f, P) - \underline{S}(f, P) < \epsilon/2$. Now consider the partition $P' = P \cup \{c - \epsilon/8M, c + \epsilon/8M\}$ of $[a, b]$. Then, $\overline{S}(f, P') \leq \overline{S}(f, P) + \frac{\epsilon}{4M}M = \overline{S}(f, P) + \epsilon/4$ and $\underline{S}(f, P') \geq \underline{S}(f, P) - \frac{\epsilon}{8M}M = \underline{S}(f, P) - \epsilon/4$. Hence, $\overline{S}(f, P') - \underline{S}(f, P') < \epsilon$. In fact, taking inf and sup over all partitions P, we find $\int_a^b f(x)dx = \inf \overline{S}(f, P) = \sup \underline{S}(f, P)$. Generalization of this argument to the case of finitely many discontinuities on $[a, b]$ is straightforward. Clearly, for the given function $\int_0^1 f(x)dx = 0$.

19.8. Let $\epsilon > 0$ and $1/n_0 \leq \epsilon$, $n_0 \in \mathcal{N}$. Then, on $[1/2n_0, 1]$, f has finitely many discontinuities. Thus there exists a partition P of $[1/2n_0, 1]$ such that $\overline{S}(f, P) - \underline{S}(f, P) < \epsilon/2$. Now consider the partition $P' = P \cup \{0, 1/2n_0\}$. Then, $\overline{S}(f, P') = \frac{1}{2n_0} + \overline{S}(f, P) < \overline{S}(f, P) + \frac{\epsilon}{2}$, $\underline{S}(f, P') = 0 + \underline{S}(f, P)$. Hence, $\overline{S}(f, P') - \underline{S}(f, P') < \epsilon$. Since $\underline{S}(f, P) = 0$ for all P it follows that $\int_0^1 f(x)dx = \sup \underline{S}(f, P) = 0$.

19.9 (i). See Example 19.2.

(ii). See Problem 19.8.

(iii). $f(x) = \left| x - \frac{1}{2} \right|$.

(iv). $f(x) = \begin{cases} 0, & 0 \le x < 1/3 \\ 1, & 1/3 \le x < 2/3 \\ 0, & 2/3 \le x \le 1. \end{cases}$

19.10. (i). f is continuous and hence integrable.

(ii). f is not bounded and hence not integrable.

(iii). Consider $[0, 1/2^n]$ and $[1/2^n, 1]$. Clearly, f is integrable on $[1/2^n, 1]$, and hence for given $\epsilon > 0$ we can choose $n \in \mathcal{N}$ such that $1/2^n < \epsilon/2$. Now there exists a partition P_2 of $[1/2^n, 1]$ such that $\overline{S}(f, P_2) - \underline{S}(f, P_2) < \epsilon/2$; also note that $\overline{S}(f, P_1) - \underline{S}(f, P_1) < \epsilon/2$ for any partition P_1 of $[0, 1/2^n]$. Let $P = P_1 \cup P_2$, then $\overline{S}(f, P) - \underline{S}(f, P) < \epsilon$.

(iv). f is not bounded and hence not integrable.

(v). f has only a finite number of discontinuities and hence integrable.

(vi). f is continuous and hence integrable.

Chapter 20

Properties of the Riemann Integral

In this chapter, we will study various properties of the Riemann integral, some of which you have seen in elementary calculus courses. Throughtout, in what follows we shall assume that if a function f is defined on a "point interval" $[c, c]$, then any Riemann sum consists of only one interval of length zero. Then, any Riemann sum is zero, and hence, the integral of f on the point interval should be zero, i.e., $\int_c^c f(x)dx = 0$. Similarly, if f is defined on a "backward interval" $[b, a]$ with $b > a$, then the length of each subinterval is negative and thus the integral over the backward interval is the negative of the integral on its forward counterpart, i.e., $\int_b^a f(x)dx = -\int_a^b f(x)dx$.

Theorem 20.1. If $f : [a, b] \to \mathcal{R}$ is integrable, then it is integrable on any subinterval of $[a, b]$.

Proof. Let $[a', b']$ be a subinterval of $[a, b]$ and $\epsilon > 0$. Since f is integrable on $[a, b]$, by Theorem 19.5 there exists a partition $P = \{a = x_0 < x_1 < \cdots < x_n = b\}$ of $[a, b]$ such that $\overline{S}(f, P) - \underline{S}(f, P) < \epsilon$. Let $P' = P \cup \{a', b'\}$. Then, by Theorem 19.3, $\overline{S}(f, P') \le \overline{S}(f, P)$ and $\underline{S}(f, P') \ge \underline{S}(f, P)$. Hence, $\overline{S}(f, P') - \underline{S}(f, P') < \epsilon$. Now, $P_1 = P' \cap [a', b']$ is a partition of $[a', b']$ and we can write $P_1 = \{a' = x_{i_1} < x_{i_1+1} < \cdots < x_{i_2} = b'\}$. Then, we have

$$
\begin{aligned}
\overline{S}(f, P_1) - \underline{S}(f, P_1) &\le \sum_{i=i_1+1}^{i_2}(M_i - m_i)\Delta x_i \\
&\le \sum_{i=1}^{n}(M_i - m_i)\Delta x_i = \overline{S}(f, P') - \underline{S}(f, P') < \epsilon.
\end{aligned}
$$

The conclusion now follows from Theorem 19.5.

Theorem 20.2. If $f : [a, b] \to \mathcal{R}$ is integrable and $c \in [a, b]$, then

$$
\int_a^b f(x)dx = \int_a^c f(x)dx + \int_c^b f(x)dx.
$$

Proof. By Theorem 20.1, f is integrable on $[a, c]$ and $[c, b]$. Let $\epsilon > 0$. Then, there exists a partition P_1 of $[a, c]$ and a partition P_2 of $[c, b]$ such that $\overline{S}(f, P_1) \le \int_a^c f(x)dx + \epsilon$ and $\overline{S}(f, P_2) \le \int_c^b f(x)dx + \epsilon$. Let $P = P_1 \cup P_2$. Then, P is a partition of $[a, b]$ and

$$
\int_a^b f(x)dx \le \overline{S}(f, P) = \overline{S}(f, P_1) + \overline{S}(f, P_2) \le \int_a^c f(x)dx + \int_c^b f(x)dx + 2\epsilon.
$$

Since ϵ is arbitrary, it follows that

$$\int_a^b f(x)dx \;\leq\; \int_a^c f(x)dx + \int_c^b f(x)dx.$$

Working similarly with lower sums, we obtain a reverse inequality. This completes the proof.

The following result shows the linearity and monotonicity of the integral.

Theorem 20.3.   Assume that $f, g : [a, b] \to \mathcal{R}$ are integrable and $\alpha \in \mathcal{R}$. Then, the following hold:

(1). $\int_a^b (f(x) + g(x))dx = \int_a^b f(x)dx + \int_a^b g(x)dx$.

(2). $\int_a^b \alpha f(x)dx = \alpha \int_a^b f(x)dx$.

(3). If $f(x) \geq g(x)$ for each $x \in [a, b]$, then $\int_a^b f(x)dx \geq \int_a^b g(x)dx$.

Proof.   (1). Let P be a partition of $[a, b]$. Then, from Problem 19.4 it follows that

$$\int_a^b (f(x) + g(x))dx \;\leq\; \overline{S}(f+g, P) \;\leq\; \overline{S}(f, P) + \overline{S}(g, P).$$

Therefore, for a given $\epsilon > 0$, one can choose a partition P such that $\overline{S}(f, P) \leq \int_a^b f(x)dx + \epsilon$ and $\overline{S}(g, P) \leq \int_a^b g(x)dx + \epsilon$. Then, we have

$$\int_a^b (f(x) + g(x))dx \;\leq\; \int_a^b f(x)dx + \int_a^b g(x)dx + 2\epsilon.$$

Working similarly with lower sums, we get the reverse inequality.

(2). The statement is obvious for $\alpha = 0$, so we assume that $\alpha \neq 0$. Let $\epsilon > 0$. Then, there exists a $\delta > 0$ such that for any partition $P = \{a = x_0 < x_1 < \cdots < x_n = b\}$ of $[a, b]$ with $\|P\| < \delta$ and for any choices $c_i \in [x_{i-1}, x_i], i = 1, 2, \cdots, n$ we have

$$\left| \sum_{i=1}^n f(c_i)\Delta x_i - \int_a^b f(x)dx \right| \;<\; \frac{\epsilon}{|\alpha|},$$

which is the same as

$$\left| \sum_{i=1}^n \alpha f(c_i)\Delta x_i - \alpha \int_a^b f(x)dx \right| \;<\; \epsilon,$$

and this means that $\int_a^b \alpha f(x)dx = \alpha \int_a^b f(x)dx$.

(3). Let $P = \{a = x_0 < x_1 < \cdots < x_n = b\}$ be a partition of $[a, b]$. For each $i = 1, 2, \cdots, n$, denote by M_i^f, M_i^g the sup values of f, g, respectively, on $[x_{i-1}, x_i]$ and observe that $M_i^f \geq M_i^g$. Hence, it follows that

$$\overline{S}(f, P) \geq \overline{S}(g, P) \geq \int_a^b g(x) dx.$$

Thus $\int_a^b g(x) dx$ is a lower bound for the upper sums of f. Thus, we have $\int_a^b f(x) dx \geq \int_a^b g(x) dx$.

Theorem 20.4. Assume that $f : [a, b] \to \mathcal{R}$ is integrable. Then, the following hold:

(1). The functions f_+, f_- defined by $f_+(x) = \max\{f(x), 0\}, f_-(x) = \min\{f(x), 0\}$ are integrable on $[a, b]$.

(2). The function $|f|$ is integrable on $[a, b]$.

(3). If $g : [a, b] \to \mathcal{R}$ is also integrable, then fg is integrable on $[a, b]$.

Proof. (1). We shall prove the integrability of f_+; the argument for f_- is similar. Let $P = \{a = x_0 < x_1 < \cdots < x_n = b\}$ be a partition of $[a, b]$. For each $i = 1, 2, \cdots, n$, denote by M_i^f, M_i^{f+} the sup values of f, f_+, respectively, on $[x_{i-1}, x_i]$. Clearly,

$$M_i^{f+} - m_i^{f+} = \begin{cases} M_i^f - m_i^f & \text{if } f \geq 0 \text{ on } [x_{i-1}, x_i], \\ 0 & \text{if } f \leq 0 \text{ on } [x_{i-1}, x_i], \\ M_i^f & \text{if } f \text{ changes sign on } [x_{i-1}, x_i]. \end{cases}$$

If f changes sign on $[x_{i-1}, x_i]$, then $M_i^f > 0$ and $m_i^f < 0$. Hence $M_i^f - m_i^f \geq M_i^f$. It follows that, in all three cases, $M_i^{f+} - m_i^{f+} \leq M_i^f - m_i^f$ for all $1 \leq i \leq n$. Thus, we have

$$\overline{S}(f_+, P) - \underline{S}(f_+, P) \leq \overline{S}(f, P) - \underline{S}(f, P). \tag{20.1}$$

The integrability of f on $[a, b]$ means that, for a given $\epsilon > 0$ we can choose a partition P such that $\overline{S}(f, P) - \underline{S}(f, P) < \epsilon$. Thus, it follows from (20.1) that $\overline{S}(f_+, P) - \underline{S}(f_+, P) < \epsilon$, which implies that f_+ is integrable.

(2). The integrability of $|f|$ follows from (1) by noticing that $|f| = f_+ - f_-$.

(3). Assume first that f, g are nonnegative on $[a, b]$. Let $P = \{a = x_0 < x_1 < \cdots < x_n = b\}$ be a partition of $[a, b]$. For each $i = 1, 2, \cdots, n, x \in [x_{i-1}, x_i]$, $f(x)g(x) \leq M_i^f g(x) \leq M_i^f M_i^g$, and hence $M_i^{fg} \leq M_i^f M_i^g$. Similarly, $m_i^f m_i^g \leq m_i^{fg}$. Hence, it follows that

$$M_i^{fg} - m_i^{fg} \leq M_i^f M_i^g - m_i^f m_i^g \leq M_i^f (M_i^g - m_i^g) \leq M^f (M_i^g - m_i^g)$$

and

$$\overline{S}(fg, P) - \underline{S}(fg, P) \leq M^f (\overline{S}(g, P) - \underline{S}(g, P)).$$

If f, g are nonpositive on $[a, b]$, then, by part (2) and the previous argument, $fg = (-f)(-g)$ is integrable. If f is nonnegative and g is nonpositive, then, by part (2) and the previous argument again $fg = -(f(-g))$ is integrable. For general f, g, the result follows from part (1) and the cases just discussed by noting that $fg = (f_+ + f_-)(g_+ + g_-) = f_+ g_+ + f_- g_+ + f_+ g_- + f_- g_-$.

Corollary 20.1. If f is integrable on $[a, b]$, then

$$\left| \int_a^b f(x)dx \right| \leq \int_a^b |f(x)|dx.$$

Now let $c \in [a, b]$ and define the function $F : [a, b] \to \mathcal{R}$ as follows:

$$F(x) = \int_c^x f(t)dt, \quad x \in [a, b]. \tag{20.2}$$

The function F is defined for $x \geq c$ as well as $x \leq c$, exists in view of Theorem 20.1, and is unique from Theorem 19.1.

Theorem 20.5. Assume that $f : [a, b] \to \mathcal{R}$ is integrable, then the function F defined by (20.2) is continuous on $[a, b]$.

Proof. Since f is integrable it is bounded on $[a, b]$, and so is $|f|$, i.e., there exists a $M > 0$ such that $|f(x)| \leq M$, $x \in [a, b]$. For $\epsilon > 0$ there exists a $\delta > 0$ such that $0 < M\delta < \epsilon$. Thus, if $x_0 \in [a, b], x_0 + h \in [a, b]$ and $|h| < \delta$, then we have

$$|F(x_0 + h) - F(x_0)| = \left| \int_c^{x_0+h} f(t)dt - \int_c^{x_0} f(t)dt \right|$$
$$= \left| \int_{x_0}^{x_0+h} f(t)dt \right| \leq \int_{x_0}^{x_0+h} |f(t)|dt$$
$$\leq M|h| < M\delta < \epsilon.$$

Theorem 20.6 (The Fundamental Theorem of Calculus). Assume that $f : [a, b] \to \mathcal{R}$ is integrable and define F by (20.2). Then, F is differentiable at any point $x_0 \in (a, b)$ at which f is continuous, and $F'(x_0) = f(x_0)$. If f is continuous from the right at a then $F'(a^+) = f(a)$ and if f is continuous from the left at b then $F'(b^-) = f(b)$.

Proof. Assume that f is continuous at $x_0 \in (a, b)$. Let $\epsilon > 0$. Then, there exists a $\delta > 0$ such that if $|x - x_0| < \delta$, then $|f(x) - f(x_0)| < \epsilon$. For sufficiently small $h, x_0 + h \in (a, b)$. Assume that $h > 0$ (the argument for $h < 0$ is similar). Using Corollary 20.1, we have

$$\left| \frac{F(x_0+h)-F(x_0)}{h} - f(x_0) \right| = \left| \frac{1}{h} \int_{x_0}^{x_0+h} f(t)dt - f(x_0) \right|$$
$$= \left| \frac{1}{h} \int_{x_0}^{x_0+h} (f(t) - f(x_0))dt \right|$$
$$\leq \frac{1}{h} \int_{x_0}^{x_0+h} |f(t) - f(x_0)|dt$$
$$\leq \frac{\epsilon}{h} \int_{x_0}^{x_0+h} dt = \epsilon.$$

Thus, F is differentiable at x_0 and $F'(x_0) = f(x_0)$.

Remark 20.1. Theorem 20.5 asserts that a continuous function is the derivative of its integral. For this reason the process of integration is viewed as an inverse operation of differentiation. Further, it reflects that the process of differentiation may be regarded as the inverse of integration.

An *antiderivative* of a function f defined on $[a, b]$ is any function F, continuous on $[a, b]$ and differentiable in (a, b) such that $F'(x) = f(x)$ for all $x \in (a, b)$. If F is an antiderivative of f on $[a, b]$, then G is also an antiderivative iff there exists a constant c such that $G(x) = F(x) + c$ for all $x \in [a, b]$. Indeed, if $G(x) = F(x) + c$, then $G'(x) = F'(x) = f(x)$, and hence G is an antiderivative of f. Conversely, suppose G is also an antiderivative of f and set $\phi(x) = G(x) - F(x)$. Then, $\phi'(x) = G'(x) - F'(x) = f(x) - f(x) = 0$, i.e., $\phi'(x) = 0$, which implies that $\phi(x) = c$. The following result which is an immediate consequence of Theorem 20.6 shows that the knowledge of antiderivative helps to evaluate the integral $\int_a^b f(x)dx$.

Theorem 20.7 (Newton-Leibniz Theorem). If f is continuous on $[a, b]$, then it has an antiderivative F on $[a, b]$, and $\int_a^b f(x)dx = F(b) - F(a)$.

The following example illustrates the importance of the continuity assumption in Theorem 20.7.

Example 20.1. Consider the function

$$F(x) = \begin{cases} x^2 \sin \frac{1}{x^2}, & 0 < x \le 1 \\ 0, & x = 0. \end{cases}$$

This function has the derivative

$$F'(x) = f(x) = \begin{cases} 2x \sin \frac{1}{x^2} - \frac{2}{x} \cos \frac{1}{x^2}, & 0 < x \le 1 \\ 0, & x = 0. \end{cases}$$

Clearly, f is not bounded in any neighborhood of 0 and hence, it cannot be integrable on $[0, 1]$. Thus, Theorem 20.7 cannot be applied for this function.

Remark 20.2. The notation $\int f(x)dx$ stands for any antiderivative of f and is usually called the indefinite integral of f, whereas the integral $\int_a^b f(x)dx$ is also called the definite integral of f. The indefinite integral is a function while the definite integral is a number.

Analogous to the mean-value theorem of differential calculus (Theorem 15.5), we have the mean-value theorem of integral calculus.

Theorem 20.8 (First Mean-Value Theorem of Integral Calculus). Let $f : [a,b] \to \mathcal{R}$ be continuous, and $g : [a,b] \to \mathcal{R}$ be integrable and nonnegative. Then, there exists a $c \in (a,b)$ such that

$$\int_a^b f(x)g(x)dx = f(c)\int_a^b g(x)dx. \tag{20.3}$$

Proof. Since f is continuous on $[a,b]$, from Theorem 12.4 there exists m and M such that $m \leq f(x) \leq M$, $x \in [a,b]$. Since $g(x) \geq 0$, $x \in [a,b]$, we have $mg(x) \leq f(x)g(x) \leq Mg(x)$, $x \in [a,b]$. Thus, from Theorem 20.3(3) it follows that

$$m\int_a^b g(x)dx \leq \int_a^b f(x)g(x)dx \leq M\int_a^b g(x)dx. \tag{20.4}$$

Again, Theorem 20.3(3) implies that $\int_a^b g(x)dx \geq 0$. If $\int_a^b g(x)dx = 0$, then from (20.4), $\int_a^b f(x)g(x)dx = 0$ and the result holds. If $\int_a^b g(x)dx > 0$, then (20.4) can be written as

$$m \leq \int_a^b f(x)g(x)dx \Big/ \int_a^b g(x)dx \leq M.$$

But, then Theorem 12.6 implies that there exists some $c \in (a,b)$ such that

$$f(c) = \int_a^b f(x)g(x)dx \Big/ \int_a^b g(x)dx.$$

Corollary 20.2 (Mean-Value Theorem of Integral Calculus). If $f : [a,b] \to \mathcal{R}$ is continuous, then there exists a point $c \in (a,b)$ such that

$$\frac{1}{b-a}\int_a^b f(x)dx = f(c).$$

Theorem 20.9 (Chain Rule for Integrals). Let $f : [a,b] \to \mathcal{R}$ be continuous, and let $g, h : [\alpha,\beta] \to [a,b]$ be differentiable functions. Define the function G on $[\alpha,\beta]$ by

$$G(x) = \int_{h(x)}^{g(x)} f(t)dt.$$

Then, G is differentiable on $[\alpha,\beta]$, and for each $x \in [\alpha,\beta]$,

$$G'(x) = f(g(x))g'(x) - f(h(x))h'(x).$$

Proof. First, assume that h is a constant function, i.e., $h(x) = c \in [a,b]$ for all $x \in [\alpha,\beta]$, and define the function F by $F(y) = \int_a^y f(t)dt$. Then,

$G(x) = F(g(x))$. Applying Theorem 15.3, we get $G'(x) = F'(g(x))g'(x)$. Since $F'(y) = f(y)$ for all $y \in [a, b]$, $F'(g(x)) = f(g(x))$. Thus, it follows that

$$G'(x) = f(g(x))g'(x). \tag{20.5}$$

For a general h, we can choose any $c \in [a, b]$ and write

$$G(x) = \int_{h(x)}^{c} f(t)dt + \int_{c}^{g(x)} f(t)dt = -\int_{c}^{h(x)} f(t)dt + \int_{c}^{g(x)} f(t)dt$$

and apply (20.5) to each part on the right of this equation.

The following example illustrates the versatility of the chain rule for integrals. We can find derivatives of functions defined by integrals even in the cases where the integrals cannot be evaluated in closed forms.

Example 20.2. From Theorem 20.9 it follows that

$$\frac{d}{dx} \int_{\cos x}^{\sin x} \frac{1}{\sqrt{1 + t^4}} dt = \frac{\cos x}{\sqrt{1 + \sin^4 x}} + \frac{\sin x}{\sqrt{1 + \cos^4 x}}.$$

Finally in this chapter, we present the theoretical basis for the two most elementary techniques of integration, namely, the change of variables and the integration by parts.

Theorem 20.10 (Change of Variables). Suppose f is continuous on $[a, b]$ and $g : [\alpha, \beta] \to [a, b]$ is differentiable with g' integrable on $[\alpha, \beta]$. Let $g(\alpha) = c$, $g(\beta) = d$. Then, $(f \circ g)g'$ is integrable on $[\alpha, \beta]$, and

$$\int_{c}^{d} f(t)dt = \int_{\alpha}^{\beta} f(g(s))g'(s)ds.$$

Proof. Since g is differentiable on $[\alpha, \beta]$, it is continuous there. Hence, $f \circ g$ is continuous on $[\alpha, \beta]$. It follows from Theorem 20.4(3) that $(f \circ g)g'$ is integrable on $[\alpha, \beta]$. Now choose an $a_1 \in [a, b]$, and define the function F on $[\alpha, \beta]$ by $F(s) = \int_{a_1}^{g(s)} f(t)dt$. Then, by Theorem 20.9, we have $F'(s) = f(g(s))g'(s)$. Thus, by Theorem 20.7, we find

$$\int_{\alpha}^{\beta} f(g(s))g'(s)ds = \int_{\alpha}^{\beta} F'(s)ds = F(\beta) - F(\alpha)$$
$$= \int_{a_1}^{g(\beta)} f(t)dt - \int_{a_1}^{g(\alpha)} f(t)dt = \int_{a_1}^{d} f(t)dt - \int_{a_1}^{c} f(t)dt$$
$$= \int_{c}^{d} f(t)dt.$$

Theorem 20.11 (Integration by Parts). If f, g are functions defined on $[a, b]$ such that f', g' are integrable on $[a, b]$, then

$$\int_{a}^{b} f(t)g'(t)dt = f(b)g(b) - f(a)g(a) - \int_{a}^{b} g(t)f'(t)dt.$$

Proof. Since f, g are differentiable on $[a, b]$, they are continuous, and hence, integrable on $[a, b]$. It follows that fg, fg' and $f'g$ are integrable on $[a, b]$. Furthermore, $(fg)' = fg' + gf'$ is integrable. Integrating this relation on both sides on $[a, b]$, we get

$$\int_a^b (f(t)g(t))'dt = \int_a^b f(t)g'(t)dt + \int_a^b g(t)f'(t)dt.$$

Applying Theorem 20.7 to the left-hand side, we get

$$f(b)g(b) - f(a)g(a) = \int_a^b f(t)g'(t)dt + \int_a^b g(t)f'(t)dt,$$

from which the result follows.

Problems

20.1. Show that if $f : [a, b] \to \mathcal{R}$ is of bounded variation, then it is integrable on $[a, b]$.

20.2. (i). Suppose that $f : [a, b] \to \mathcal{R}$ is continuous and $f(x) \geq 0$ for all $x \in [a, b]$. If $\int_a^b f(x)dx = 0$, prove that $f(x) = 0$ for all $x \in [a, b]$.

(ii). Give an example to show that in (i), the assumption that f is continuous cannot be omitted.

(iii). Suppose that $f : [a, b] \to \mathcal{R}$ is continuous and $\int_a^b f(x)g(x)dx = 0$ for every integrable function g. Show that $f(x) = 0$ for all $x \in [a, b]$.

20.3. (i). Give an example to show that the conclusion of Theorem 20.8 may not hold without assuming that g is nonnegative on $[a, b]$. What happens if $f(x) = 1$ for all $x \in [a, b]$?

(ii). Apply Theorem 20.8 to show that

$$\frac{1}{32} \leq \int_0^1 \frac{100x^{99}}{(1+x^2)^5}dx \leq 1.$$

20.4. Prove or disprove the following:

(i). If $|f|$ is integrable on $[a, b]$, then so is f.

(ii). If g and h are integrable on $[a, b]$, then so are $\max(g, h)$ and $\min(g, h)$.

(iii). If f is integrable on $[a, b]$, then given any $\epsilon > 0$ there exists a $\delta > 0$ such that $\int_c^d |f(x)|dx < \epsilon$ when $[c, d] \subset [a, b], d - c < \delta$.

20.5. Let $f : [a, b] \to \mathcal{R}$ be integrable. Show that the function F defined by (20.2) is Lipschitz continuous on $[a, b]$ with Lipschitz constant $k = \sup_{[a,b]} |f(x)|$.

20.6. (i). Let f be a positive and continuous function on $[a, b]$. Show that for the function F defined in (20.2), F^{-1} exists and continuous. Moreover, if F^{-1} is differentiable (indeed this is true), find its derivative.

(ii). If f is nondecreasing on $[a, b]$ and $x_0 \in (a, b)$, show that F is differentiable iff f is continuous at x_0.

20.7. Let the conditions of Theorem 16.1 be satisfied. Show that Taylor's formula (16.1) for all $x \in [a, b]$ can be written as

$$f(x) = \sum_{k=0}^{n} (x - a)^k \frac{f^{(k)}(a)}{k!} + R_{n+1}, \qquad (20.6)$$

where R_{n+1} is the remainder term

$$R_{n+1} = \frac{1}{n!} \int_a^x (x - t)^n f^{(n+1)}(t) dt.$$

Further, from R_{n+1} deduce the Lagrange form of the remainder term in (16.2).

20.8 (Weierstrass's Mean-Value Theorem). Suppose $f, g : [a, b] \to \mathcal{R}$ are integrable with f increasing and g nonnegative on $[a, b]$. If $m \le f(a+0)$ and $M \ge f(b-0)$, then show that there exists an $x_0 \in [a, b]$ such that

$$\int_a^b f(x)g(x) dx = m \int_a^{x_0} g(x) dx + M \int_{x_0}^b g(x) dx. \qquad (20.7)$$

In particular, if f is also nonnegative on $[a, b]$, then there exists an $x_0 \in [a, b]$ such that

$$\int_a^b f(x)g(x) dx = M \int_{x_0}^b g(x) dx.$$

This relation is known as *Bonnet's Mean-Value Theorem*.

20.9. Show that

(i). If an odd function $f : [-a, a] \to \mathcal{R}$ is integrable, then $\int_{-a}^a f(x) dx = 0$.

(ii). If an even function $f : [-a, a] \to \mathcal{R}$ is integrable, then $\int_{-a}^a f(x) dx = 2 \int_0^a f(x) dx$.

(iii). If a function $f : \mathcal{R} \to \mathcal{R}$ is periodic with period w, then $\int_0^a f(x) dx = \int_w^{a+w} f(x) dx$.

(iv). If $f : \mathcal{R} \to \mathcal{R}$ is continuously differentiable, then $\left(\int_0^x f(t) dt \right)' \ne \int_0^x f'(t) dt$.

20.10. Let $f, g : (a, b) \to \mathcal{R}$ be piecewise continuous.

(i). Verify that

$$\frac{1}{2} \int_a^b \int_a^b [f(x)g(y) - g(x)f(y)]^2 dx dy = \|f\|^2 \|g\|^2 - (f \cdot g)^2,$$

where $(f \cdot g) = \int_a^b f(x)g(x)dx$ and $\|f\|^2 = (f \cdot f)$.

(ii). Deduce the following Cauchy-Schwarz inequality $|(f \cdot g)| \le \|f\|\|g\|$.

(iii). Deduce from part (ii) the following triangle inequality, $\|f + g\| \le \|f\| + \|g\|$.

(iv). Use the triangle inequality to show that

$$\int_0^1 \left(\sqrt{7 + 2t} + \sqrt{\pi \sin \pi t} \right)^2 dt \le 18.$$

Answers or Hints

20.1. From Theorem 14.8, $f = g - h$, where g and h are nondecreasing functions on $[a, b]$. From Problem 19.5 both the functions g and h are integrable on $[a, b]$. Thus f is integrable on $[a, b]$ now follows from Theorem 20.3(1).

20.2. (i). Suppose there exists an $x' \in [a, b]$ such that $f(x') > 0$. Since f is continuous at x', there exists a $\delta > 0$ such that $|f(x) - f(x')| < f(x')/2$ if $0 < |x - x'| < \delta$, i.e., $f(x) > f(x')/2$ for $|x - x'| < \delta$. Now consider any partition $P : a = x_0 < x_1 < x_2 < \cdots < x_n = b$ of $[a, b]$ with $x_i - x_{i-1} < \delta/2$. Then, there exists j's such that $\cup_j[x_j, x_{j+1}] \supset (x' - \delta/2, x' + \delta/2)$, and $\underline{S}(f, P) > (\delta/2)(f(x_0)/2)$. Thus, $\int_a^b f(x)dx \ge f(x_0)\delta/4 > 0$, which contradicts $\int_a^b f(x)dx = 0$.

(ii). On the interval $[0, 1]$ consider the function $f(x) = \begin{cases} 0, & x \ne 1/2 \\ 1, & x = 1/2. \end{cases}$

(iii). Since f is continuous, f is integrable on $[a, b]$. Take $g = f$. Then $fg = f^2$ is continuous, so integrable, $f^2 \ge 0$, and $\int_a^b f^2(x)dx = 0$. But, then by (i) $f^2 \equiv 0$.

20.3. (i). Take $f(x) = g(x) = \sin x, x \in [0, 2\pi]$. Clearly, $\int_0^{2\pi} \sin^2 xdx = \pi, \int_0^{2\pi} \sin xdx = 0$. When $f(x) = 1$, then $\int_a^b f(x)g(x)dx = \int_a^b g(x)dx = f(c)\int_a^b g(x)dx$ regardless what g is, so long as g is integrable.

(ii). Let $f(x) = 1/(1 + x^2)^5, g(x) = 100x^{99}$, so that $\int_0^1 g(x)dx = 1$ and $(1/32) \le f(c) \le 1$.

20.4. (i). False, let $f(x) = \begin{cases} 1, & x \in [a, b] \cap \mathcal{Q} \\ -1, & x \in [a, b] \backslash \mathcal{Q}, \end{cases}$ then for any partition P, $\overline{S}(f, P) = (b - a)$, $\underline{S}(f, P) = -(b - a)$. Hence, f is not integrable, whereas $|f| = 1$ is integrable.

(ii). True, since $\max(g, h) = (g + h + |g - h|)/2$, $\min(g, h) = -\max(-g, -h)$ (see Chapter 6), and g, h integrable implies $|g - h|$ is integrable.

(iii). f integrable implies $|f|$ is integrable and $|f| < M$ on $[a, b]$. For a given $\epsilon > 0$ take $\delta = \epsilon/M$, then $\int_c^d |f(x)| dx \leq \int_c^d M dx < \epsilon$ if $d - c < \delta$, $[c, d] \subset [a, b]$.

20.5. For $a \leq x_1 \leq x_2 \leq b$, from (20.2) it follows that $F(x_2) - F(x_1) = \int_{x_1}^{x_2} f(x) dx$, and hence from Corollary 20.1, we find $|F(x_2) - F(x_1)| \leq \int_{x_1}^{x_2} |f(x)| dx \leq k(x_2 - x_1)$.

20.6. (i). Since f is positive $\int_{x_1}^{x_2} f(t) dt > 0$ for $x_1, x_2 \in [a, b], x_2 > x_1$. Hence, $F(x)$ is strictly monotone increasing. Thus, F^{-1} exists. Moreover, from Theorem 20.6, F' exists, thus by the chain rule it follows that $1 = (F^{-1} \circ F)'(x) = (F^{-1})'(F(x))F'(x)$, and hence $(F^{-1})'(F(x)) = 1/F'(x)$, which implies that $(F^{-1})'(x) = 1/F'(F^{-1}(x)) = 1/f(F^{-1}(x))$.

(ii). Since f is monotone, $\lim_{x \to x_0^-} f(x)$ and $\lim_{x \to x_0^+} f(x)$ exist. Now similar to that in Theorem 20.6 one can show that $\lim_{x \to x_0^-} f(x) = F'_-(x_0)$ and $\lim_{x \to x_0^+} f(x) = F'_+(x_0)$.

20.7. From Theorem 20.11 it follows that
$R_{n+1} = \frac{1}{n!} \left[0 - (x - a)^n f^{(n)}(a) - \int_a^x n(x - t)^{n-1}(-1)f^{(n)}(t) dt \right]$.

Thus, (20.6) is the same as $f(x) = \sum_{k=0}^{n-1} (x-a)^k \frac{f^{(k)}(a)}{k!} + R_n$. Now repeated application of Theorem 20.11 proves (20.6). Next, note that $x - t \geq 0$. Hence, from Theorem 20.8, we have $R_{n+1} = \frac{1}{n!} f^{(n+1)}(x_0) \int_a^x (x - t)^n dt$, $x_0 \in (a, x)$.

20.8. Observe that $F(x) = m \int_a^x g(t) dt + M \int_x^b g(t) dt$ is continuous on $[a, b]$, and $mg(x) \leq f(x)g(x) \leq Mg(x)$, $x \in (a, b)$. Thus, $F(b) = m \int_a^b g(t) dt \leq \int_a^b f(t)g(t) dt \leq M \int_a^b g(t) dt = F(a)$. Hence, there exists an $x_0 \in (a, b)$ such that $F(x_0) = \int_a^b f(t)g(t) dt$. For the remaining part take $m = 0$ in (20.7).

20.9. (i). $\int_{-a}^a f(x) dx = \int_{-a}^0 f(x) dx + \int_0^a f(x) dx$. In the first integral use the substitution $x = -t$ and the fact that f is odd, to obtain $\int_{-a}^0 f(x) dx = \int_a^0 f(-t)(-dt) = \int_a^0 f(t) dt = -\int_0^a f(t) dt$.

(ii). Same as (i) except that now f is even.

(iii). Use the substitution $x = t - \omega$ and the fact that f is periodic of period ω.

(iv). Let $f(x) = 1$.

20.10. (i). $\int_a^b \int_a^b [f(x)g(y) - g(x)f(y)]^2 dx dy$
$= \int_a^b \int_a^b \{f^2(x)g^2(y) - 2f(x)g(y)g(x)f(y) + g^2(x)f^2(y)\} dx dy$
$= \left(\int_a^b f^2(x) dx \right) \left(\int_a^b g^2(y) dy \right) - 2 \left(\int_a^b f(x)g(x) dx \right) \left(\int_a^b f(y)g(y) dy \right)$
$\quad + \left(\int_a^b g^2(x) dx \right) \left(\int_a^b f^2(y) dy \right)$
$= \|f\|^2 \|g\|^2 - 2(f \cdot g)(f \cdot g) + \|g\|^2 \|f\|^2 = 2\|f\|^2 \|g\|^2 - 2(f \cdot g)^2$.

(ii). In (i) the left-hand side is nonnegative, and hence $\|f\|^2 \|g\|^2 \geq (f \cdot g)^2$.

(iii). $\|f+g\|^2 = (f+g \cdot f+g) = (f \cdot f) + (f \cdot g) + (g \cdot f) + (g \cdot g) \leq \|f\|^2 + 2\|f\|\|g\| + \|g\|^2.$

(iv). Let $f(t) = \sqrt{7+2t}$ and $g(t) = \sqrt{\pi \sin \pi t}$ so that $\int_0^1 (\sqrt{7+2t} + \sqrt{\pi \sin \pi t})^2 dt = \|f+g\|^2 \leq (\|f\| + \|g\|)^2 \leq (\sqrt{\int_0^1 (7+2t)dt} + \sqrt{\int_0^1 \pi \sin \pi t \, dt})^2 \leq (\sqrt{8} + \sqrt{2})^2 = 18.$

Chapter 21

Improper Integrals

In Chapters 19 and 20, we studied the Riemann integration of bounded functions on bounded intervals. If we extend the interval or the function to be unbounded at a point(s), then the integral is called *improper*. Improper integrals are defined as the limit of Riemann integrals. We begin with the following definition.

Assume that $f : [a, \infty) \to \mathcal{R}$ is integrable on $[a,b]$ for every $b \geq a$, and $\lim_{b\to\infty} \int_a^b f(x)dx = \int_a^\infty f(x)dx$ exists and is finite, then the improper integral $\int_a^\infty f(x)dx$ is said to converge, otherwise it diverges. If $f : (-\infty, b] \to \mathcal{R}$ is integrable on $[a,b]$ for every $b \geq a$, and $\lim_{a\to-\infty} \int_a^b f(x)dx = \int_{-\infty}^b f(x)dx$ exists and is finite, then the improper integral $\int_{-\infty}^b f(x)dx$ is said to converge, otherwise it diverges. Similarly, if $f : \mathcal{R} \to \mathcal{R}$ is integrable on $[a,b]$ for every $b \geq a$, and $\lim_{a\to-\infty} \left(\lim_{b\to\infty} \int_a^b f(x)dx \right) = \lim_{b\to\infty} \left(\lim_{a\to-\infty} \int_a^b f(x)dx \right) = \int_{-\infty}^\infty f(x)dx$ exists and is finite, then the improper integral $\int_{-\infty}^\infty f(x)dx$ is said to converge, otherwise it diverges. In what follows we shall mainly discuss the results for the improper integral $\int_a^\infty f(x)dx$. A similar theory holds for the other two improper integrals.

Example 21.1. Since for $a > 0$,

$$\int_a^b x^{-\alpha}dx = \begin{cases} \frac{b^{1-\alpha}-a^{1-\alpha}}{1-\alpha}, & \alpha \neq 1 \\ \ln b/a, & \alpha = 1 \end{cases}$$

it follows that

$$\int_a^\infty x^{-\alpha}dx = \begin{cases} 1/(\alpha-1)a^{\alpha-1}, & \alpha > 1 \\ \infty, & \alpha \leq 1. \end{cases}$$

The following result follows from Theorem 20.3(1) and (2), and the definition of the improper integral.

Theorem 21.1. If functions $f, g : [a, \infty)$ are improper integrals, and c_1 and c_2 are constants, then $c_1 f + c_2 g$ is an improper integral and

$$\int_a^\infty (c_1 f(x) + c_2 g(x))dx = c_1 \int_a^\infty f(x)dx + c_2 \int_a^\infty g(x)dx.$$

The following result follows from Theorem 20.3(3) and the definition of the improper integral.

Theorem 21.2. Suppose that functions $f, g : [a, \infty) \to \mathcal{R}$ are integrable on $[a, b]$ for every $b \geq a$ and $0 \leq f(x) \leq g(x)$ for all $x \geq a$. Then, the following hold:

(1). If $\int_a^\infty g(x)dx$ converges, then $\int_a^\infty f(x)dx$ converges, and $\int_a^\infty f(x)dx \leq \int_a^\infty g(x)dx$.

(2). If $\int_a^\infty f(x)dx$ diverges, then $\int_a^\infty g(x)dx$ diverges.

Example 21.2. Since for all $x \geq 1$, $1/\sqrt{x^6 + 9} \leq 1/x^3$, in view of Example 21.1 and Theorem 21.2(1) the improper integral $\int_1^\infty dx/\sqrt{x^6 + 9}$ converges. For all $x \geq 1$, $1/x \leq (1 + |\sin x|)/\sqrt{x}$, in view of Example 21.1 and Theorem 21.2(2) the improper integral $\int_1^\infty (1 + |\sin x|)/\sqrt{x}dx$ diverges.

The function $f : [a, \infty) \to \mathcal{R}$ is said to be *absolutely integrable* on $[a, \infty)$ if $|f|$ is improperly integrable on $[a, \infty)$. Further, it is said to be *conditionally integrable* on $[a, \infty)$ if f is improperly integrable but not absolutely integrable.

Theorem 21.3. If $f : [a, \infty) \to \mathcal{R}$ is *absolutely integrable* on $[a, \infty)$, then f is improperly integrable on $[a, \infty)$, and $|\int_a^\infty f(x)dx| \leq \int_a^\infty |f(x)|dx$.

Proof. Since $0 \leq |f(x)| + f(x) \leq 2|f(x)|$, from Theorem 21.2 it follows that $|f(x)| + f(x)$ is improperly integrable on $[a, \infty)$. Now from Theorem 21.1, we find $f = (|f| + f) - |f|$ is improperly integrable on $[a, \infty)$. For the required inequality it suffices to note that for every $b \geq a$, $|\int_a^b f(x)dx| \leq \int_a^b |f(x)|dx$.

The following classical example shows that the converse of Theorem 21.3 does not hold.

Example 21.3. From integration by parts, we have

$$\int_1^b \frac{\sin x}{x}dx = \cos 1 - \frac{\cos b}{b} - \int_1^b \frac{\cos x}{x^2}dx.$$

Since $|\cos x|/x^2 \leq 1/x^2$ and $1/x^2$ is absolutely integrable on $[1, \infty)$ from Example 21.1, it follows that $\cos x/x^2$ is absolutely integrable on $[1, \infty)$. Thus, $\sin x/x$ is improperly integrable on $[1, \infty)$, and

$$\int_1^\infty \frac{\sin x}{x}dx = \cos 1 - \int_1^\infty \frac{\cos x}{x^2}dx.$$

Now we shall show that $\sin x/x$ is not absolutely integrable on $[1, \infty)$. For this, it suffices to note that

$$\int_1^{n\pi} \frac{|\sin x|}{x}dx \geq \sum_{k=2}^n \int_{(k-1)\pi}^{k\pi} \frac{|\sin x|}{x}dx \geq \sum_{k=2}^n \frac{1}{k\pi} \int_{(k-1)\pi}^{k\pi} |\sin x|dx$$
$$= \sum_{k=2}^n \frac{2}{k\pi} = \frac{2}{\pi}\sum_{k=2}^n \frac{1}{k} \to \infty \text{ as } n \to \infty \text{ (see Example 9.2)}.$$

Theorem 21.4 (Riemann's Integral Test for Sequences). Let $f : [1, \infty) \to \mathcal{R}$ be continuous, nonnegative, and nonincreasing, and let $a_k = f(k)$, $k \in \mathcal{N}$. Then, $\sum_{k=1}^{\infty} a_k$ converges iff $\int_1^{\infty} f(x)dx$ converges.

Proof. Since f is nonincreasing $f(k+1) \leq f(x) \leq f(k)$ for all $x \in [k, k+1]$, $k \in \mathcal{N}$. Further, since f is continuous, it is Riemann integrable on $[k, k+1]$. Thus, from Theorem 20.3(3) it follows that

$$\int_k^{k+1} f(k+1)dx \leq \int_k^{k+1} f(x)dx \leq \int_k^{k+1} f(k)dx,$$

which is the same as

$$a_{k+1} = f(k+1) \leq \int_k^{k+1} f(x)dx \leq f(k) = a_k.$$

Hence, we have $s_{n+1} - a_1 \leq \int_1^{n+1} f(x)dx \leq s_n$, where $s_n = \sum_{k=1}^n a_k$. Since all the terms in this inequality are nondecreasing as n increases, the result follows from Theorem 21.2.

Example 21.4. From Example 21.1 and Theorem 21.4 it follows that the p-series $\sum_{k=1}^{\infty} 1/k^p$ converges if $p > 1$ and diverges if $0 < p \leq 1$. The function $f(x) = 1/(x \ln x)$ is nonnegative and nonincreasing in $[3, \infty)$. Since $F'(x) = f(x)$, where $F(x) = \ln \ln x$,

$$\int_3^b f(x)dx = \ln \ln b - \ln \ln 3,$$

which implies that $\int_3^{\infty} f(x)dx$ diverges. Thus, from Theorem 21.4 the series $\sum_{k=3}^{\infty} 1/(k \ln k)$ diverges.

Now we shall define the improper integrals for unbounded functions. Assume that the unbounded function $f : [a, b) \to \mathcal{R}$ is integrable on $[a, c]$ for every $c \in [a, b)$ and $\lim_{c \to b-} \int_a^c f(x)dx = \int_a^b f(x)dx$ exists and is finite, then the improper integral $\int_a^b f(x)dx$ is said to converge, otherwise it diverges. If the unbounded function $f : (a, b] \to \mathcal{R}$ is integrable on $[c, b]$ for every $c \in (a, b]$ and $\lim_{c \to a+} \int_c^b f(x)dx = \int_a^b f(x)dx$ exists, then the improper integral $\int_a^b f(x)dx$ is said to converge, otherwise it diverges. Similarly, if an unbounded function $f : [a, c) \cup (c, b] \to \mathcal{R}$ is integrable on $[a, s] \cup [t, b]$ for all $s \in [a, c)$ and $t \in (c, b]$ and both the limits $\lim_{s \to c-} \int_a^s f(x)dx = \int_a^c f(x)dx$ and $\lim_{t \to c+} \int_t^b f(x)dx = \int_c^b f(x)dx$ exist, then $\int_a^b f(x)dx = \int_a^c f(x)dx + \int_c^b f(x)dx$ is said to converge, otherwise diverges. In general, if the improper integral $\int_a^b f(x)dx$ can be broken into several subintegrals and if each of these is convergent, then $\int_a^b f(x)dx$ is a convergent improper integral. However, if one of the subintegrals is divergent, then $\int_a^b f(x)dx$ is a divergent improper integral

Example 21.5. From Example 21.1 it follows that

$$\int_0^1 x^{-\alpha} dx = \begin{cases} 1/(1-\alpha), & \alpha < 1 \\ \infty, & \alpha \geq 1. \end{cases}$$

Now since $|\sin x|/\sqrt{x} \leq 1/\sqrt{x}$, $0 < x \leq 1$, it follows that $\int_0^1 \sin x/\sqrt{x}\,dx$ converges absolutely.

Example 21.6. If $0 < \epsilon < 1$, then $\int_0^{1-\epsilon}(1/\sqrt{1-x})dx = 2 - 2\sqrt{\epsilon}$, and hence

$$\int_0^1 \frac{1}{\sqrt{1-x}}dx = \lim_{\epsilon \to 0} \int_0^{1-\epsilon} \frac{1}{\sqrt{1-x}}dx = 2,$$

i.e., $\int_0^1(1/\sqrt{1-x})dx$ is absolutely convergent. Now since, for $0 \leq x < 1$,

$$\frac{1}{\sqrt{1-x^2}} = \frac{1}{\sqrt{1-x}\sqrt{1+x}} \leq \frac{1}{\sqrt{1-x}}$$

it follows that $\int_0^1 1/\sqrt{1-x^2}\,dx$ is absolutely convergent.

Example 21.7. Let $f : [0, \infty) \to \mathcal{R}$ be defined by

$$f(x) = \begin{cases} 1 & \text{if } n - \frac{1}{n^2} \leq x \leq n, \ n \in \mathcal{N} \\ 0 & \text{otherwise.} \end{cases}$$

Fix $c > 1$. Note that f is integrable on $[0, c]$, since f is almost everywhere continuous on $[0, c]$ (see Theorem 22.1). Clearly, $\int_0^c f(x)dx$ is a monotone increasing function of c, and

$$0 < \int_0^c f(x)dx < \sum_{n=1}^{\infty} \frac{1}{n^2},$$

and hence $\int_0^c f(x)dx$ is bounded above. Thus, $\lim_{c \to \infty} \int_0^c f(x)dx$ exists, i.e., $\int_0^\infty f(x)dx$ converges. It is interesting to note that for this function $\lim_{x \to \infty} f(x)$ does not exist, in fact, $\lim_{n \to \infty} f(n) = 1$, but $\lim_{n \to \infty} f(n + 1/2) = 0$.

Example 21.8. Let $f : (3, \infty) \to \mathcal{R}$ be defined by

$$f(x) = \frac{1}{\sqrt{|(x-3)(x-5)(x-7)|}},$$

which is unbounded at $x = 3, 5, 7$. Thus, we can write

$$\int_3^\infty \cdot = \int_3^4 \cdot + \int_4^5 \cdot + \int_5^6 \cdot + \int_6^7 \cdot + \int_7^8 \cdot + \int_8^\infty \cdot .$$

Clearly, each integral in the right side converges, and hence $\int_3^\infty f(x)dx$ is convergent.

Remark 21.1. Let $c \in (a, b)$ be a point where the function $f : [a, b] \to \mathcal{R}$ is unbounded. If $\int_a^b f(x)dx$ is divergent, but

$$\lim_{h \to 0+} \left(\int_a^{c-h} f(x)dx + \int_{c+h}^b f(x)dx \right) = A$$

exists, then the value A is called *Cauchy's Principal Value* (C.P.V.) of the divergent integral. For the infinite integral $\int_{-\infty}^{\infty} f(x)dx$ C.P.V. is defined as $\lim_{h \to \infty} \int_{-h}^{h} f(x)dx$.

Example 21.9. Consider the improper integral $I = \int_0^5 1/(x-2)^3 dx$. Since both the associated integrals

$$\lim_{h \to 0+} \int_0^{2-h} \frac{1}{(x-2)^3} dx \quad \text{and} \quad \lim_{k \to 0+} \int_{2+k}^5 \frac{1}{(x-2)^3} dx$$

diverge, the integral I is divergent. Now

$$
\begin{aligned}
\text{C.P.V. } (I) &= \lim_{h \to 0+} \left(\int_0^{2-h} \frac{1}{(x-2)^3} dx + \int_{2+h}^5 \frac{1}{(x-2)^3} dx \right) \\
&= \lim_{h \to 0+} \left(\left[-\frac{1}{2h^2} + \frac{1}{8} \right] + \left[-\frac{1}{18} + \frac{1}{2h^2} \right] \right) = \frac{5}{72}.
\end{aligned}
$$

Example 21.10. The infinite integral $\int_{-\infty}^{\infty} x dx$ does not exist, but since

$$\int_{-h}^h x dx = \left. \frac{x^2}{2} \right|_{-h}^h = 0,$$

the C.P.V. of the integral is zero.

Problems

21.1. Let $f, g : [a, \infty) \to \mathcal{R}$ be such that f is bounded on $[a, \infty)$ and integrable on every $[a, b]$, $b \geq a$, and $|g|$ is improperly integrable on $[a, \infty)$. Show that $|fg|$ is improperly integrable on $[a, \infty)$.

21.2. Show that the product of two improper integrable functions may not be improperly integrable.

21.3. Let $f : [a, b) \to \mathcal{R}$ be bounded and monotonic, and let $g : [a, b) \to \mathcal{R}$ be improper integrable. Show that $\int_a^b f(x)g(x)$ is convergent. In particular, show that $\int_0^{\pi/2} (\cos x / \sqrt{x})dx$ is convergent.

21.4. Let $f : [a, \infty) \to \mathcal{R}$ be improperly integrable. Show that if $A = \lim_{x \to \infty} f(x)$ exists, then $A = 0$.

21.5. Let $f, g : [a, b) \to \mathcal{R}$ be nonnegative and integrable on $[a, c]$ for all $c \in [a, b)$, and $A = \lim_{x \to b-} f(x)/g(x)$ exists. Show that

(i). If $0 \leq A < \infty$ and g is improperly integrable on $[a, b)$, then f is also improperly integrable on $[a, b)$.

(ii). If $0 \leq A < \infty$ and f is not improperly integrable on $[a, b)$, then g is also not improperly integrable on $[a, b)$.

21.6. Let $f, g : [a, \infty) \to \mathcal{R}$ be continuous, and let g be positive and nondecreasing. Also, let f be nonnegative and periodic of period w, and $\int_0^w f(x)dx > 0$. Show that the improper integral $\int_0^\infty dx/g(x)$ and the integral $\int_0^\infty f(x)/g(x)dx$ simultaneously converge or diverge.

21.7. Let $f : [1, \infty) \to \mathcal{R}$ be absolutely integrable on $[1, \infty)$. Show that $\lim_{n \to \infty} \int_1^\infty f(x^n)dx = 0$.

21.8. Give an example of a function $f : [1, \infty) \to \mathcal{R}$ which is nonnegative on $[1, \infty)$ and integrable on $[1, c]$ for any $c > 1$, for which

(i). $\sum_{k=1}^\infty f(k)$ converges, but $\int_1^\infty f(x)dx$ diverges.

(ii). $\int_1^\infty f(x)dx$ converges, but $\sum_{k=1}^\infty f(k)$ diverges.

21.9. Let $f : \mathcal{R} \to \mathcal{R}$ and $\int_{-\infty}^\infty f(x)dx$ converges to A. Show that

$$\text{C.P.V.} \int_{-\infty}^\infty f(x)dx = A.$$

21.10. Discuss the convergence or divergence of the following improper integrals.

(i). $\int_0^\infty \frac{1}{x^2 + \sqrt{x}} dx$.

(ii). $\int_0^\infty \frac{x}{\sqrt{x^4 + 1}} dx$.

(iii). For each $p > 0$, $\int_1^\infty \frac{\sin x}{x^p} dx$.

(iv). For each $p > 0$, $\int_e^\infty \frac{\cos x}{\ln^p x} dx$.

(v). $\int_1^\infty \frac{\sin 1/x}{x} dx$.

(vi). $\int_0^1 \frac{\cos x}{\sqrt{x - x^2}} dx$.

(vii). $\int_0^\infty \frac{x^{p-1}}{1+x} dx$.

(viii). $\int_0^1 x^{n-1} \ln x dx$, $n \leq 1$.

(ix). $\int_0^\infty x^{a-1} e^{-x} dx$, $a > 0$.

(x). $\int_0^1 x^{m-1}(1 - x)^{n-1} dx$, $m > 0, n > 0$.

Answers or Hints

21.1. Let $M = \sup_{x \in [a,\infty)} |f(x)|$. Then, $0 \le |f(x)g(x)| \le M|g(x)|$ for all $x \in [a, \infty)$. Now use Theorem 21.3.

21.2. Let $f(x) = 1/x^\alpha$, $x \in (0,1)$, $\alpha \in \mathcal{R}$. We will show that f is improperly integrable iff $\alpha < 1$. For $\alpha \ne 1$, we have
$\int_0^1 \frac{1}{x^\alpha} dx = \lim_{\epsilon \to +0} \int_\epsilon^1 \frac{1}{x^\alpha} dx = \lim_{\epsilon \to +0} \frac{x^{-\alpha+1}}{-\alpha+1}\Big|_\epsilon^1 = \lim_{\epsilon \to +0} \left(\frac{1}{-\alpha+1} - \frac{\epsilon^{-\alpha+1}}{-\alpha+1} \right).$
This limit is finite iff $-\alpha + 1 > 0$, thus $\alpha < 1$. For $\alpha = 1$, we have
$\int_0^1 \frac{1}{x} dx = \lim_{\epsilon \to +0} \int_\epsilon^1 \frac{1}{x} dx = \lim_{\epsilon \to +0}(\ln 1 - \ln \epsilon) = +\infty.$
Thus, $1/x$ is not improperly integrable on $(0,1)$. Since $1/x^{1/2} \cdot 1/x^{1/2} = 1/x$ we find that the product of two improperly integrable functions is not an improperly integrable function.

21.3. For simplicity we assume that $f(x) \ge 0$ and $g(x) \ge 0$. We shall show that $\lim_{\epsilon \to b-} \int_a^\epsilon f(x)g(x)dx$ exists. Let $0 \le m \le f(x) \le M$ for all $x \in [a, b)$. It follows that $f(x)g(x) \le Mg(x)$. Thus, $\lim_{\epsilon \to b-} \int_a^\epsilon f(x)g(x)dx \le M \lim_{\epsilon \to b-} \int_a^\epsilon g(x)dx$, which in view of Theorem 21.2 converges. For the remaining part, let $f(x) = \cos x$ and $g(x) = 1/\sqrt{x}$.

21.4. Suppose that $A > 0$. Then, there exists large b such that for all $x \ge b$, $f(x) \ge A/2$. But, then $\lim_{x \to \infty} \int_b^x f(x)dx > \lim_{x \to \infty} \int_b^x A/2 dx = (A/2)\lim_{x \to \infty}(x - b) = \infty$.

21.5. (i). For $\epsilon > 0$, there is $\mu > 0$ such that $|f(x)/g(x) - A| < \epsilon$ for all x with $|x - b| < \mu$. We assume that $A + \epsilon > 0$, then $\lim_{\tau \to -b} \int_a^\tau f(x)dx < \lim_{\tau \to -b} \int_a^\tau g(x)(A + \epsilon)dx < \infty$. This means that f is improperly integrable on $[a, b)$.

(ii). Following part (i), we have $\infty = \lim_{\tau \to b-} \int_a^\tau f(x)/(A + \epsilon)dx < \lim_{\tau \to b-} \int_a^\tau g(x)dx$, thus g is not improperly integrable on $[a, b)$.

21.6. Since $\int_{k\omega}^{(k+1)\omega} f(x)dx = \int_0^\omega f(x)dx = \mu > 0$. It follows that $\mu/g((k+1)\omega) \le \int_{k\omega}^{(k+1)\omega} f(x)/g(x)dx \le \mu/g(k\omega)$, $k = 0, 1, \cdots$. Thus, $\int_0^\infty f(x)/g(x)dx$ converges iff $\sum_{k=0}^\infty 1/g(k\omega)$ converges. But, from Theorem 21.4, $\sum_{k=0}^\infty 1/g(k\omega)$ converges iff $\int_0^\infty dx/g(x)$ converges.

21.7. Let $t = x^n > 1$, so that $dt = nx^{n-1}dx$ and $dx = dt/nt^{\frac{n-1}{n}}$. It follows that
$\int_1^\infty f(x^n)dx \le \left| \int_1^\infty f(t)\frac{dt}{nt^{\frac{n-1}{n}}} \right| < \int_1^\infty |f(t)|\frac{dt}{nt^{\frac{n-1}{n}}} < \frac{1}{n}\int_1^\infty |f(t)|dt.$
Since f is absolutely integrable on $[1, \infty)$, we have $\int_1^\infty |f(t)|dt < \infty$, therefore $\lim_{n \to \infty} \int_1^\infty f(x^n)dx = 0$.

21.8. (i). Consider $g(x) = 0$ if $x \in [n - 1/n^2, n + 1/n^2]$, $n \in \mathcal{N}\setminus\{1\}$ and $g(x) = 1$, otherwise. Let $f(x) = g(x) + 1/x^2$. The function f for all $c > 1$ is almost everywhere continuous on $[1, c]$, and hence in view of Theorem 21.1 it is integrable on $[1, c]$, and we have $\int_1^\infty f(x)dx = \infty$. However, $\sum_{k \in \mathcal{N}} f(k) = \sum 1/k^2 < \infty$.

(ii). Consider the function f from Example 21.7 and $g(x) = f(x) + 1/x^2$. Clearly, $\lim_{n \to \infty} \int_1^n g(x)dx$ exists, whereas $\lim_{k \to \infty} g(k)$ does not exist, thus from Theorem 9.1, the series is divergent.

21.9. From the definition, we have $\lim_{h \to \infty} \int_{-h}^h f(x)dx = \int_{-\infty}^\infty f(x)dx =$ A.

21.10. (i). Convergent.
(ii). Divergent.
(iii). Convergent.
(iv). Convergent.
(v). Convergent.
(vi). Convergent.
(vii). Convergent iff $0 < p < 1$.
(viii). Convergent if $0 < n \leq 1$ and divergent if $n \leq 0$.
(ix). Convergent.
(x). Convergent.

Chapter 22

Riemann-Lebesgue Theorem

In Example 19.2 and Problem 19.8, the functions we considered have infinitely many discontinuities. The function in Problem 19.8 is Riemann integrable, whereas the function in Example 19.2 is not Riemann integrable. The natural question which we will answer in this chapter is what type of infinitely many discontinuities a function may have to remain Riemann integrable. For this, we need the following definitions and elementary results.

For an open interval $I = (a, b)$ the (Lebesgue) *measure* is defined as the length of the interval, i.e., $b - a$ and denoted as $m(I)$. The measure of infinite intervals $(-\infty, a), (a, \infty)$ and $(-\infty, \infty)$ is infinite. A set $E \subset \mathcal{R}$ is said to be of *measure zero* if for every $\epsilon > 0$ there exists a countable collection of open intervals $\{I_1, I_2, \cdots\}$ which covers E such that $\sum_{k=1}^{\infty} m(I_k) \leq \epsilon$. From the definition it is clear that if the measure of E is zero, then the measure of every subset of E is also zero. It is also clear that the measure of a finite set as well as of a countable set $E = \{x_1, x_2, \cdots\}$ is zero. Indeed, each of the points x_k of E can be covered by an interval I_k of length less than $\epsilon/2^k$, i.e., $I_k = (x_k - \epsilon 2^{-k-1}, x_k + \epsilon 2^{-k-1})$, and then

$$\sum_{k=1}^{\infty} m(I_k) = \sum_{k=1}^{\infty} \frac{\epsilon}{2^k} = \epsilon.$$

In what follows we shall need the following result.

Lemma 22.1. The union of a countable collection of sets of measure zero is a set of measure zero.

Proof. We arrange these sets as $E_1, E_2, \cdots$, and let $E = \cup_{k=1}^{\infty} E_k$. Let $\epsilon > 0$. Clearly, for each $k \in \mathcal{N}$ we can choose a collection of intervals $\{I_j^{(k)}\}$, $j \in \mathcal{N}$ which covers E_k such that

$$\sum_{j=1}^{\infty} m\left(I_j^{(k)}\right) \leq \frac{\epsilon}{2^k}.$$

Then, the collection $\{I_j^{(k)}\}$, $j, k \in \mathcal{N}$ is countable (see Chapter 5, P3), covers

E, and

$$\sum_{k=1}^{\infty}\sum_{j=1}^{\infty} m\left(I_j^{(k)}\right) \leq \sum_{k=1}^{\infty}\frac{\epsilon}{2^k} = \epsilon.$$

Remark 22.1. There are even uncountable subsets of $\mathcal{R}$ which have measure zero. For example, the Cantor set C (see Problem 5.6) is of measure zero. Of course, a nonempty open interval (no matter how small) is never of measure zero. From Problem 22.1, the set of irrational numbers in $[a, b]$ is not of measure zero.

An assertion is said to hold *almost everywhere* in $[a, b]$ if the set of points at which the assertion does not hold is of measure zero. In particular, a function $f : [a, b] \to \mathcal{R}$ is said to be *almost everywhere continuous* on $[a, b]$ if the set of points $E \subset [a, b]$ where f is discontinuous is of measure zero.

Example 22.1. On the interval $[0, 1]$ the functions considered in (12.4), and in Problems 19.7 and 19.8 are almost everywhere continuous, whereas the functions in (12.3) and in Problem 11.3(i) are not almost everywhere continuous. The function $f : [0, 1] \to \mathcal{R}$ defined as

$$f(x) = \begin{cases} 1, & \text{if } x \in C \\ 0, & \text{if } x \in [0, 1]\backslash C, \end{cases} \tag{22.1}$$

where C is the Cantor set, is almost everywhere continuous.

Let $f : [a, b] \to \mathcal{R}$ be a bounded function. The *oscillation* of f on a nonempty set $A \subseteq [a, b]$ is defined by

$$\Omega_f(A) = \sup_{x \in A} f(x) - \inf_{x \in A} f(x). \tag{22.2}$$

Clearly, $\Omega_f(A)$ is a nonnegative finite number, and if $B \subseteq A$, then $\Omega_f(B) \leq \Omega_f(A)$. The *oscillation* of f *at a point* $x \in [a, b]$ is defined by

$$\omega_f(x) = \lim_{h \to 0+} \Omega_f((x - h, x + h)), \tag{22.3}$$

where necessary changes are made when x is an endpoint of $[a, b]$.

Example 22.2. For the greatest integer function $f(x) = [x]$ (see Chapter 6), $\omega_f(x) = \begin{cases} 1, & \text{if } x \text{ is an integer} \\ 0, & \text{otherwise.} \end{cases}$ For the function $f(x) = \begin{cases} \sin(1/x), & x \neq 0 \\ 0, & x = 0 \end{cases}$ it is clear from Example 13.3 that $\omega_f(0) = 2$.

Lemma 22.2. Let $f : [a, b] \to \mathcal{R}$ be a bounded function, and let E be the set of points of discontinuity of f on $[a, b]$. Then, the following holds

$$E = \cup_{k=1}^{\infty}\left\{x \in [a, b] : \omega_f(x) \geq \frac{1}{k}\right\}.$$

Proof. From (22.2) and (22.3) it is clear that f is continuous at $x \in [a,b]$ iff $\omega_f(x) = 0$. Thus, $x \in E$ iff $\omega_f(x) > 0$. But, then from the Archimedean Property (Theorem 3.3), $\omega_f(x) > 0$ iff $\omega_f(x) > 1/k$ for some $k \in \mathcal{N}$.

Lemma 22.3. Let $f : [a,b] \to \mathcal{R}$ be a bounded function. Then, for each $\epsilon > 0$, the set $G = \{x \in [a,b] : \omega_f(x) \geq \epsilon\}$ is compact.

Proof. Assume that G is not compact, so that it is nonempty. Since G is bounded, it cannot be closed. Thus, there exists points $x_i \in G$ such that $\lim_{i \to \infty} x_i = x$ but $x \notin G$, and hence $\omega_f(x) < \epsilon$. This means that there exists an $h_0 > 0$ such that $\Omega_f((x - h_0, x + h_0)) < \epsilon$. Since $x_i \to x$, we can find $N \in \mathcal{N}$ such that $(x_N - h_0/2, x_N + h_0/2) \subset (x - h_0, x + x_0)$, but then $\Omega_f((x_N - h_0/2, x_N + h_0/2)) < \epsilon$. Thus, $\omega_f(x_N) < \epsilon$, which contradicts our assumption that $x_N \in G$.

Lemma 22.4. Let $f : I \to \mathcal{R}$ be bounded, where I is a closed and bounded interval. If for $\epsilon > 0$, $\omega_f(x) < \epsilon$ for all $x \in I$, then there exists a $\delta > 0$ such that $\Omega_f(J) < \epsilon$ for all closed intervals $J \subseteq I$ which satisfy $m(J) < \delta$.

Proof. For each $x \in I$, let δ_x be such that $\Omega_f((x - \delta_x, x + \delta_x)) < \epsilon$. Since I is closed and bounded, from the open cover $\{(x - \delta_x, x + \delta_x)\}_{x \in I}$ of I, we can choose $x_1, \cdots, x_N$ such that $I \subseteq \cup_{k=1}^{N}(x_k - \delta_{x_k}/2, x_k + \delta_{x_k}/2)$. Now let $\delta = \min_{1 \leq k \leq N} \delta_{x_k}/2$. If $J \subseteq I$, then there exists some $k \in \{1, \cdots, N\}$ such that $J \cap (x_k - \delta_{x_k}/2, x_k + \delta_{x_k}/2) \neq \emptyset$. If J also satisfies $m(J) < \delta$, then in view of $2\delta \leq \delta_{x_k}$ it follows that $J \subseteq (x_k - \delta_{x_k}, x_k + \delta_{x_k})$ and this from our assumption implies that $\Omega_f(J) \leq \Omega_f((x_k - \delta_{x_k}, x_k + \delta_{x_k})) < \epsilon$.

We are now in the position to prove the main result of this chapter.

Theorem 22.1 (Riemann-Lebesgue Theorem). Let $f : [a,b] \to \mathcal{R}$ be bounded. Then, f is Riemann integrable on $[a,b]$ iff f is almost everywhere continuous on $[a,b]$, i.e., iff the set of discontinuities E of f on $[a,b]$ has measure zero.

Proof. Suppose that $m(E) = 0$. Let $M = \sup_{a \leq x \leq b} f(x)$ and $m = \inf_{a \leq x \leq b} f(x)$. Given $\epsilon > 0$, we choose $k_0 \in \mathcal{N}$ such that $(M - m + b - a)/k_0 < \epsilon$. Since $m(E) = 0$, the measure of the set $H = \{x \in [a,b] : \omega_f(x) \geq 1/k_0\}$ is also zero. This means there exists a collection of intervals which covers H whose total sum of lengths is less than $1/(2k_0)$. Now we enlarge these intervals slightly and assume that there exists open intervals $I_1, I_2, \cdots$ which cover H such that $\sum_{j=1}^{\infty} m(I_j) < 1/k_0$. Thus Lemma 22.3 implies that there exists an $N \in \mathcal{N}$ such that $\{I_1, \cdots, I_N\}$ covers H and

$$\sum_{k=1}^{N} m(I_k) < \frac{1}{k_0}. \tag{22.4}$$

Now we shall find a partition P such that $\overline{S}(f, P) - \underline{S}(f, P) < \epsilon$. The endpoints of $I_k's$ will be included in this partition. For the rest of the points we will further divide the part of $[a, b]$ which is not covered by the $I_k's$. Let $D \subseteq [a, b] \backslash (\cup_{k=1}^{N} I_k)$. Since $I_k's$ cover H, $\omega_f(x) < 1/k_0$ for all $x \in D$. Thus, Lemma 22.4 implies that there exists a $\delta > 0$ such that if $J \subseteq D$ satisfies $m(J) < \delta$, then $\Omega_f(J) < 1/k_0$. We subdivide $[a, b] \backslash (\cup_{k=1}^{N} I_k)$ into intervals J_ℓ, $\ell = 1, \cdots, r$ such that $m(J_\ell) < \delta$. From this, it follows that

$$\Omega_f(J_\ell) < \frac{1}{k_0}, \quad \ell = 1, \cdots, r. \tag{22.5}$$

Let $P = \{x_1, \cdots, x_n\}$, where each x_i is either an endpoint of some I_k or J_ℓ. Clearly, if $(x_{i-1}, x_i) \cap H \neq \emptyset$, then x_{i-1} and x_i are endpoints of some I_k. Thus, from (22.4) it follows that

$$\sum_{(x_{i-1}, x_i) \cap H \neq \emptyset} (M_i(f) - m_i(f))(x_i - x_{i-1}) \leq \frac{M - m}{k_0}.$$

Now if $(x_{i-1}, x_i) \cap H = \emptyset$, then x_{i-1} and x_i are endpoints of some J_ℓ, and hence from (22.5), we have

$$\sum_{(x_{i-1}, x_i) \cap H = \emptyset} (M_i(f) - m_i(f))(x_i - x_{i-1}) \leq \frac{b - a}{k_0}.$$

From the above two inequalities, we find

$$\overline{S}(f, P) - \underline{S}(f, P) = \sum_{i=1}^{n} (M_i(f) - m_i(f))(x_i - x_{i-1}) \leq \frac{M - m + b - a}{k_0} < \epsilon,$$

which implies that f is integrable on $[a, b]$.

Conversely, suppose that f is integrable on $[a, b]$, but $m(E) \neq 0$. Then, in view of Lemmas 22.1 and 22.2 there exists a $k_0 \in \mathcal{N}$ such that the measure of the set $H = \{x \in [a, b] : \omega_f(x) \geq 1/k_0\}$ is not zero. This means that there exists an $\epsilon_0 > 0$ such that if $\{I_j\}$, $j \in \mathcal{N}$ is any open covering of H, then

$$\sum_{j=1}^{\infty} m(I_j) \geq \epsilon_0. \tag{22.6}$$

Now let $P = \{x_0, \cdots, x_n\}$ be a partition of $[a, b]$. Clearly, if $(x_{i-1}, x_i) \cap H \neq \emptyset$, then by definition, $M_i(f) - m_i(f) \geq 1/k_0$. Thus, it follows that

$$\begin{aligned} \overline{S}(f, P) - \underline{S}(f, P) &= \sum_{i=1}^{n} (M_i(f) - m_i(f))(x_i - x_{i-1}) \\ &\geq \sum_{(x_{i-1}, x_i) \cap H \neq \emptyset} (M_i(f) - m_i(f))(x_i - x_{i-1}) \\ &\geq \frac{1}{k_0} \sum_{(x_{i-1}, x_i) \cap H \neq \emptyset} (x_i - x_{i-1}). \end{aligned}$$

Finally, since $\{[x_{i-1}, x_i] : (x_{i-1}, x_i) \cap H \neq \emptyset\}$ is a collection of intervals which covers H, from (22.6) it follows that

$$\overline{S}(f, P) - \underline{S}(f, P) \geq \frac{\epsilon_0}{k_0} > 0,$$

and hence f cannot be integrable on $[a, b]$.

Example 22.3. On the interval $[0, 1]$ the functions considered in (12.4), in Problems 19.7 and 19.8, and (22.1) are integrable, whereas the functions in (12.3) and in Problem 11.3(i) are not integrable.

Problems

22.1. Let A be a set of nonzero measures and $B \subset A$ be of measure zero. Show that $A - B$ is not of measure zero.

22.2. Let $f, g : [a, b] \to \mathcal{R}$ be almost everywhere continuous on $[a, b]$. Show that $f \pm g$ and fg are almost everywhere continuous on $[a, b]$. Thus, from Theorem 22.1 if f and g are Riemann integrable on $[a, b]$, then $f \pm g$ and fg are Riemann integrable on $[a, b]$.

22.3. Find $\omega_f(x)$, where $f(x) = \begin{cases} 1, & \text{if } x \text{ is irrational} \\ -1, & \text{if } x = 0 \\ 0, & \text{if } x \text{ is rational.} \end{cases}$

22.4. Let $f : [a, b] \to \mathcal{R}$ be an integrable function on $[a, b]$, and let $g(x) = f(x)$, $x \in [a, b] \backslash S$, where the set $S \subset [a, b]$ has a finite number of points. Show that the function g is integrable on $[a, b]$ and $\int_a^b f(x)dx = \int_a^b g(x)dx$. (Note that f need not be defined at points in S.) What happens when the subset S is infinite?

22.5. If $f : [a, b] \to \mathcal{R}$ is an integrable function on $[a, b]$, show that f^α, $\alpha > 0$ is also integrable on $[a, b]$.

22.6. Let $f : [a, b] \to \mathcal{R}$ be an integrable function and let $m = \inf_{x \in [a,b]} f(x)$, $M = \sup_{x \in [a,b]} f(x)$. Further, let the function $g : [m, M] \to \mathcal{R}$ be continuous. Show that $g \circ f$ is integrable on $[a, b]$. Further, show that this result does not hold if g is discontinuous even at a single point.

Answers or Hints

22.1. Follows from Lemma 22.1.

22.2. Follows from Lemma 22.1 and $fg = [(f + g)^2 - f^2 - g^2]/2$.

22.3. $\omega_f(0) = 2$ and for $x \neq 0$, $\omega_f(x) = 0$.

22.4. Let $\phi(x) = f(x) - g(x)$. Clearly, $\phi(x)$ is zero everywhere except at points in S. Thus, $\phi(x)$ is integrable, and its integral over $[a, b]$ is zero. Hence, $g(x) = f(x) - \phi(x)$ is integrable, and $\int_a^b f(x)dx = \int_a^b g(x)dx$. If the subset S is infinite, then $g(x)$ may not be integrable. For this consider $f(x) = 1, x \in [0, 1]$ which is integrable, whereas $g(x) = \begin{cases} 1, & x \in [0, 1] \backslash \mathcal{Q} \\ 0, & x \in \mathcal{Q}, \end{cases}$ is not integrable.

22.5. From Theorem 22.1 we have that f is continuous almost everywhere on $[a, b]$. As a result f^α is continuous almost everywhere on $[a, b]$ (see Theorem 12.3). Now Theorem 22.1 guarantees that f^α is integrable on $[a, b]$.

22.6. Essentially the same reasoning as in 22.5 guarantees that $g \circ f$ in integrable on $[a, b]$. To show that the result does not hold if g is discontinuous even at a single point, consider $f(x) = x^2, g(x) = x^{-1/2}, x \in [0, 1]$. Clearly, $(g \circ f)(x) = x^{-1}$ is not integrable on $[0, 1]$.

Chapter 23

Riemann-Stieltjes Integral

The Riemann-Stieltjes integral is a generalization of the Riemann integral. Instead of dealing with just one function f (integrand), it deals with two functions f (integrand) and g (integrator). In what follows we will assume that all functions are defined on the interval $[a, b]$ and are bounded. We will begin with the assumption that the function g is nondecreasing, but later consider the general function. In this chapter, we will use the same notations of Chapter 19.

Let $f : [a, b] \to \mathcal{R}$ be a bounded function and $g : [a, b] \to \mathcal{R}$ be a nondecreasing function on $[a, b]$, and let $P = \{a = x_0 < x_1 < \cdots < x_n = b\}$ be a partition of $[a, b]$. We write $\Delta g(x_i) = g(x_i) - g(x_{i-1})$, $i = 1, 2, \cdots, n$. It is clear that $\Delta g(x_i) \geq 0$. The *upper Riemann-Stieltjes sum* $\overline{S}(f, g, P)$ corresponding to f, g, and P is defined by

$$\overline{S}(f, g, P) = \sum_{i=1}^{n} M_i \Delta g(x_i),$$

where $M_i = \sup_{x \in [x_{i-1}, x_i]} f(x)$, $i = 1, 2, \cdots, n$. The *lower Riemann-Stieltjes sum* $\underline{S}(f, g, P)$ corresponding to f, g and P is defined by

$$\underline{S}(f, g, P) = \sum_{i=1}^{n} m_i \Delta g(x_i),$$

where $m_i = \inf_{x \in [x_{i-1}, x_i]} f(x)$, $i = 1, 2, \cdots, n$. The *upper Riemann-Stieltjes integral* $\overline{\int_a^b} f(x) dg$ of f, g is defined by

$$\overline{\int_a^b} f(x) dg = \inf_P \overline{S}(f, g, P).$$

The *lower Riemann-Stieltjes integral* $\underline{\int_a^b} f(x) dg$ of f, g is defined by

$$\underline{\int_a^b} f(x) dg = \sup_P \underline{S}(f, g, P).$$

The following results are, respectively, analogs of Theorem 19.3, Remark 19.1, and Corollary 19.1.

Theorem 23.1. Let $f : [a,b] \to \mathcal{R}$ be a bounded function and $g : [a,b] \to \mathcal{R}$ be a nondecreasing function on $[a,b]$, and P_1 and P_2 are partitions of $[a,b]$ such that $P_1 \subseteq P_2$. Then

$$\overline{S}(f,g,P_1) \geq \overline{S}(f,g,P_2) \quad \text{and} \quad \underline{S}(f,g,P_1) \leq \underline{S}(f,g,P_2).$$

Remark 23.1. If in Theorem 23.1, P_1 and P_2 are arbitrary partitions of $[a,b]$, then we have

$$\underline{S}(f,g,P_1) \leq \underline{S}(f,g,P_1 \cup P_2) \leq \overline{S}(f,g,P_1 \cup P_2) \leq \overline{S}(f,g,P_2).$$

Corollary 23.1. Let $f : [a,b] \to \mathcal{R}$ be a bounded function and $g : [a,b] \to \mathcal{R}$ be a nondecreasing function on $[a,b]$. Then, $\underline{\int_a^b} f(x)dg \leq \overline{\int_a^b} f(x)dg$.

If $\underline{\int_a^b} f(x)dg = \overline{\int_a^b} f(x)dg$, then f is said to be *Riemann-Stieltjes integrable* with respect to g on $[a,b]$, and we write

$$\int_a^b f(x)dg = \underline{\int_a^b} f(x)dg = \overline{\int_a^b} f(x)dg.$$

The proof of the following result is the same as in Theorem 19.5.

Theorem 23.2. Let $f : [a,b] \to \mathcal{R}$ be a bounded function and $g : [a,b] \to \mathcal{R}$ be a nondecreasing function on $[a,b]$. Then, f is Riemann-Stieltjes integrable with respect to g on $[a,b]$, iff for each $\epsilon > 0$ there exists a partition P of $[a,b]$ such that

$$|\overline{S}(f,g,P) - \underline{S}(f,g,P)| < \epsilon.$$

Example 23.1. Consider the functions $f(x) = x$, $x \in [0,2]$, $g(x) = \begin{cases} 0, & x \in [0,1] \\ 1, & x \in (1,2]. \end{cases}$ For a given $\epsilon > 0$, we choose a partition P, such that $\|P\| < \epsilon$. Clearly, if $\Delta g(x_k) \neq 0$, then $[1,\delta) \subseteq [x_{k-1}, x_k]$, where $\delta > 1$. Thus, there is exactly one k for which $\Delta g(x_k) \neq 0$, and for this k, we have

$$\begin{aligned} |\overline{S}(f,g,P) - \underline{S}(f,g,P)| &\leq |\textstyle\sum_{i=1}^n M_i \Delta g(x_i) - \sum_{i=1}^n m_i \Delta g(x_i)| \\ &= M_k \Delta g(x_k) - m_k \Delta g(x_k) \\ &= (M_k - m_k)(1-0) = x_{k+1} - x_k \leq \|P\| < \epsilon. \end{aligned}$$

Hence, from Theorem 23.2, f is Riemann-Stieltjes integrable with respect to g on $[0,2]$, and its value is 1.

Example 23.2. Consider the functions $f(x) = \begin{cases} 0, & x \in [0,1] \\ 1, & x \in (1,2] \end{cases}$ and

$g(x) = \begin{cases} 0, & x \in [0,1) \\ 1, & x \in [1,2]. \end{cases}$ We choose a partition P, which includes the number $x_k = 1$ as a partition point. Then, as in Example 23.1 it follows that $\overline{S}(f,g,P) = M_k = 0$ and $\underline{S}(f,g,P) = m_k = 0$. Hence, f is Riemann-Stieltjes integrable with respect to g on $[0,2]$, and its value is 0. Here, the choice of the partition (includes the number 1 as a partition point) is important. In fact, if we choose a partition P' which does not include the point 1 as a partition point, then it follows that $\overline{S}(f,g,P') = 1$ and $\underline{S}(f,g,P') = 0$.

Example 23.3. Consider the functions $f(x) = g(x) = \begin{cases} 0, & x \in [0,1] \\ 1, & x \in (1,2]. \end{cases}$ For any partition P of $[0,2]$ there exists a refinement P' for which there exists a subinterval $[x_{k-1}, x_k]$ which contains the point 1 and a point greater than 1. Thus, as in Example 23.1 it follows that $\overline{S}(f,g,P') = M_k = 1$ and $\underline{S}(f,g,P') = m_k = 0$. Hence, f is not Riemann-Stieltjes integrable with respect to g on $[0,2]$.

Example 23.4. Let $f : [0,n] \to \mathcal{R}$ be continuous. Then, from the definition of Riemann-Stieltjes integral it follows that

$$\int_0^n f(x)d[x] = \sum_{i=1}^n f(x_i).$$

We shall now state results which are analogs of Problems 19.5 and 19.6.

Theorem 23.3. If $f : [a,b] \to \mathcal{R}$ is monotonic and $g : [a,b] \to \mathcal{R}$ is nondecreasing continuous on $[a,b]$, then f is Riemann-Stieltjes integrable with respect to g on $[a,b]$.

Theorem 23.4. If $f : [a,b] \to \mathcal{R}$ is continuous and $g : [a,b] \to \mathcal{R}$ is nondecreasing on $[a,b]$, then f is Riemann-Stieltjes integrable with respect to g on $[a,b]$.

Now we state and prove a result which relates the Riemann integral and Riemann-Stieltjes integral.

Theorem 23.5. Let $f : [a,b] \to \mathcal{R}$ be bounded, $g : [a,b] \to \mathcal{R}$ be nondecreasing and differentiable on $[a,b]$, f and g' be both Riemann integrable on $[a,b]$. Then, f is Riemann-Stieltjes integrable with respect to g on $[a,b]$, and

$$\int_a^b f(x)dg = \int_a^b f(x)g'(x)dx. \qquad (23.1)$$

Proof. Since f and g' are Riemann integrable from Problem 22.2 the function fg' is Riemann integrable on $[a,b]$. Let P be any partition of $[a,b]$.

Then, it follows that

$$\overline{S}(f, g, P) - \underline{S}(f, g, P) = \sum_{i=1}^{n} (M_i - m_i) \Delta g(x_i).$$

Since g' exists on $[a, b]$, from the mean-value Theorem 15.5, we have $\Delta g(x_i) = g'(t_i)(x_i - x_{i-1})$ where $t_i \in (x_{i-1}, x_i)$. Now since g' is Riemann integrable, it is bounded, i.e., there exists a positive number G such that $|g'(x)| \leq G$ for $x \in [a, b]$. Thus, it follows that

$$|\overline{S}(f, g, P) - \underline{S}(f, g, P)| \leq G \sum_{i=1}^{n} (M_i - m_i)(x_i - x_{i-1}) = G|\overline{S}(f, P) - \underline{S}(f, P)|.$$

Let $\epsilon > 0$. Since f is Riemann integrable on $[a, b]$, there exists a partition P such that $|\overline{S}(f, P) - \underline{S}(f, P)| < \epsilon/G$. For this partition, we have $|\overline{S}(f, g, P) - \underline{S}(f, g, P)| < \epsilon$. In conclusion both integrals $\int_a^b f(x)dg$ and $\int_a^b f(x)g'(x)dx$ exist.

We shall now show that equality (23.1) holds. Let $\epsilon > 0$ be given and let a partition P be such that

$$\int_a^b f(x)g'(x)dx - \epsilon < \sum_{i=1}^{n} (fg')(s_i)\Delta x_i < \int_a^b f(x)g'(x)dx + \epsilon$$

for any $s_i \in [x_{i-1}, x_i]$. Now using the fact that g nondecreasing implies $g'(x) \geq 0$, we have

$$\overline{S}(f, g, P) = \sum_{i=1}^{n} M_i \Delta g(x_i) = \sum_{i=1}^{n} M_i g'(t_i)(x_i - x_{i-1})$$
$$\geq \sum_{i=1}^{n} f(t_i)g'(t_i)\Delta x_i > \int_a^b f(x)g'(x)dx - \epsilon,$$

and hence

$$\int_a^b f(x)dg \geq \int_a^b f(x)g'(x)dx. \tag{23.2}$$

Similarly, we have

$$\underline{S}(f, g, P) = \sum_{i=1}^{n} m_i \Delta g(x_i) = \sum_{i=1}^{n} m_i g'(t_i)(x_i - x_{i-1})$$
$$\leq \sum_{i=1}^{n} f(t_i)g'(t_i)\Delta x_i < \int_a^b f(x)g'(x)dx + \epsilon,$$

and hence

$$\int_a^b f(x)dg \leq \int_a^b f(x)g'(x)dx. \tag{23.3}$$

From (23.2) and (23.3) the required equality (23.1) follows.

The following results are analogs to Theorems 20.2 to 20.4, and Corollary 20.1.

Theorem 23.6. Let $f : [a, b] \to \mathcal{R}$ be bounded. Then, f is Riemann-Stieltjes integrable with respect to the nondecreasing function $g : [a, b] \to \mathcal{R}$, iff, for every $c \in (a, b)$, f is Riemann-Stieltjes integrable with respect to g on $[a, c]$ and $[c, b]$. Further, it follows that

$$\int_a^b f(x)dg = \int_a^c f(x)dg + \int_c^b f(x)dg.$$

Theorem 23.7. (1). Let $f_1, f_2 : [a, b] \to \mathcal{R}$ be Riemann-Stieltjes integrable with respect to the nondecreasing function $g : [a, b] \to \mathcal{R}$ on $[a, b]$. Then, for any real numbers α and β, $\alpha f_1 + \beta f_2$ is Riemann-Stieltjes integrable with respect g on $[a, b]$, and

$$\int_a^b (\alpha f_1(x) + \beta f_2(x))dg = \alpha \int_a^b f_1(x)dg + \beta \int_a^b f_2(x)dg.$$

Further, if $f_1(x) \leq f_2(x)$ on $[a, b]$, then

$$\int_a^b f_1(x)dg \leq \int_a^b f_2(x)dg.$$

(2). Let $f : [a, b] \to \mathcal{R}$ be Riemann-Stieltjes integrable with respect to the nondecreasing functions $g_1, g_2 : [a, b] \to \mathcal{R}$ on $[a, b]$. Then, for any nonnegative real numbers α and β, f is Riemann-Stieltjes integrable with respect $\alpha g_1 + \beta g_2$ (which is nondecreasing) on $[a, b]$, and

$$\int_a^b f(x)(\alpha dg_1 + \beta dg_2) = \alpha \int_a^b f(x)dg_1 + \beta \int_a^b f(x)dg_2. \qquad (23.4)$$

Proof. We shall sketch the proof of part 2 only. Let P be a partition of $[a, b]$, then from the definition it immediately follows that

$$\overline{S}(f, \alpha g_1 + \beta g_2, P) = \alpha \overline{S}(f, g_1, P) + \beta \overline{S}(f, g_2, P)$$

and

$$\underline{S}(f, \alpha g_1 + \beta g_2, P) = \alpha \underline{S}(f, g_1, P) + \beta \underline{S}(f, g_2, P).$$

Now let $\epsilon > 0$. Since f is Riemann-Stieltjes integrable with respect to functions g_1, g_2 on $[a, b]$, there exists partitions P_1 and P_2 such that

$$\overline{S}(f, g_1, P_1) - \underline{S}(f, g_1, P_1) < \frac{\epsilon}{2(\alpha + 1)}$$

and

$$\overline{S}(f, g_2, P_2) - \underline{S}(f, g_2, P_2) < \frac{\epsilon}{2(\beta + 1)}.$$

Let $P = P_1 \cup P_2$. Then, the above relations lead to

$$\overline{S}(f, \alpha g_1 + \beta g_2, P) - \underline{S}(f, \alpha g_1 + \beta g_2, P) < \frac{\alpha \epsilon}{2(\alpha + 1)} + \frac{\beta \epsilon}{2(\beta + 1)} < \epsilon,$$

which implies that f is Riemann-Stieltjes integrable with respect to $\alpha g_1 + \beta g_2$ on $[a, b]$. From the preceding relations, (23.4) is also clear.

Theorem 23.8. Let $f_1, f_2 : [a, b] \to \mathcal{R}$ be Riemann-Stieltjes integrable with respect to the nondecreasing function $g : [a, b] \to \mathcal{R}$ on $[a, b]$. Then, $f_1 f_2$ is Riemann-Stieltjes integrable with respect to g.

Theorem 23.9. Let $f : [a, b] \to \mathcal{R}$ be Riemann-Stieltjes integrable with respect to the nondecreasing function $g : [a, b] \to \mathcal{R}$ on $[a, b]$. Then, $|f|$ is Riemann-Stieltjes integrable with respect to g, and $\left| \int_a^b f(x)dg \right| \leq \int_a^b |f(x)|dg$.

Next, we state results which are analogs to Theorems 20.8, 20.10, and 20.11.

Theorem 23.10. (1). Let $f : [a, b] \to \mathcal{R}$ be continuous and $g : [a, b] \to \mathcal{R}$ strictly increasing on $[a, b]$. Then, there exists a $c \in (a, b)$ such that

$$\int_a^b f(x)dg = f(c)[g(b) - g(a)].$$

(2). Let $f : [a, b] \to \mathcal{R}$ be strictly increasing and $g : [a, b] \to \mathcal{R}$ continuous on $[a, b]$. Then, there exists a $c \in (a, b)$ such that

$$\int_a^b f(x)dg = f(a)[g(c) - g(a)] + f(b)[g(b) - g(c)].$$

Theorem 23.11. Let $f : [a, b] \to \mathcal{R}$ be Riemann-Stieltjes integrable with respect to the function $g : [a, b] \to \mathcal{R}$ on $[a, b]$. Let $\phi : [c, d] \to \mathcal{R}$ be a strictly increasing continuous function on $[c, d]$ with $a = \phi(c)$, $b = \phi(d)$. Let $h = f \circ \phi$ and $\beta = g \circ \phi$. Then, h is Riemann-Stieltjes integrable with respect to the function β on $[c, d]$, and

$$\int_a^b f(x)dg = \int_c^d f(\phi(t))dg(\phi(t)) = \int_c^d h(t)d\beta.$$

Example 23.5. Let $x = \sqrt{t}$, from Theorem 23.11, we find

$$\int_0^9 (x^3 + [\sqrt{x}])d\sqrt{x} = \int_0^3 t^6 dt + \int_0^3 [t]dt = \frac{3^7}{7} + 3 = \frac{2208}{7}.$$

Theorem 23.12. Suppose $f : [a, b] \to \mathcal{R}$ and $g : [a, b] \to \mathcal{R}$ are nondecreasing on $[a, b]$. Then, f is Riemann-Stieltjes integrable with respect to g iff g is Riemann-Stieltjes integrable with respect to f. Further, in this case

$$\int_a^b f(x)dg = f(b)g(b) - f(a)g(a) - \int_a^b g(x)df.$$

Example 23.6. For the functions $f(x) = x$ and $g(x) = x + [x]$, Example 23.4 gives

$$\int_0^2 f(x)dg = \int_0^2 xdx + \int_0^2 f(x)d[x] = 2 + f(1) + f(2) = 5,$$

whereas Theorem 23.10 gives the same answer, but differently

$$\int_0^2 f(x)dg = f(2)g(2) - f(0)g(0) - \int_0^2 g(x)df$$
$$= 2 \times 4 - 0 \times 0 - \int_0^2 (x + [x])dx = 8 - 2 - 1 = 5.$$

Remark 23.2. Suppose $g : [a, b] \to \mathcal{R}$ is a nonincreasing function and assume that $f : [a, b] \to \mathcal{R}$ is Riemann-Stieltjes integrable with respect to $-g$. Then, the Riemann-Stieltjes integration of f with respect to g is defined by

$$\int_a^b f(x)dg = -\int_a^b f(x)d(-g).$$

Similarly, if $f : [a, b] \to \mathcal{R}$ is Riemann-Stieltjes integrable with respect to the nondecreasing functions $g_1, g_2 : [a, b] \to \mathcal{R}$ on $[a, b]$, then the Riemann-Stieltjes integration of f with respect to $g_1 - g_2$ is defined by

$$\int_a^b f(x)dg = \int_a^b f(x)dg_1 - \int_a^b f(x)dg_2.$$

When the function g is of bounded variation on $[a, b]$, Remark 23.2 can be employed to define the Riemann-Stieltjes integral. For this, we consider two cases:

(a). When f is continuous and g is of bounded variation on $[a, b]$.

(b). When f and g both are of bounded variation and continuous on $[a, b]$.

For (a), we recall from Theorem 14.8 that the function g can be written as the difference of two nondecreasing functions. In fact, $g(x) = v(x) - (v(x) - g(x))$, where $v(x) = V_f(a, x)$. Thus, from Remark 23.2 it follows that

$$\int_a^b f(x)dg = \int_a^b f(x)dv - \int_a^b f(x)d(v - g).$$

For (b), we notice that f and g can be written as $f = f_1 - f_2$ and $g = g_1 - g_2$, where f_1, f_2 and g_1, g_2 are nondecreasing, and thus it follows that

$$\int_a^b f(x)dg = \int_a^b f_1(x)dg_1 - \int_a^b f_1(x)dg_2 - \int_a^b f_2(x)dg_1 + \int_a^b f_2(x)dg_2.$$

We conclude this chapter with the following general definition of Riemann-Stieltjes integration: Let $f, g : [a, b] \to \mathcal{R}$ be bounded functions and P be a partition of $[a, b]$. The sum $S(f, g, P) = \sum_{i=1}^{n} f(t_i)\Delta g(x_i)$, $t_i \in [x_{i-1}, x_i]$ is called a Riemann-Stieltjes sum. We say f is Riemann-Stieltjes integrable with respect to g on $[a, b]$ if there exists a real number A such that given $\epsilon > 0$ there exists a $\delta > 0$ such that any partition P of $[a, b]$ with $\|P\| < \delta$ and any Riemann-Stieltjes sum $S(f, g, P)$ (i.e., for any choices $t_i \in [x_{i-1}, x_i], i = 1, 2, \cdots, n$) satisfies $|S(f, g, P) - A| < \epsilon$. The number A, if it exists, is called the *Riemann-Stieltjes integral* of f with respect to g.

The above definition is applicable to a larger class of functions. For example, Theorem 23.7 can be improved to the following result.

Theorem 23.13. (1). Let $f_1, f_2 : [a, b] \to \mathcal{R}$ be Riemann-Stieltjes integrable with respect to the function $g : [a, b] \to \mathcal{R}$ on $[a, b]$. Then, for any real numbers α and β, $\alpha f_1 + \beta f_2$ is Riemann-Stieltjes integrable with respect g on $[a, b]$, and

$$\int_a^b (\alpha f_1(x) + \beta f_2(x))dg = \alpha \int_a^b f_1(x)dg + \beta \int_a^b f_2(x)dg.$$

(2). Let $f : [a, b] \to \mathcal{R}$ be Riemann-Stieltjes integrable with respect to functions $g_1, g_2 : [a, b] \to \mathcal{R}$ on $[a, b]$. Then, for any real numbers α and β, f is Riemann-Stieltjes integrable with respect to $\alpha g_1 + \beta g_2$ on $[a, b]$, and

$$\int_a^b f(x)(\alpha dg_1 + \beta dg_2) = \alpha \int_a^b f(x)dg_1 + \beta \int_a^b f(x)dg_2.$$

Chapter 24

Sequences of Functions

In this chapter, we will study the convergence of sequences of functions. We will introduce pointwise and uniform convergences, show that pointwise convergence lacks several important properties, provide some necessary and sufficient criteria for the uniform convergence, and prove that the deficiencies of the pointwise convergence are regained by the uniform convergence.

Let S be a nonempty subset of $\mathcal{R}$. A sequence of real-valued functions $\{f_n\}$, where for each $n \in \mathcal{N}$, $f_n : S \to \mathcal{R}$ is said to *converge pointwise* on S to a function f iff for each $x_0 \in S$ the sequence $\{f_n(x_0)\}$ converges to a real number $f(x_0)$. Thus, $\{f_n\}$ converges pointwise on S to the function f iff given $\epsilon > 0$ and $x_0 \in S$ there is an $N \in \mathcal{N}$ (depending on ϵ and the point x_0) such that $n > N$ implies $|f_n(x_0) - f(x_0)| < \epsilon$. The function f is called the *pointwise limit of the sequence* $\{f_n\}$ on S. Clearly, the pointwise limit of a sequence, if it exists, it is unique.

Example 24.1. Consider $f_n(x) = nx$, $x \in \mathcal{R}^+$. Since $\lim_{n \to \infty} f_n(x) = \infty$ for any $x > 0$, the sequence $\{f_n\}$ does not converge pointwise on $\mathcal{R}^+$. Consider $f_n(x) = \cos^n x$, $-\pi/2 \le x \le \pi/2$. Clearly, the sequence $\{f_n\}$ converges pointwise to the function $f(x) = \begin{cases} 0, & x \in [-\pi/2, 0) \cup (0, \pi/2] \\ 1, & x = 0. \end{cases}$
Consider $f_n(x) = nx(1-x)^n$, $x \in [0,1]$. Since $f_n(0) = f_n(1) = 0$ and for $0 < x < 1$, $\lim_{n \to \infty} f_n(x) = x \lim_{n \to \infty} e^{n \ln(1-x)} = 0$, the sequence $\{f_n\}$ converges pointwise to the function $f(x) = 0$, $x \in [0,1]$.

The following examples illustrate that certain basic properties of the functions f_n do not carry over to the pointwise limit function f.

Example 24.2. Consider $f_n(x) = n/(nx + 1)$, $x \in (0,1)$. Clearly, the sequence $\{f_n\}$ converges pointwise to the function $f(x) = 1/x$, $x \in (0,1)$. Since $|f_n(x)| < n$, $x \in (0,1)$ each f_n is bounded on (0,1), but the pointwise limiting function $f(x) = 1/x$ is not bounded on $(0,1)$. Hence, the pointwise limit function of bounded functions is not necessarily bounded.

Example 24.3. Consider

$$f_n(x) = \begin{cases} 2n^2x, & 0 \le x \le 1/(2n) \\ 2n^2(1/n - x), & 1/(2n) < x < 1/n \\ 0, & 1/n \le x \le 1. \end{cases}$$

Clearly, the sequence $\{f_n\}$ converges pointwise to the function $f(x) = 0$, $x \in [0, 1]$. Note that $\max f_n(x) = n \to \infty$ as $n \to \infty$. Hence, a pointwise convergent sequence need not be bounded.

Example 24.4. Consider $f_n(x) = x^n$ and $f(x) = \begin{cases} 0, & 0 \le x < 1 \\ 1, & x = 1. \end{cases}$
Clearly, $f_n \to f$ pointwise on $[0, 1]$. Here each f_n is continuous and differentiable on $[0, 1]$, but f is neither differentiable nor continuous at $x = 1$. Hence, the pointwise limit function of continuous (differentiable) functions is not necessarily continuous (differentiable).

Example 24.5. Consider $f_n(x) = \begin{cases} 1, & x = p/n \in \mathcal{Q} \text{ (least form)} \\ 0, & \text{otherwise} \end{cases}$
and $f(x) = \begin{cases} 1, & x \in \mathcal{Q} \\ 0, & \text{otherwise.} \end{cases}$ Clearly, $f_n \to f$ pointwise on $[0, 1]$. Here each f_n is integrable on $[0, 1]$ with integral zero (see Problem 19.7), but from Example 19.2, f is not integrable on $[0, 1]$. Hence, the pointwise limit function of integrable functions is not necessarily integrable.

Example 24.6. Consider the sequence $\{f_n\}$ of continuous functions
with $f_n(x) = \begin{cases} n - n^2x, & 0 < x < 1/n \\ 0, & 1/n \le x \le 1 \end{cases}$ and $f(x) = 0$. Clearly, $f_n \to f$
pointwise on $[0, 1]$. Here for each $n \in \mathcal{N}$, $\int_0^1 f_n(x)dx = 1/2$, and $\int_0^1 f(x)dx = 0$. Hence, there exists Riemann integrable functions f_n and f such that $f_n \to f$ pointwise on S, but

$$\lim_{n \to \infty} \int_S f_n(x)dx \ne \int_S \lim_{n \to \infty} f_n(x)dx. \qquad (24.1)$$

Example 24.7. Consider $f_n(x) = x^n/n$ and $f(x) = 0$. Clearly, $f_n \to f$ pointwise on $[0, 1]$. Here each f_n and f are differentiable on $[0, 1]$, but $0 = f'(1) \ne \lim_{n \to \infty} f'_n(1) = 1$. Hence, there exists differentiable functions f_n and f such that $f_n \to f$ pointwise on S, but

$$\left(\lim_{n \to \infty} f_n(x) \right)' \ne \lim_{n \to \infty} f'_n(x) \qquad (24.2)$$

at least for some $x \in S$.

From the preceding examples it is clear that for a given sequence of functions pointwise convergence is only of limited importance, and therefore, we need a stronger concept which we introduce now.

A sequence of functions $\{f_n\}$ is said to *converge uniformly* on S to the function f iff given $\epsilon > 0$ there is an $N \in \mathcal{N}$ (depending only on ϵ) such that $n > N$ implies $|f_n(x) - f(x)| < \epsilon$ for all $x \in S$. Clearly, if on S, $|f_n(x) - f(x)| < M_n$, where M_n is independent of x, and if $M_n \to 0$, then $f_n \to f$ uniformly on S. The uniform limit of a sequence, if it exists, is unique. It is also clear that the uniform convergence of $\{f_n\}$ to f on S implies that $\{f_n\}$ also converges pointwise to f on S. However, the following examples show that the converse is not true.

Example 24.8. Consider $f_n(x) = x^n$ and $f(x) = 0$. We shall show that $f_n \to f$ uniformly on $[0, a]$ for any $a < 1$, and pointwise but not uniformly, on $[0, 1)$. For $0 < a < 1$, let $\epsilon > 0$ and $N > [\ln \epsilon / \ln a]$, so that $x \in [0, a]$ and $n > N$ implies that

$$|x^n - 0| \leq a^n < a^N < a^{\ln \epsilon / \ln a} = \epsilon.$$

Hence, $x^n \to 0$ uniformly on $[0, a]$, $a < 1$. Now suppose that $x^n \to 0$ uniformly on $[0, 1)$. Then, for $\epsilon = 1/8$ there exists $N \in \mathcal{N}$ such that for all $n > N$ and $x \in [0, 1)$, $|x^n - 0| < 1/8$. But, for $x = 1 - 1/n$ this implies $(1 - 1/n)^n < 1/8$, and hence as $n \to \infty$, $e^{-1} < 1/8$, which is a contradiction.

Example 24.9. Consider $f_n(x) = nxe^{-nx^2}$ and $f(x) = 0$. Clearly, $f_n \to f$ pointwise on $[0, 1]$. However, since

$$\left| f_n \left(\frac{1}{\sqrt{n}} \right) - f \left(\frac{1}{\sqrt{n}} \right) \right| = \sqrt{n} e^{-1} \to \infty$$

the convergence is not uniform on $[0, 1]$.

We note that if $\{f_n\}$ converges pointwise to f on S, and if $\{f_n\}$ converges uniformly on S, then $\{f_n\}$ converges uniformly to f on S. Thus, to test the uniform convergence by the above definition, it becomes necessary to find the pointwise limit function (if it exists).

Example 24.10. Consider $f_n(x) = (\sin nx)/n$, $x \in \mathcal{R}$. The sequence $\{f_n\}$ converges pointwise to $f(x) = 0$, $x \in \mathcal{R}$. Now since $|f_n(x) - 0| = |\sin nx|/n \leq 1/n < \epsilon$ for all $x \in \mathcal{R}$, $n > 1/\epsilon$, the convergence is in fact uniform. It is interesting to note that the sequence $\{f_n'\}$ of derivatives $f_n'(x) = \cos nx$ does not converge pointwise on $\mathcal{R}$. For this, it suffices to note that $f_n'(\pi) = (-1)^n$, which diverges as $n \to \infty$. Hence, there exists differentiable functions f_n and f such that $f_n \to f$ uniformly, but $\lim_{n \to \infty} f_n'(x)$ does not exist at least for some x, i.e., (24.2) holds.

Example 24.11. Consider $f_n(x) = xe^{-nx} + ((n+1)/n)\sin x$, $x \in [0, \infty)$. The sequence $\{f_n\}$ converges pointwise to $f(x) = \sin x$, $x \in [0, \infty)$.

Now since

$$\begin{aligned}
|f_n(x) - f(x)| &= \left|xe^{-nx} + \frac{n+1}{n}\sin x - \sin x\right| \\
&= \left|xe^{-nx} + \frac{1}{n}\sin x\right| \\
&\leq \frac{1}{ne} + \frac{1}{n} < \frac{2}{n} < \epsilon
\end{aligned}$$

provided $n > 2/\epsilon$, the convergence is uniform.

An analog of Theorem 7.7 for the sequences of functions, we have the following result.

Theorem 24.1 (Cauchy's Criterion for Uniform Convergence). A sequence of functions $\{f_n\}$ converges uniformly on S iff given $\epsilon > 0$ there is an $N \in \mathcal{N}$ such that $m, n > N$ implies

$$|f_n(x) - f_m(x)| < \epsilon \quad \text{for all} \quad x \in S. \tag{24.3}$$

Proof. Suppose that $\{f_n\}$ converges uniformly to f on S. Then, for given $\epsilon > 0$ there exists an $N \in \mathcal{N}$ such that for all $n > N$ and $x \in S$, $|f_n(x) - f(x)| < \epsilon/2$. But, then for all $m, n > N$ and $x \in S$, $|f_n(x) - f_m(x)| \leq |f_n(x) - f(x)| + |f(x) - f_m(x)| < \epsilon/2 + \epsilon/2 = \epsilon$. Hence, condition (24.3) is necessary.

Conversely, suppose that (24.3) holds. Then, for each fixed $x \in S$, the sequence of real numbers $\{f_n(x)\}$ is a Cauchy sequence. Hence, in view of Theorem 7.7, $\lim_{n\to\infty} f_n(x)$ exists for each $x \in S$. Now we define $f(x) = \lim_{n\to\infty} f_n(x)$, $x \in S$ and claim that this convergence is also uniform. For this, it suffices to keep n fixed in (24.3) and let $m \to \infty$, to obtain $|f_n(x) - f(x)| < \epsilon$ for all $n > N$ and $x \in S$.

Example 24.12. Consider $f_n(x) = \sqrt{x^2 + 1/n}$, $x \in \mathcal{R}$. The sequence $\{f_n\}$ converges pointwise to $f(x) = |x|$, $x \in \mathcal{R}$. Now since

$$\begin{aligned}
|f_n(x) - f(x)| &= \left|\sqrt{x^2 + 1/n} - \sqrt{x^2}\right| \cdot \frac{|\sqrt{x^2+1/n}+\sqrt{x^2}|}{|\sqrt{x^2+1/n}+\sqrt{x^2}|} \\
&= \frac{(x^2+1/n)-x^2}{\sqrt{x^2+1/n}+\sqrt{x^2}} < \frac{1}{\sqrt{n}} < \epsilon
\end{aligned}$$

provided $n > 1/\epsilon^2$, the convergence is in fact uniform. As in Example 24.10, here also each f_n is continuously differentiable; however, the limit function $|x|$ is not differentiable at $x = 0$. Now for $n \geq m > 1/\epsilon^2$, we have

$$|f_n(x) - f_m(x)| = \frac{(x^2 + 1/n) - (x^2 + 1/m)}{\sqrt{x^2 + 1/n} + \sqrt{x^2 + 1/m}} < \frac{1}{\sqrt{n}} - \frac{1}{\sqrt{m}} < \frac{1}{\sqrt{n}} < \epsilon.$$

Thus, the uniform convergence of the sequence $\{f_n\}$ also follows from Theorem 24.1.

Our next result provides sufficient conditions so that a given sequence of continuous functions converges uniformly.

Theorem 24.2 (Dini's Theorem). Let $\{f_n\}$ be a sequence of continuous functions on the compact set S such that

$$f_1(x) \leq f_2(x) \leq \cdots \leq f_n(x) \leq \cdots, \qquad x \in S. \qquad (24.4)$$

If $\{f_n\}$ converges pointwise on S to the continuous function f, then $\{f_n\}$ converges uniformly to f on S.

Proof. For each $n \in \mathcal{N}$, we define $g_n = f - f_n$. Then, from (24.4) it follows that

$$g_1(x) \geq g_2(x) \geq \cdots \geq g_n(x) \geq \cdots \geq 0, \qquad x \in S. \qquad (24.5)$$

Also, since $\{f_n\}$ converges to f on S, we have

$$\lim_{n \to \infty} g_n(x) = 0, \qquad x \in S. \qquad (24.6)$$

We shall show that $\{g_n\}$ converges uniformly to 0 on S. For this, let $\epsilon > 0$. If $x \in S$, then (24.6) implies that there is an $N(x) \in \mathcal{N}$ such that $g_{N(x)}(x) < \epsilon/2$. Now since $g_{N(x)}$ is continuous at x, there is an open interval I_x such that $g_{N(x)}(y) < \epsilon$, $y \in I_x$. The I_x for all $x \in S$ forms an open covering of S. But, then from Theorem 4.4 a finite number of the I_x, say, $I_{x_1}, I_{x_2}, \cdots, I_{x_k}$ also cover S. Let $N = \max\{N(x_1), N(x_2), \cdots, N(x_k)\}$. If $y \in S$, then $y \in I_{x_j}$ for some $j = 1, 2, \cdots, k$. Hence, $g_{N(x_j)}(y) < \epsilon$. But, since $N(x_j) \leq N$, (24.5) implies that $g_N(y) \leq g_{N(x_j)}(y)$, and therefore, $0 \leq g_N(y) < \epsilon$ for all $y \in S$. Finally, (24.5) shows that $0 \leq g_n(y) < \epsilon$ for all $n > N$ and $y \in S$, and hence $\{g_n\}$ converges uniformly to 0 on S.

It is clear that Theorem 24.2 remains true if the inequality signs in (24.4) are all reversed. Indeed then we define $g_n = f_n - f$ and proceed as above.

Example 24.13. Consider again the sequence $\{f_n\}$ with $f_n(x) = \sqrt{x^2 + 1/n}$, $x \in [-1, 1]$. Clearly, these continuous functions satisfy (24.4) with the inequality signs reversed, and $\{f_n\}$ converges pointwise to a continuous function $|x|$, $x \in [-1, 1]$. Thus, from Theorem 24.2 this sequence $\{f_n\}$ converges uniformly to $|x|$, $x \in [-1, 1]$.

Now let the sequences $\{f_n\}$ and $\{g_n\}$ converge uniformly on S to f and g, respectively. Then, it is easily seen that the sequences $\{f_n \pm g_n\}$ converge uniformly to $f \pm g$ on S. Further, if $\{f_n\}$ converges uniformly to f on S_1 and S_2 (nonempty subsets of $\mathcal{R}$), then $\{f_n\}$ converges uniformly to f on $S_1 \cup S_2$. Moreover, if $\{f_n\}$ converges uniformly on S to f and each f_n is increasing, then f is also increasing. However, the following example shows that the sequence $\{f_n g_n\}$ need not converge uniformly, while the pointwise convergence exists.

Example 24.14. The sequences $\left\{\frac{n+x+1}{(n+1)x}\right\}$ and $\left\{\frac{(n+1)x^2}{1+n^2x^2}\right\}$ converge uniformly on $(0,1)$. For this it suffices to note that

$$\left|\frac{n+x+1}{(n+1)x} - \frac{1}{x}\right| = \frac{1}{n+1} \to 0$$

and

$$\left|\frac{(n+1)x^2}{1+n^2x^2} - 0\right| < \frac{n+1}{n^2} \to 0.$$

We shall show that the product of these sequences, i.e., the sequence $\left\{\frac{nx+x^2+x}{1+n^2x^2}\right\}$ does not converge uniformly. Since

$$\left|\frac{nx+x^2+x}{1+n^2x^2} - 0\right| < \frac{3nx}{1+n^2x^2}, \quad x \in (0,1)$$

from Problem 25.2, the sequence $\{f_n\}$ converges to 0 pointwise in $(0,1)$. Now assume that for $\epsilon = 1/4$ there exists $N \in \mathcal{N}$ such that for all $n > N$ and $x \in (0,1)$,

$$\left|\frac{nx+x^2+x}{1+n^2x^2} - 0\right| < \frac{1}{4}.$$

But, then for $x = 1/n$, we have

$$\left|\frac{1+(1/n)+(1/n^2)}{1+1}\right| \to \frac{1}{2},$$

which is a contradiction.

Chapter 25

Sequences of Functions (Contd.)

The purpose of this chapter is to prove several results which show that the deficiencies of pointwise convergence listed in Examples 24.2 to 24.7 are regained by uniform convergence.

We begin with the following result which deals with the uniform convergence of bounded functions.

Theorem 25.1. Let $\{f_n\}$ be a sequence of bounded functions which converge uniformly on S to the function f. Then, f is bounded on S.

Proof. In the definition of uniform convergence of $\{f_n\}$ on S let $\epsilon = 1$, so that there exists an $N \in \mathcal{N}$ such that $|f_n(x) - f(x)| < 1$ for all $x \in S$, $n \geq N$. Thus, in particular, we have $|f_N(x) - f(x)| < 1$ for all $x \in S$, which implies that $|f(x)| \leq |f(x) - f_N(x)| + |f_N(x)| < 1 + |f_N(x)|$. Now since each f_n is bounded on S, the result follows.

Remark 25.1. A sequence of functions $\{f_n\}$ is said to be *uniformly bounded* on S iff there exists a positive constant K such that $|f_n(x)| \leq K$ for all $x \in S$ and $n \in \mathcal{N}$. It is clear that if uniformly bounded sequence $\{f_n\}$ converges uniformly on S to the function f, then f is bounded by the same constant. From this and the triangle inequality it immediately follows that if $\{f_n\}$ and $\{g_n\}$ are uniformly bounded, and converge uniformly on S to f and g, then the sequence $\{f_n g_n\}$ converges uniformly to fg on S.

Remark 25.2. In Theorem 25.1 uniform convergence as well as boundedness of the sequence $\{f_n\}$ are only sufficient conditions, not necessary conditions. For this, in Example 24.3 we have discussed a sequence which is unbounded and converges pointwise to a bounded function. As another example, consider $f_n(x) = n^2 x (1 - x^2)^n$, $x \in [0, 1]$. Since $\max f_n(x) = (n^2/\sqrt{2n+1})(1 + 1/2n)^{-n} \to \infty$ as $n \to \infty$ the sequence $\{f_n\}$ is unbounded, but converges pointwise to a bounded function $f(x) = 0$, $x \in [0, 1]$.

Remark 25.3. In Theorem 8.3 we proved that every bounded sequence has a convergent subsequence. However, in the case of sequences of functions, uniform boundedness does not imply that there exists a subsequence which converges uniformly. For this, we consider the sequence $\{f_n\}$ with

$$f_n(x) = \frac{x^2}{x^2 + (1 - nx)^2}, \quad x \in [0, 1].$$

Clearly, $|f_n(x)| \leq 1$, $x \in [0, 1]$ and $\lim_{n \to \infty} f_n(x) = 0 = f(x)$ pointwise. However, since $f_n(1/n) = 1$, there cannot be a subsequence of $\{f_n\}$ which converges uniformly on $[0, 1]$.

Example 25.1. Consider again the sequence discussed in Example 24.2. Clearly, this sequence is bounded but converges pointwise to an unbounded function. Thus, the contrapositive form of Theorem 25.1 implies that the convergence cannot be uniform.

Our next result shows that the uniform limit of a sequence of continuous functions is continuous.

Theorem 25.2. Let $\{f_n\}$ be a sequence of continuous functions which converge uniformly on S to the function f. Then, f is continuous on S.

Proof. Since the convergence is uniform, for any given $\epsilon > 0$ there is an $N \in \mathcal{N}$ such that $n > N$ implies $|f_n(x) - f(x)| < \epsilon/3$ for all $x \in S$. Let $x_0 \in S$ be an arbitrary point. Then, in particular, we have $|f_{N+1}(x_0) - f(x_0)| < \epsilon/3$. Now since f_{N+1} is continuous at x_0, there exists a $\delta > 0$ such that $|x - x_0| < \delta$, $x \in S$ implies that $|f_{N+1}(x) - f_{N+1}(x_0)| < \epsilon/3$. Thus, for all $x \in S$ with $|x - x_0| < \delta$, we have

$$\begin{aligned}
|f(x) - f(x_0)| &\leq |f(x) - f_{N+1}(x)| + |f_{N+1}(x) - f_{N+1}(x_0)| \\
&\quad + |f_{N+1}(x_0) - f(x_0)| \\
&< \tfrac{\epsilon}{3} + \tfrac{\epsilon}{3} + \tfrac{\epsilon}{3} = \epsilon.
\end{aligned}$$

Hence, f is continuous at x_0.

Remark 25.4. In Theorem 25.2 uniform convergence is a sufficient condition but not a necessary condition. For this, consider the sequence of continuous functions with $f_n(x) = nx/(1 + n^2 x^2)$, $x \in \mathcal{R}$, which converges pointwise to a continuous function $f(x) = 0$, $x \in \mathcal{R}$ (see Problem 25.2). Since $f_n(1/n) = 1/2$, $n \geq 1$ the convergence is not uniform.

Example 25.2. Consider the sequence $\{f_n\}$ of continuous functions defined by $f_n(x) = \sin^n x$, $x \in [0, \pi]$. Clearly, this sequence converges pointwise to the function $f(x) = \begin{cases} 0, & x \in [0, \pi/2) \cup (\pi/2, \pi] \\ 1, & x = \pi/2. \end{cases}$ Since $f(x)$ is not continuous at $\pi/2$ the contrapositive form of Theorem 25.2 implies that the convergence cannot be uniform. Similarly, the sequence of continuous functions considered in Example 24.4 converges to a discontinuous function, and hence the convergence cannot be uniform.

Next we shall show that the uniform limit of a sequence of Riemann integrable functions is Riemann integrable.

Theorem 25.3. Let $\{f_n\}$ be a sequence of Riemann integrable functions on $[a, b]$, which converge uniformly to the function f. Then, f is Riemann integrable on $[a, b]$.

Proof. Since Riemann integrable functions on $[a, b]$ are bounded and the convergence of the sequence $\{f_n\}$ on $[a, b]$ is uniform, from Theorem 25.1 it follows that the limiting function f is also bounded on $[a, b]$.

Now let for each $n \in \mathcal{N}$, E_n be the set of points of $[a, b]$ at which f_n is not continuous, and let $E = \cup_{n=1}^{\infty} E_n$. In view of Theorem 22.1 each set E_n is of measure zero. Thus, from Lemma 22.1 it follows that E is also of measure zero. Further, if $x \in [a, b] \backslash E$, then $x \notin E_n$, $n \geq 1$, and hence every f_n is continuous at x. Hence, in view of Theorem 25.2 the function f is also continuous at x. Therefore, f is almost everywhere continuous on $[a, b]$. Since f is bounded on $[a, b]$, Theorem 22.1 ensures that f is Riemann integrable on $[a, b]$.

Remark 25.5. In Theorem 25.3 uniform convergence is a sufficient condition but not a necessary condition. For this, consider the sequence discussed in Example 24.4. Clearly the limiting function f is Riemann integrable on $[0, 1]$.

Example 25.3. For the sequence of Riemann integrable functions considered in Example 24.5 the limiting function is not Riemann integrable, and hence the contrapositive form of Theorem 25.3 implies that the convergence cannot be uniform.

In our next result we shall provide sufficient conditions so that equality holds in (24.1).

Theorem 25.4. Let $\{f_n\}$ be a sequence of Riemann integrable functions on $[a, b]$, which converge uniformly to the function f. Then, f is Riemann integrable on $[a, b]$, and

$$\lim_{n \to \infty} \int_a^b f_n(x)dx = \int_a^b \lim_{n \to \infty} f_n(x)dx = \int_a^b f(x)dx. \qquad (25.1)$$

Proof. Since the sequence $\{f_n\}$ converges uniformly to f, there exists $N \in \mathcal{N}$ such that for all $x \in [a, b]$,

$$|f_n(x) - f(x)| < \frac{\epsilon}{b - a}, \quad n \geq N. \qquad (25.2)$$

In view of Theorem 25.3, the function f is Riemann integrable on $[a, b]$. Thus, together with Theorem 20.3(1) the functions f_n, f, and $f_n - f$ are Riemann integrable on $[a, b]$. Hence, from (25.2), for all $x \in [a, b]$, it follows that

$$\left| \int_a^b f_n(x)dx - \int_a^b f(x)dx \right| = \left| \int_a^b [f_n(x) - f(x)]dx \right|$$

$$\leq \int_a^b |f_n(x) - f(x)|dx < \int_a^b \frac{\epsilon}{b-a}dx = \epsilon,$$

i.e., (25.1) holds.

Remark 25.6. In Theorem 25.4 uniform convergence is a sufficient condition but not a necessary condition. For this, again we consider the sequence of continuous functions with $f_n(x) = nx/(1 + n^2x^2)$, $x \in [0, 1]$, which converges pointwise to a continuous function $f(x) = 0$, $x \in [0, 1]$, but not uniformly to $f(x) = 0$, $x \in [0, 1]$. For this sequence, we have

$$\lim_{n\to\infty} \int_0^1 f_n(x)dx = \lim_{n\to\infty} \int_0^1 \frac{nx}{1+n^2x^2}dx$$

$$= \lim_{n\to\infty} \frac{1}{2n} \ln(1 + n^2) = 0 = \int_0^1 f(x)dx.$$

Example 25.4. Consider again the sequence $\{f_n\}$, where $f_n(x) = n^2x(1 - x^2)^n$, $x \in [0, 1]$ whose limiting function is $f(x) = 0$, $x \in [0, 1]$. Clearly, $\int_0^1 n^2x(1 - x^2)^n = n^2/2(n + 1) \to \infty$, whereas $\int_0^1 f(x)dx = 0$. Thus, the contrapositive form of Theorem 25.4 implies that the convergence cannot be uniform. Similarly, for the sequence of Riemann integrable functions considered in Example 24.6 the convergence cannot be uniform.

The following result gives sufficient conditions for the equality in (24.2).

Theorem 25.5. Let $\{f_n\}$ be a sequence of functions on $[a, b]$ such that each f_n is continuously differentiable on $[a, b]$. Suppose that $\{f_n\}$ converges pointwise on $[a, b]$ to f, and $\{f_n'\}$ converges uniformly on $[a, b]$ to g, then f is continuously differentiable on $[a, b]$, further $\{f_n\}$ converges uniformly to f and $\{f_n'\}$ converges uniformly to f' on $[a, b]$, i.e., $\lim_{n\to\infty} f_n'(x) = f'(x) = g(x)$, $x \in [a, b]$.

Proof. Since $\{f_n'\}$ converges uniformly to g on $[a, b]$, Theorem 25.2 implies that the function g is continuous on $[a, b]$. Thus, for each $x \in [a, b]$, Theorem 25.4 implies that

$$\lim_{n\to\infty} \int_a^x f_n'(t)dt = \int_a^x g(t)dt.$$

Now, from Theorem 20.7, we have

$$\lim_{n\to\infty} [f_n(x) - f_n(a)] = \int_a^x g(t)dt,$$

which in view of $\lim_{n\to\infty} f_n(x) = f(x)$ and $\lim_{n\to\infty} f_n(a) = f(a)$ implies that

$$f(x) - f(a) = \int_a^x g(t)dt, \quad x \in [a, b].$$

But, now from Theorem 20.6, we find $f'(x) = g(x)$, $x \in [a, b]$.

Finally, in this chapter we shall prove the following result which weakens the hypotheses of Theorem 25.5.

Theorem 25.6. Let $\{f_n\}$ be a sequence of functions on (a, b) which converges at some $x_0 \in (a, b)$. Suppose that each f_n is differentiable on (a, b), and $\{f'_n\}$ converges uniformly on (a, b), then $\{f_n\}$ converges uniformly on (a, b) and

$$\lim_{n \to \infty} f'_n(x) = \left(\lim_{n \to \infty} f_n(x) \right)'$$

for each $x \in (a, b)$.

Proof. Let $c \in (a, b)$ be fixed. Then, for $n \in \mathcal{N}$ and $x \in (a, b)$, we have

$$f_n(x) = f_n(c) + (x - c)g_n(x), \tag{25.3}$$

where

$$g_n(x) = \begin{cases} (f_n(x) - f_n(c))/(x - c), & x \neq c \\ f'_n(c), & x = c. \end{cases}$$

First, we shall show that for any $c \in (a, b)$, the sequence $\{g_n\}$ converges uniformly on (a, b). For this, let $\epsilon > 0$, $n, m \in \mathcal{N}$, and $x \in (a, b) \backslash \{c\}$. Then, by Theorem 15.5 there exists a $\hat{x}$ between x and c such that

$$g_n(x) - g_m(x) = \frac{[f_n(x) - f_m(x)] - [f_n(c) - f_m(c)]}{x - c} = f'_n(\hat{x}) - f'_m(\hat{x}).$$

Since $\{f'_n\}$ converges uniformly on (a, b), it follows that there exists an $N \in \mathcal{N}$ such that $|g_n(x) - g_m(x)| < \epsilon$ for all $n, m \geq N$ and $x \in (a, b) \backslash \{c\}$. Further, since $g_n(c) = f'_n(c)$, this also holds for $x = c$. Now, since g_n converges uniformly and (25.3) holds for $c = x_0$, and $\{f_n(x_0)\}$ converges, it follows from (25.3) that $\{f_n\}$ converges uniformly on (a, b).

Again, let $c \in (a, b)$ be fixed. We let $f(x) = \lim_{n \to \infty} f_n(x)$ and $g(x) = \lim_{n \to \infty} g_n(x)$. To complete the proof, we need to show that

$$f'(c) = \lim_{n \to \infty} f'_n(c). \tag{25.4}$$

Since each g_n is continuous at c, from Theorem 25.2 it follows that g is also continuous at c. Now, since $g_n(c) = f'_n(c)$, the right side of (25.4) can be written as

$$\lim_{n \to \infty} f'_n(c) = \lim_{n \to \infty} g_n(c) = g(c) = \lim_{x \to c} g(x). \tag{25.5}$$

Next, if $x \neq c$, then we have

$$\frac{f(x) - f(c)}{x - c} = \lim_{n \to \infty} \frac{f_n(x) - f_n(c)}{x - c} = \lim_{n \to \infty} g_n(x) = g(x).$$

Thus, the left side of (25.4) reduces to

$$f'(c) \;=\; \lim_{x \to c} \frac{f(x) - f(c)}{x - c} \;=\; \lim_{x \to c} g(x). \tag{25.6}$$

The required equality (25.4) now follows from (25.5) and (25.6).

Problems

25.1. Show that the given sequence of functions $\{f_n\}$ converges pointwise to f on S, where

(i). $f_n(x) = x^n e^{-nx}$, $f(x) = 0$, $S = [0, \infty)$.

(ii). $f_n(x) = x/(1 + nx)$, $f(x) = 0$, $S = [0, \infty)$.

(iii). $f_n(x) = x + 1/(1 + nx)$, $f(x) = \begin{cases} 1, & x = 0 \\ x, & x > 0, \end{cases}$ $S = [0, \infty)$.

(iv). $f_n(x) = \begin{cases} 2n, & 1/n \le x \le 2/n \\ 0, & \text{for all other } x \in [0, 1], \end{cases}$ $f(x) = 0$, $S = [0, 1]$.

(v). $f_n(x) = \sin(nx + 1)/\sqrt{n + 2}$, $f(x) = 0$, $S = \mathcal{R}$.

(vi). $f_n(x) = x^2/\sqrt{x^2 + 1/n}$, $f(x) = |x|$, $S = \mathcal{R}$.

(vii). $f_n(x) = (nx^2 + 1)/(nx + 1)$, $f(x) = x$, $S = [1, 2]$.

(viii). $f_n(x) = (n + \cos x)/(ne^x + \sin x)$, $f(x) = e^{-x}$, $S = \mathcal{R}$.

25.2. Consider $f_n(x) = nx/(1 + n^2 x^2)$ and $f(x) = 0$. Show that

(i). $f_n \to f$ pointwise on $\mathcal{R}$, however the convergence is not uniform.

(ii). $f_n \to f$ uniformly on $\{x : |x| > k > 0\}$.

25.3. Determine which sequences given in Problem 25.1 converge uniformly.

25.4. Let g be a continuous function on $[0, 1]$ and $g(1) = 0$. Show that the sequence of functions $\{g(x)x^n\}_{n=1}^{\infty}$ converges uniformly on $[0, 1]$.

25.5. Let $\{f_n\}$ be a sequence of continuous functions which converges uniformly to the continuous function f on $\mathcal{R}$. Show that $\lim_{n \to \infty} f_n(x + 1/n) = f(x)$, $x \in \mathcal{R}$.

25.6. Let $\{f_n\}$ be a sequence of bounded functions which converges uniformly to f on a nonempty set S. Show that $\{(f_1(x) + \cdots + f_n(x))/n\}$ also converges uniformly to f on S. Is the converse true?

25.7. Let $\{f_n\}$ be a sequence of Lipschitz continuous functions on $[a, b]$ with the same Lipschitz constant L, i.e., for all $n \in \mathcal{N}$ and $x, y \in [a, b]$, $|f_n(x) -$

$f_n(y)| \le L|x - y|$. Further, let $\{f_n\}$ converges pointwise to f on $[a, b]$. Show that $\{f_n\}$ converges uniformly to f on $[a, b]$.

25.8. A sequence $\{f_n\}$ defined on a set S is said to be *equicontinuous* if for every $\epsilon > 0$ there exists a $\delta = \delta(\epsilon) > 0$ such that for all $x, y \in S$ and each $n \in \mathcal{N}$, $|f_n(x) - f_n(y)| < \epsilon$, whenever $|x - y| < \delta$. It is clear that every function of an equicontinuous sequence is uniformly continuous. It also follows that the sequence considered in Remark 25.3 is not equicontinuous. Show that if the sequence $\{f_n\}$ of continuous functions converges uniformly on a compact subset $D \subseteq S$, then $\{f_n\}$ is equicontinuous on D.

25.9. Show that Theorem 24.2 does not hold if

(i). The set S is not compact.

(ii). The function f is not continuous.

(iii). The sequence $\{f_n\}$ is not monotonic.

(iv). Functions f_n are not continuous.

25.10. Let a sequence $\{f_n\}$ be recursively defined as follows: $f_1(x) = \sqrt{x}$, and for $n > 1$, $f_{n+1}(x) = \sqrt{x + f_n(x)}$. Show that $\{f_n\}$ converges uniformly on $[a, b]$, where $0 < a < b < \infty$. Is the convergence uniform on $[0, 1]$?

25.11. Let $\{f_n\}$ be a sequence of functions that converges uniformly to f on $[a, b] \backslash \{x_0\}$ where $x_0 \in [a, b]$. Suppose that for each $n \in \mathcal{N}$, $\lim_{x \to x_0} f_n(x)$ exists. Show that

$$\lim_{x \to x_0} \left(\lim_{n \to \infty} f_n(x) \right) = \lim_{n \to \infty} \left(\lim_{x \to x_0} f_n(x) \right).$$

25.12. Let $\{f_n\}$ be a sequence of continuous functions which converges uniformly to the continuous function f on $[a, b]$. Further, let g be a continuous function on $[a, b]$. Show that

$$\lim_{n \to \infty} \int_a^b f_n(x)g(x)dx = \int_a^b f(x)g(x)dx.$$

25.13. (i). Suppose that $\{f_n\}$ is a sequence of nonnegative integrable functions which satisfies $\lim_{n \to \infty} \int_a^b f_n(x)dx = 0$. Show that if $g : [a, b] \to \mathcal{R}$ is continuous on $[a, b]$, then $\lim_{n \to \infty} \int_a^b g(x)f_n(x)dx = 0$.

(ii). Prove that if f is continuous on $[0, 1]$, then $\lim_{n \to \infty} \int_0^1 x^n f(x)dx = 0$.

25.14. Suppose $\{f_n\}$ is a sequence of Riemann integrable functions on $[a, b]$ and there exists a $M > 0$ with $|f_n(x)| \le M$ for all $x \in [a, b]$ and $n \in \mathcal{N}$. Also assume there exists a Riemann integrable function f on $[a, b]$ with $|f(x)| \le M$ for $x \in [a, b]$ and with the following property: For any

$\epsilon > 0$ there exists a c_ϵ, $a < c_\epsilon < b$, and a $n_\epsilon \in \mathcal{N}$ with $|c_\epsilon - a| < \frac{\epsilon}{4M}$ and $|f_n(x) - f(x)| < \frac{\epsilon}{2(b-a)}$ for $x \in [c_\epsilon, b]$ and $n \geq n_\epsilon$. Show that

$$\lim_{n \to \infty} \left(\int_a^b f_n(x)dx \right) = \int_a^b f(x)dx.$$

25.15. Suppose that $\{f_n\}$ is a sequence of functions convex on an interval I and $f(x) = \lim_{n \to \infty} f_n(x)$ exists for each $x \in I$. Prove that f is convex on I.

Answers or Hints

25.1. (i). $|f_n(x)| \leq e^{-n}$, $x \in [0, \infty)$.
(ii). $f_n(0) = 0$ and for $x > 0$, $0 < f_n(x) < x/nx = 1/$, $x \in [0, \infty)$.
(iii). Similar as part (ii).
(iv). $f_n(0) = 0$ and if $x > 0$, then $f_N(x) = f_{N+1}(x) = \cdots = 0$ for $2/N < x$, $x \in [0, 1]$.
(v). $-1/\sqrt{n+2} \leq f_n(x) \leq 1/\sqrt{n+2}$, $x \in \mathcal{R}$.
(vi). $f_n(0) = 0$, and if $x \neq 0$, $\lim_{n \to \infty} x^2/\sqrt{x^2 + 1/n} = x^2/|x| = |x|$, $x \in \mathcal{R}$.
(vii). $\lim_{n \to \infty}(nx^2 + 1)/(nx + 1) = \lim_{n \to \infty}(x^2 + 1/n)/(x + 1/n) = x$, $x \in [1, 2]$.
(viii). $\lim_{n \to \infty}(n + \cos x)/(ne^x + \sin x) = e^{-x}$, $x \in \mathcal{R}$.

25.2. (i). Clearly, $f_n(0) = 0$ and for $x \neq 0$, $f_n(x) = (1/nx)/[1 + (1/n^2x^2)] \to 0$. Thus, $f_n \to f$ pointwise on $\mathcal{R}$. If $\delta > 0$ there exists $N \in \mathcal{N}$ such that $1/n \in (-\delta, \delta)$ for all $n > N$, and then $f_n(1/n) = 1/2$ for all $n > N$. Therefore, for $\epsilon = 1/4$, $|f_n(1/n) - 0| = 1/2 > 1/4$ for all $n > N$. Hence, the convergence is not uniform.
(ii). If $x : |x| > k > 0$, then for any $\epsilon > 0$,

$$|f_n(x) - 0| = \left| \frac{nx}{1 + n^2x^2} \right| < \frac{1}{|nx|} < \frac{1}{nk} < \epsilon, \ n > [1/n\epsilon].$$

Hence, $f_n \to f$ uniformly on $\{x : |x| > k > 0\}$.

25.3. (i). $|f_n(x) - f(x)| \leq e^{-n}$, $x \in [0, \infty)$ converges uniformly.
(ii). $|f_n(x) - f(x)| \leq 1/n$, $x \in [0, \infty)$ converges uniformly.
(iii). Note that $f_n(x) - f(x) = 1/(1 + nx) = 1/2$ for $x = 1/n$. Hence, the convergence is not uniform.
(iv). Note that $f_n(x) - f(x) = 2n$ for $1/n \leq x \leq 2/n$. Hence, the convergence is not uniform.
(v). Since $|f_n(x) - f(x)| \leq 1/\sqrt{n+2}$, $x \in \mathcal{R}$ the convergence is uniform.
(vi). $\left| \frac{x^2}{\sqrt{x^2+1/n}} - |x| \right| \leq \left(\sqrt{x^2 + 1/n} - \sqrt{x^2} \right) < \epsilon$ for $n > \epsilon^{-2}$ (see Example 24.12). Hence the convergence is uniform.

(vii). $\left|\frac{nx^2+1}{nx+1} - x\right| \le \frac{1+|x|}{nx+1} \le \frac{3}{1+n}$, $x \in [1,2]$. Hence the convergence is uniform.

(viii). Since $\left|\frac{n+\cos x}{ne^x+\sin x} - e^{-x}\right| \le \frac{2}{n}$, $x \in \mathcal{R}$ the convergence is uniform.

25.4. Let $M = \sup_{x \in [0,1]} |g(x)|$. If $M = 0$, then g is the constant zero function and the result is obvious. So, let $M > 0$. Since $\lim_{x \to 1-} g(x) = g(1) = 0$, given $\epsilon > 0$ there exists a $\delta \in [0,1)$ such that $|g(x)| < \epsilon$ for all $x \in (\delta, 1]$. Since $\delta < 1$, $\lim_{n \to \infty} \delta^n = 0$. Hence there exists a positive integer N such that $\delta^n < \epsilon/M$ for all $n \ge N$. Thus for all $n \ge N$ and for all $x \in [0,1]$, if $x \in (\delta, 1]$, then $|g(x)x^n| \le |g(x)| < \epsilon$ and if $x \in [0,\delta]$, then $|g(x)x^n| \le Mx^n \le M\delta^n < M(\epsilon/M) = \epsilon$. Hence the sequence of functions $\{g(x)x^n\}_{n=1}^{\infty}$ converges to 0 uniformly on $[0,1]$.

25.5. For each $\epsilon > 0$ there exists $n_1 \in \mathcal{N}$ such that $n > n_1$ implies $|f_n(x+1/n) - f(x+1/n)| < \epsilon/2$ for all $x \in \mathcal{R}$. Since f is continuous, there is $n_2 \in \mathcal{N}$ such that for $n > n_2$ we have $|f(x+1/n) - f(x)| < \epsilon/2$. Thus, for $n > n_3 = \max\{n_1, n_2\}$ it follows that $|f_n(x+1/n) - f(x)| \le |f_n(x+1/n) - f(x+1/n)| + |f(x+1/n) - f(x)| < \epsilon$. Therefore, $\lim_{n \to \infty} f_n(x+1/n) = f(x)$, $x \in \mathcal{R}$.

A general version of this problem can be stated as follows: Let $\{f_n\}$ be a sequence of continuous functions which converges uniformly to the continuous function f on $\mathcal{R}$. Then, $\lim_{n \to \infty} f_n(x_n) = f(x)$ for all sequences $\{x_n\}$ in $\mathcal{R}$ converging to x.

25.6. Let $\epsilon > 0$. Then there exists $n_0 \in \mathcal{N}$ with $|f_{n+1}(x) - f(x)| < \epsilon$ for all $n \ge n_0 \in \mathcal{N}$ and $x \in \mathcal{R}$. Let $F_n(x) = \sum_{k=1}^{n} f_k(x)$ and $g_n(x) = n$. Clearly, $\left|\frac{F_{n+1}(x)-F_n(x)}{g_{n+1}(x)-g_n(x)} - f(x)\right| = |f_{n+1}(x) - f(x)| < \epsilon$ for all $n \ge n_0 \in \mathcal{N}$ and $x \in \mathcal{R}$. Thus, it follows that
$$\left(f(x) - \tfrac{\epsilon}{2}\right)(g_{n+1}(x) - g_n(x)) < F_{n+1}(x) - F_n(x) < (g_{n+1}(x) - g_n(x))\left(f(x) + \tfrac{\epsilon}{2}\right),$$
which on summing from n_0 to $n-1$, gives
$$\left(f(x) - \tfrac{\epsilon}{2}\right)(g_n(x) - g_{n_0}(x)) < F_n(x) - F_{n_0}(x) < (g_n(x) - g_{n_0}(x))\left(f(x) + \tfrac{\epsilon}{2}\right)$$
for all $n \ge n_0$ and $x \in \mathcal{R}$. Since $g_n(x) = n$, we can divide this by $g_n(x)$, to obtain
$$\left(f(x) - \tfrac{\epsilon}{2}\right)\left(1 - \frac{g_{n_0}(x)}{g_n(x)}\right) < \frac{F_n(x)}{g_n(x)} - \frac{F_{n_0}(x)}{g_n(x)} < \left(1 - \frac{g_{n_0}(x)}{g_n(x)}\right)\left(f(x) + \tfrac{\epsilon}{2}\right).$$
From this it follows that there exists $m_0 > n_0$ such that $f(x) - \frac{\epsilon}{2} < \frac{F_n(x)}{g_n(x)} < f(x) + \frac{\epsilon}{2}$ for all $n \ge m_0$ and $x \in \mathcal{R}$, which means that $\left|\frac{f_1(x)+\cdots+f_n(x)}{n} - f(x)\right| = \left|\frac{F_n(x)}{g_n(x)} - f(x)\right| < \frac{\epsilon}{2}$.

The converse is not true. For this, let $\{f_n\}$ be a sequence with $f_n(x) = 1$ for n odd, and $f_n(x) = 0$ for n even. It follows that $\{(f_1(x)+\cdots+f_n(x))/n\}$ is equal to $\frac{n+1}{2n}$ or $\frac{n}{2n}$, according as n is odd or even, which converges uniformly to $\frac{1}{2}$, but $\{f_n(x)\}$ diverges.

25.7. For $x, y \in [a,b]$, $\epsilon > 0$, with $\epsilon < L|x-y|$ there exist n_x and n_y such that $|f_n(x) - f(x)| < \epsilon$ for $n \ge n_x$ and $|f_n(y) - f(y)| < \epsilon$ for $n \ge n_y$. Thus, for

$n \geq \max\{n_x, n_y\}$, we have $|f(x) - f(y)| \leq |f_n(x) - f(x)| + |f_n(x) - f_n(y)| + |f_n(y) - f(y)| \leq 3L|x - y|$, and hence f is Lipschitz continuous, which implies that f is continuous. We consider $F_n(x) = f_n(x) + Lx$. For $x \geq y$, it follows that $F_n(x) - F_n(y) = f_n(x) - f_n(y) + L(x - y) \geq -L(x - y) + L(x - y) = 0$, thus $F_n(x)$ is an increasing function. We will prove that $F_n(x)$ converges uniformly to $F(x) = f(x) + Lx$. Since f is continuous, for a given $\epsilon > 0$ there exists $\delta > 0$ such that $|x - y| < \delta$ implies $|F(x) - F(y)| < \epsilon/2$. For $x \in [a, b]$, clearly $\{(x - \delta/2, x + \delta/2)\}$ is an open cover for $[a, b]$, therefore there is a finite subcover cover such that $[a, b] \subset (x_1 - \delta/2, x_1 + \delta/2) \cup \cdots \cup (x_s - \delta/2, x_s + \delta/2)$, where we assume that $x_1 < \cdots < x_s$. Hence, we have $x_i - x_{i-1} < \delta$ for $i \in \{2, \cdots, s\}$. For $i \in \{2, \cdots, s\}$ let n_i be such that $n \geq n_i$ implies $|F_n(x_i) - F(x_i)| < \epsilon/2$. Let $n' = \max\{n_1, \cdots, n_s\}$ and $n \geq n'$. For $x \in [a, b]$, we have an i such that $x \in [x_{i-1}, x_i]$. Finally, since F_n is increasing, $F_n(x_{i-1}) - F(x) \leq F_n(x) - F(x) \leq F_n(x_i) - F(x)$. Hence, $|F_n(x) - F(x)| \leq \max\{|F_n(x_{i-1}) - F(x)|, |F_n(x_i) - F(x)|\}$. But, $|F_n(x_{i-1}) - F(x)| \leq |F_n(x_{i-1}) - F(x_{i-1})| + |F(x_{i-1}) - F(x)| \leq \epsilon$, and hence $|F_n(x) - F(x)| \leq \epsilon$. From this, we have $|F_n(x) - F(x)| = |f_n(x) - f(x)| \leq \epsilon$.

25.8. Let $\epsilon > 0$ be given. In view of uniform convergence there exist an $N \in \mathcal{N}$ and a $\delta > 0$ such that for all $x \in D$ and $n \geq N$, $|f_n(x) - f_N(x)| < \epsilon/3$. Further, since continuity on the compact set implies uniform continuity, we have $|f_i(x) - f_i(y)| < \epsilon/3$, $1 \leq i \leq N$, $x, y \in D$ provided $|x - y| < \delta$. Now if $x, y \in D$, $|x - y| < \delta$ and $n > N$, we find $|f_n(x) - f_n(y)| \leq |f_n(x) - f_N(x)| + |f_N(x) - f_N(y)| + |f_N(y) - f_n(y)| < \epsilon$.

25.9. (i). Consider $\{f_n\}$, where $f_n(x) = x^n$, $S = (0, 1)$ and $f_n(x) = $
$$\begin{cases} 0, & x < n \\ (x - n)/n, & n \leq x < 2n \\ 1, & x \geq 2n, \end{cases} \quad S = \mathcal{R}.$$

(ii). Consider $\{f_n\}$, where $f_n(x) = \begin{cases} 1 - 2nx, & 0 \leq x < 1/2n \\ 0, & 1/2n \leq x \leq 1, \end{cases}$ $S = [0, 1]$.

(iii). Consider $\{f_n\}$, where $f_n(x) = \begin{cases} 0, & 0 \leq x < 1/2n \\ 4nx - 2, & 1/2n \leq x < 3/4n \\ 4 - 4nx, & 3/4n \leq x < 1/n \\ 1, & 1/n \leq x \leq 1, \end{cases}$ $S = [0, 1]$.

(iv). Consider $\{f_n\}$, where $f_n(x) = \begin{cases} 0, & 0 \leq x < 1 - 1/n \\ 1, & 1 - 1/n \leq x < 1 \\ 0, & x = 1, \end{cases}$ $S = [0, 1]$.

25.10. Since for each $x_0 \in [a, b]$, $f_n(x_0) \geq f_{n-1}(x_0)$, there exists $L = \lim_{n \to \infty} f_n(x_0)$. This limit satisfies the recurrence relation $L^2 = x_0 + L$. Thus, we have $L = \frac{1 + \sqrt{1 + 4x_0}}{2}$. Let $f(x) = \frac{1 + \sqrt{1 + 4x}}{2}$. From this, and Theorem 24.2, we find $\{f_n\}$ converges uniformly to f on $[a, b]$, where $0 < a < b < \infty$. Since $|f_n(0) - f(0)| = 1$ the convergence is not uniform on $[0, 1]$.

25.11. Assume that $\lambda_n = \lim_{x \to x_0} f_n(x)$. For the convergence of $\{\lambda_n\}$ we shall show that it is a Cauchy sequence. Let $\epsilon > 0$. Since $\{f_n\}$ converges uniformly to f on $[a,b] \setminus \{x_0\}$, Theorem 24.1 implies that $\{f_n\}$ is uniformly Cauchy on $[a,b] \setminus \{x_0\}$. Thus, there exists an $N \in \mathcal{N}$, such that for $n, m > N$

$$|f_n(x) - f_m(x)| < \epsilon/3 \quad \text{for every} \ x \in [a,b] \setminus \{x_0\}. \tag{25.7}$$

Since $\lim_{x \to x_0} f_n(x) = \lambda_n$, for each $k \in \mathcal{N}$ there exists a number $\delta_k > 0$ such that

$$|\lambda_k - f_k(x)| < \epsilon/3 \quad \text{if} \ \ 0 < |x - x_0| < \delta_k \quad \text{and} \quad x \in [a,b]. \tag{25.8}$$

Now, from (25.7) and (25.8), for $n, m > N$, $0 < |x - x_0| < \delta = \min\{\delta_n, \delta_m\}$ and $x \in [a,b]$ it follows that

$$|\lambda_n - \lambda_m| \le |\lambda_n - f_n(x)| + |f_n(x) - f_m(x)| + |f_m(x) - \lambda_m| < \epsilon,$$

and hence $\{\lambda_n\}$ is a Cauchy sequence and we let $\lambda_0 = \lim_{n \to \infty} \lambda_n$. Next, we note that

$$|f(x) - \lambda_0| \le |f(x) - f_n(x)| + |f_n(x) - \lambda_n| + |\lambda_n - \lambda_0|. \tag{25.9}$$

We shall show that as x is sufficiently close to x_0 each term of the right-hand side of (25.9) can be made arbitrarily small. For this, from the uniform convergence of $\{f_n\}$ to f on $[a,b] \setminus \{x_0\}$, there exists an $N_1 \in \mathcal{N}$ such that if $n > N_1$, then $|f(x) - f_n(x)| < \epsilon/3$ for every $x \in [a,b] \setminus \{x_0\}$. Now choose $N_2 \in \mathcal{N}$ so that if $n > N_2$, then $|\lambda_n - \lambda_0| < \epsilon/3$. Let $N = \max\{N_1, N_2\}$, then $|f(x) - \lambda_0| < \epsilon/3 + |f_N(x) - \lambda_N| + \epsilon/3$. Finally, there exists a $\delta_N > 0$ such that $|f_N(x) - \lambda_N| < \epsilon/3$ if $0 < |x - x_0| < \delta_N$ and $x \in [a,b]$. Thus, we have $|f(x) - \lambda_0| < \epsilon$ if $0 < |x - x_0| < \delta_N$ and $x \in [a,b]$. This implies that $\lim_{x \to x_0} (\lim_{n \to \infty} f_n(x)) = \lim_{x \to x_0} f(x) = \alpha_0 = \lim_{n \to \infty} \alpha_n = \lim_{n \to \infty} (\lim_{x \to x_0} f_n(x))$.

25.12. Since g is continuous on $[a,b]$, there exists $M > 0$ such that $|g(x)| \le M$, $x \in [a,b]$. Since $\{f_n\}$ converges uniformly to f, for a given $\epsilon > 0$ there exists $N \in \mathcal{N}$ such that for $n > N$, $|f_n(x) - f(x)| < \epsilon/M$, $x \in [a,b]$. Thus, it follows that $|f_n(x)g(x) - f(x)g(x)| < \epsilon$ for all $n > N$, i.e., the sequence of continuous functions $\{f_n(x)g(x)\}$ converges uniformly to $f(x)g(x)$, $x \in [a,b]$. The result now follows from Theorem 25.4.

25.13. (i). There exists M such that $|g(x)| \le M$, $x \in [a,b]$. Let $\epsilon > 0$ and let $n_0 \in N$ be such that for $n > n_0$, $|\int_a^b f_n(x)dx| < \epsilon/M$. Then, $|\int_a^b g(x)f_n(x)dx| \le \int_a^b |g(x)f_n(x)|dx \le M \int_a^b f_n(x)dx \le \epsilon$.

(ii). Let $g(x) = f(x)$ and $f_n(x) = x^n$ and we apply (i).

25.14. Let $\epsilon > 0$. Let n_ϵ and c_ϵ be as in the statement of the problem. Then for $n \ge n_\epsilon$ we have $\left| \int_a^b f_n(x)dx - \int_a^b f(x)dx \right| \le \int_a^b |f_n(x) - f(x)|dx \le \int_a^{c_\epsilon} |f_n(x) - f(x)|dx + \int_{c_\epsilon}^b |f_n(x) - f(x)|dx < \frac{\epsilon}{2} + \frac{\epsilon}{2} = \epsilon$.

A general version of this problem can be stated as follows: Let $\{f_n\}$ be a uniformly bounded sequence of Riemann integrable functions converging pointwise to a Riemann integrable function f on $[a,b]$. Show that

$$\lim_{n \to \infty} \left(\int_a^b f_n(x)dx \right) = \int_a^b f(x)dx.$$

25.15. From the definition, we have $f_n((1-t)x_0+tx_1) \leq (1-t)f_n(x_0) + tf_n(x_1)$, $0 \leq t \leq 1$, $x_0, x_1 \in I$, and for all $n \in \mathcal{N}$. Letting $n \to \infty$, it immediately follows that $f((1-t)x_0+tx_1) \leq (1-t)f(x_0)+tf(x_1)$, $0 \leq t \leq 1$, i.e., f is convex on I.

Chapter 26

Series of Functions

Just as the convergence of a series of numbers is defined in Chapter 9 to mean the convergence of the sequence of partial sums, the convergence of a series of functions is also defined in terms of the sequence of partial sums. Thus, every result established in Chapters 24 and 25 for sequences of functions can be directly extended for series of functions.

Let $\{f_n\}$ be a sequence of functions defined on $S \subseteq \mathcal{R}$. We introduce a new sequence $\{s_n\}$, where $s_n(x) = \sum_{k=1}^{n} f_k(x)$, $x \in S$. Here, $s_n(x)$ is called the nth *partial sum* of the series $\sum_{n=1}^{\infty} f_n(x)$. This series is said to *converge pointwise (uniformly)* on S to $f(x)$ if $\{s_n\}$ converges pointwise (uniformly) to $f(x)$ on S as $n \to \infty$. Further, the series $\sum_{n=1}^{\infty} f_n(x)$ is said to *converge absolutely* to $\hat{f}(x)$ on S if $\sum_{n=1}^{\infty} |f_n(x)|$ converges for each $x \in S$. The series $\sum_{n=1}^{\infty} f_n(x)$ is *uniformly bounded* on S iff there exists a real constant L such that $\sum_{k=1}^{n} |f_k(x)| \leq L$ for all $x \in S$ and $n \in \mathcal{N}$.

Example 26.1. Consider the series $\sum_{n=1}^{\infty} f_n(x)$, where

$$f_n(x) = \frac{nx}{1 + n^2 x^2} - \frac{(n-1)x}{1 + (n-1)^2 x^2}, \quad x \in \mathcal{R}.$$

Clearly, $\sum_{n=1}^{\infty} f_n(x)$ is a series of continuous functions $f_n(x)$ on $\mathcal{R}$ and since its nth partial sum is $s_n(x) = nx/(1+n^2 x^2)$ from Problem 25.2 it follows that this series converges pointwise to $f(x) = 0$ on $\mathcal{R}$, but the convergence is not uniform.

Example 26.2. Similarly, for the series $\sum_{n=1}^{\infty} f_n(x)$ on $\mathcal{R}$ defined by $f_1(x) = 1/(x^2 + 1)$, and for $n > 1$

$$f_n(x) = \frac{x^{2n-2}}{x^{2n} + 1} - \frac{x^{2n-4}}{x^{2n-2} + 1}$$

the nth partial sum is $s_n(x) = x^{2n-2}/(x^{2n} + 1)$. Thus, this series converges pointwise on $\mathcal{R}$ to $f(x) = \begin{cases} 0, & |x| < 1 \\ 1/2, & x = \pm 1 \\ 1/x^2, & |x| > 1. \end{cases}$ The convergence is not uniform.

Theorem 26.1 (Cauchy's Criterion for Uniform Convergence). A series of functions $\sum_{n=1}^{\infty} f_n(x)$ converges uniformly on S iff given $\epsilon > 0$ there is an $N \in \mathcal{N}$ such that $m > n > N$ implies

$$|s_n(x) - s_m(x)| \; = \; |f_{n+1}(x) + f_{n+2}(x) + \cdots + f_m(x)| \; < \; \epsilon \quad \text{for all } \; x \in S.$$

Remark 26.1. If series $\sum_{n=1}^{\infty} f_n(x)$ converges uniformly on S, then Theorem 26.1 implies that $|f_{n+1}(x)| < \epsilon$ for all $n > N$. Thus, if there exists an $\epsilon > 0$ and a sequence $\{x_n\}$ in S such that $|f_n(x_n)| \geq \epsilon$, for all $n \in \mathcal{N}$, then $\sum_{n=1}^{\infty} f_n(x)$ does not converge uniformly on S. The series $\sum_{n=1}^{\infty} nx/(1+n^2x^2)$ does not converge uniformly on $S = [0, 1]$. For this, we note that $f_n(1/n) = 1/2$.

Theorem 26.2 (Dini's Theorem). Let $\sum_{n=1}^{\infty} f_n(x)$ be a series of nonnegative and continuous functions on the compact set S which converges pointwise on S to the continuous function f, then the convergence is uniform to f on S.

Theorem 26.3. Let $\sum_{n=1}^{\infty} f_n(x)$ be a series of continuous functions which converge uniformly on S to the function f. Then, f is continuous on S.

Theorem 26.4 (Term-by-Term Integration). Let $\sum_{n=1}^{\infty} f_n(x)$ be a series of Riemann integrable functions on $[a, b]$, which converge uniformly to the function f. Then, f is Riemann integrable on $[a, b]$, and

$$\int_a^b \sum_{n=1}^{\infty} f_n(x) dx \; = \; \sum_{n=1}^{\infty} \int_a^b f_n(x) dx.$$

Theorem 26.5 (Term-by-Term Differentiation). Let $\sum_{n=1}^{\infty} f_n(x)$ be a series of functions on (a, b) which converges at some $x_0 \in (a, b)$. Suppose that each f_n is differentiable on (a, b), and $\sum_{n=1}^{\infty} f_n'(x)$ converges uniformly on (a, b), then $\sum_{n=1}^{\infty} f_n(x)$ converges uniformly to a differentiable function $f(x)$ on (a, b), and

$$\left(\sum_{n=1}^{\infty} f_n(x) \right)' \; = \; \sum_{n=1}^{\infty} f_n'(x)$$

for each $x \in (a, b)$.

Now we shall discuss three commonly used tests for uniform convergence of series of functions.

Theorem 26.6 (Weierstrass M-Test). If $|f_n(x)| \leq M_n$ for all $x \in S$ and $n \in \mathcal{N}$, and the series of nonnegative constants $\sum_{n=1}^{\infty} M_n$ converges, then the series $\sum_{n=1}^{\infty} f_n(x)$ converges uniformly and absolutely on S.

Proof. According to Cauchy's criterion $\sum_{n=1}^{\infty} M_n$ converges if for a given $\epsilon > 0$ there exists an $N \in \mathcal{N}$ such that for all $m > n > N$, $\sum_{k=n+1}^{m} M_k < \epsilon$. Now since

$$|s_n(x) - s_m(x)| = \left| \sum_{j=n+1}^{m} f_j(x) \right| \leq \sum_{j=n+1}^{m} |f_j(x)| \leq \sum_{j=n+1}^{m} M_j < \epsilon$$

the uniform convergence of $\sum_{n=1}^{\infty} f_n(x)$ on S follows from Theorem 26.1. The absolute convergence follows from the fact that $\sum_{k=n+1}^{\infty} |f_k(x)| \leq \sum_{k=n+1}^{\infty} M_k < \epsilon$ for $n > N$.

Example 26.3. Since for all $n \in \mathcal{N}$ and $x \in [1, \infty)$, $|n^2 e^{-nx}| \leq n^2 e^{-n}$, and by Theorem 9.8, $\sum_{n=1}^{\infty} n^2 e^{-n}$ converges, Theorem 26.6 ensures that the series $\sum_{n=1}^{\infty} n^2 e^{-nx}$ converges uniformly and absolutely on $[1, \infty)$.

Example 26.4. The series $\sum_{n=1}^{\infty} \frac{(-1)^{n-1}}{n} \sin \frac{x}{n}$ converges at $x = 0$. Term-by-term differentiation of this series gives $\sum_{n=1}^{\infty} \frac{(-1)^{n-1}}{n^2} \cos \frac{x}{n}$, which in view of Theorem 26.6 converges uniformly on $\mathcal{R}$. Thus, Theorem 26.5 guarantees that the series $\sum_{n=1}^{\infty} \frac{(-1)^{n-1}}{n} \sin \frac{x}{n}$ converges uniformly to a differentiable function $f(x)$ on $\mathcal{R}$, and $\sum_{n=1}^{\infty} \frac{(-1)^{n-1}}{n^2} \cos \frac{x}{n} = f'(x)$, $x \in \mathcal{R}$.

Theorem 26.7 (Abel's Test). Let $\sum_{n=1}^{\infty} f_n(x)$ be uniformly convergent on S, and let the sequence $\{g_n\}$ be uniformly bounded on S and for each $x \in S$, $\{g_n(x)\}$ be monotonic. Then, the series $\sum_{n=1}^{\infty} f_n(x) g_n(x)$ is uniformly convergent on S.

Proof. We write $\sum_{n=1}^{\infty} f_n = s_k + t_k$, where s_k is the kth partial sum, and $t_k = \sum_{n=k+1}^{\infty} f_n$. It follows that

$$f_k g_k = -t_k[g_k - g_{k+1}] + t_{k-1} g_k - t_k g_{k+1}.$$

Summing this from $k = n + 1$ to $n + m$, yields

$$\sum_{k=n+1}^{n+m} f_k g_k = \sum_{k=n+1}^{n+m} [-t_k(g_k - g_{k+1})] + t_n g_{n+1} - t_{n+m} g_{n+m+1}$$

and hence

$$\left| \sum_{k=n+1}^{n+m} f_k g_k \right| \leq \sum_{k=n+1}^{n+m} |t_k| |g_k - g_{k+1}| + |t_n| |g_{n+1}| + |t_{n+m}| |g_{n+m+1}|. \quad (26.1)$$

Since $\{g_n\}$ is uniformly bounded, there is a $M > 0$ such that $|g_n(x)| \leq M$, $x \in S$ and for all $n \in \mathcal{N}$. Further, since the sequence $\{g_n\}$ is monotone, for each $x \in S$, we have

$$\sum_{k=n+1}^{n+m} |g_k(x) - g_{k+1}(x)| = |g_{n+1}(x) - g_{n+m+1}(x)| < 2M.$$

Now, since $\sum_{n=1}^{\infty} f_n(x)$ is uniformly convergent on S, for a given $\epsilon > 0$, Theorem 26.1 ensures that there exists an $N \in \mathcal{N}$ such that for all $n > N$, $|t_n| < \epsilon/(4M)$. Thus, from (26.1) it follows that

$$\left| \sum_{k=n+1}^{n+m} f_k g_k \right| < \frac{\epsilon}{4M}(2M) + \frac{\epsilon}{4M}M + \frac{\epsilon}{4M}M = \epsilon.$$

The result now follows from Theorem 26.1.

Theorem 26.8 (Dirichlet's Test). Let $\{f_n\}$ and $\{g_n\}$ be such that $\sum_{n=1}^{\infty} f_n(x)$ be uniformly bounded on S, the sequence $\{g_n\}$ be uniformly convergent on S, and let for each $x \in S$, the sequence $\{g_n(x)\}$ be monotonic and tend to zero, then $\sum_{n=1}^{\infty} f_n(x)g_n(x)$ is uniformly convergent on S.

Proof. We begin with the identity

$$f_k g_k = s_k[g_k - g_{k+1}] - s_{k-1}g_k + s_k g_{k+1},$$

where again s_k is the kth partial sum. Summing from $k = n+1$ to $n+m$, we have

$$\left| \sum_{k=n+1}^{n+m} f_k g_k \right| \le \sum_{k=n+1}^{n+m} |s_k||g_k - g_{k+1}| + |s_{n+m}||g_{n+m+1}| + |s_n||g_{n+1}|. \quad (26.2)$$

Now, since $\sum_{n=1}^{\infty} f_n(x)$ is uniformly bounded on S, there is a $L > 0$ such that $|s_n(x)| \le L$ for all $x \in S$ and all $n \in \mathcal{N}$. By the uniform convergence of $\{g_n\}$ and the fact that for each $x \in S$ the sequence $\{g_n(x)\}$ tends to zero, there exists an $N \in \mathcal{N}$ such that for all $x \in S$ and $n > N$, $|g_n(x)| < \epsilon/(4L)$. Hence, for $m > 0$, $n > N$ and $x \in S$, from (26.2) we have

$$
\begin{aligned}
\left| \sum_{k=n+1}^{n+m} f_k g_k \right| &< L\sum_{k=n+1}^{n+m} |g_k - g_{k+1}| + 2L\frac{\epsilon}{4L} \\
&< L|g_{n+1}(x) - g_{n+m+1}(x)| + \frac{\epsilon}{2} \le \epsilon.
\end{aligned}
$$

The result now follows from Theorem 26.1.

Example 26.5. For both of the series

$$\sum_{n=1}^{\infty} \frac{\cos nx}{n^p} \quad \text{and} \quad \sum_{n=1}^{\infty} \frac{\sin nx}{n^p},$$

the Weierstrass M-test confirms uniform and absolute convergence for $p > 1$ on $\mathcal{R}$ (see Problem 26.1(iii)). However, this test fails for $0 < p \le 1$. By using Dirichlet's test, we shall show that both of these series converge uniformly on $S = [\mu, 2\pi - \mu]$, $0 < \mu < 2\pi - \mu$. For this, we note that the respective partial sums of the series

$$\frac{1}{2} + \cos x + \cos 2x + \cdots \quad \text{and} \quad \sin x + \sin 2x + \cdots$$

are

$$s_n(x) = \frac{\sin(n+1/2)x}{2\sin(x/2)} \quad \text{and} \quad S_n(x) = \frac{\cos x/2 - \cos(n+1/2)x}{2\sin(x/2)}, \quad n = 1, 2, \cdots.$$

These partial sums are uniformly bounded on $[\mu, 2\pi - \mu]$. In fact, we have

$$|s_n(x)| \leq \frac{1}{2\sin(\mu/2)} \quad \text{and} \quad |S_n(x)| \leq \frac{1}{\sin(\mu/2)}, \quad n = 1, 2, \cdots.$$

We also note that $n^{-p} \geq (n+1)^{-p}$ and $n^{-p} \to 0$, and $n \to \infty$.

Problems

26.1. Show that

(i). $\sum_{n=1}^{\infty} x^{n-1}$ converges pointwise to $1/(1-x)$ iff $|x| < 1$.

(ii). $\sum_{n=1}^{\infty} x/(x+1)^{n-1}$ converges pointwise to $f(x) = \begin{cases} 0, & x = 0 \\ x+1, & 0 < x \leq 1. \end{cases}$

(iii). $\sum_{n=1}^{\infty} (\cos nx)/n^p$, $p > 1$ converges uniformly and absolutely on $\mathcal{R}$.

(iv). $\sum_{n=1}^{\infty} (xe^{-x})^n$ converges uniformly for $x \in [0, 2]$.

(v). $\sum_{n=1}^{\infty} x^n/(n^2+1)$ converges uniformly and absolutely on $[-1, 1]$.

(vi). $\sum_{n=1}^{\infty} 1/(x^4 + n^4)$ converges uniformly on $\mathcal{R}$.

(vii). $\sum_{n=1}^{\infty} x^2/(1+x^2)^n$ converges uniformly in any interval not including $x = 0$.

26.2. Suppose that $\sum_{n=1}^{\infty} f_n(x)$ and $\sum_{n=1}^{\infty} g_n(x)$ converge uniformly to functions $f(x)$ and $g(x)$ on S. Show that $\sum_{n=1}^{\infty} (f_n(x) \pm g_n(x))$ converges uniformly to $f(x) \pm g(x)$ on S.

26.3. Suppose that both of the series $\sum_{n=1}^{\infty} a_n$ and $\sum_{n=1}^{\infty} b_n$ converge absolutely. Show that the series $\sum_{n=1}^{\infty} [a_n \cos(nx) + b_n \sin(nx)]$ converges uniformly and absolutely on $\mathcal{R}$ to some continuous function $f(x)$. Further, show that under the stronger hypothesis that both of the series $\sum_{n=1}^{\infty} (na_n)$ and $\sum_{n=1}^{\infty} (nb_n)$ converge absolutely, the function $f(x)$ is continuously differentiable on $\mathcal{R}$ and $f'(x) = \sum_{n=1}^{\infty} [-na_n \sin(nx) + nb_n \cos(nx)]$.

26.4. If $\sum_{n=1}^{\infty} f_n(x)$ converges uniformly on S, show that the sequence of functions $\{f_n\}$ converges uniformly to the zero function. Is the converse true?

26.5. Suppose that $\{f_n\}$ is a decreasing sequence of nonnegative functions which converges uniformly to 0 on S. Show that the alternative series $\sum_{n=1}^{\infty} (-1)^{n-1} f_n(x)$ converges uniformly on S.

26.6. Show that absolute convergence of $\sum_{n=1}^{\infty} f_n(x)$ on S implies convergence of $\sum_{n=1}^{\infty} f_n(x)$ on S. Is the converse true?

26.7. (i). Give an example of a series which is uniformly convergent but not absolutely convergent S.

(ii). Give an example of a series which is absolutely convergent but not uniformly convergent on S.

26.8. Show that the converse of the Weierstrass M-test does not hold, i.e., $\sum_{n=1}^{\infty} f_n(x)$ may converge uniformly and absolutely on S, but $\sum_{n=1}^{\infty} M_n$ diverges, where $M_n = \sup\{f_n(x) : x \in S\}$.

26.9. Show that the series $\sum_{n=0}^{\infty} x^n/n!$

(i). Converges pointwise, but not uniformly in $\mathcal{R}$.

(ii). Converges absolutely in $\mathcal{R}$.

(iii). Converges uniformly on $[-a, a]$, where a is any fixed number.

(iv). Converges to a continuous function $f(x)$ on $\mathcal{R}$.

26.10. Let $f_n(x) = ne^{-nx}$, $x \in \mathcal{R}$.

(i). Find the set of all x for which the series $\sum_{n=1}^{\infty} f_n(x)$ converges pointwise.

(ii). Does the series $\sum_{n=1}^{\infty} f_n(x)$ converge uniformly on $[r, \infty)$ where $r > 0$? On $(0, \infty)$? Justify your answers.

(iii). Is the sum function $f(x) = \sum_{n=1}^{\infty} f_n(x)$ continuous?

(iv). Evaluate $\int_1^2 f(x)dx$.

Answers or Hints

26.1. (i). $s_n(x) = (1 - x^n)/(1 - x)$.
(ii). $s_n(0) = 0$ and for $0 < x \leq 1$, $s_n(x) = x(1-(x+1)^{-n})/(1-(x+1)^{-1})$.
(iii). $\sum_{n=1}^{\infty} |(\cos nx)/n^p| \leq \sum_{n=1}^{\infty} 1/n^p$, $x \in \mathcal{R}$.
(iv). $\sum_{n=1}^{\infty} |(xe^{-x})^n| \leq \sum_{n=1}^{\infty} e^{-n}$, $x \in [0, 2]$.
(v). $\sum_{n=1}^{\infty} x^n/(n^2 + 1) \leq \sum_{n=1}^{\infty} 1/n^2$, $x \in [-1, 1]$.
(vi). $\sum_{n=1}^{\infty} 1/(x^4 + n^4) \leq \sum_{n=1}^{\infty} 1/n^4$, $x \in \mathcal{R}$.
(vii). $s_n(x) = 1 - 1/(1 + x^2)^n$.

26.2. Follows from the definition.

26.3. $|a_n \cos(nx)| \leq |a_n|$ and $|b_n \sin(nx)| \leq |b_n|$. Use Theorems 26.3 and 26.5.

26.4. The first part of the problem follows from Remark 26.2. The converse is not true; see Example 26.2 for $|x| < 1$.

26.5. Since $\sum_{n=1}^{\infty} f_n(x)$ converges uniformly to 0, for $\epsilon > 0$, there is $N \in \mathcal{N}$ such that for all $m > n > N$, $|f_n(x)| < \epsilon/(m - n)$, $x \in S$. Now from Theorem 26.1 it follows that $|s_n(x) - s_m(x)| = |(-1)^n f_{n+1}(x) + \cdots + (-1)^{m-1} f_m(x)| \leq |f_{n+1}(x)| + \cdots + |f_m(x)| < (m-n)\epsilon/(m-n) = \epsilon$, which implies the uniform convergence of $\sum_{n=1}^{\infty} (-1)^{n-1} f_n(x)$.

26.6. Note that $\left|\sum_{k=n+1}^{m} f_k(x)\right| \leq \sum_{k=n+1}^{m} |f_k(x)|$.

26.7. (i). $\sum_{n=1}^{\infty} (-1)^{n-1}/(n+x^2)$, $x \in \mathcal{R}$.
(ii). On $\mathcal{R}$ consider the series $\sum_{n=1}^{\infty} f_n(x)$, where

$$f_n(x) = \begin{cases} 0, & x < 1/(n+1) \\ \sin^2(\pi/x), & 1/(n+1) \leq x \leq 1/n \\ 0, & x > 1/n. \end{cases}$$

26.8. Consider the function $g(x) = \begin{cases} 2x, & x \in [0, 1/2] \\ -2x+2, & x \in [1/2, 1] \\ 0, & x \in \mathcal{R}\backslash[0,1]. \end{cases}$ Let $f_n(x)$
$= g(x-n)/n$. The series $\sum_{n=1}^{\infty} f_n(x)$ converges uniformly for all x, since from Theorem 26.1 there exists an $N \in \mathcal{N}$ such that for all $m > n > N$, $|f_{n+1}(x) + \cdots + f_m(x)| < 1/n < \epsilon$. Clearly, $M_n = \sup f_n(x) = 1/n$ and thus $\sum_{n=1}^{\infty} M_n = \sum_{n=1}^{\infty} 1/n$ diverges.

26.9. (i). For pointwise convergence apply Ratio test. For uniform convergence Cauchy criterion fails.
(ii). Apply the ratio test for the series $\sum_{n=0}^{\infty} |x|^n/n!$.
(iii). Since $\sum_{n=0}^{\infty} |x|^n/n! \leq \sum_{n=0}^{\infty} a^n/n!$ and $\sum_{n=0}^{\infty} a^n/n!$ converges, uniform converges on $[-a, a]$ follows from Theorem 26.6.
(iv). From part (iii) and Theorem 26.3 it follows that the limiting function $f(x)$ on $[-a, a]$ is continuous. Since $a > 0$ is arbitrary, the function $f(x)$ is continuous on R.

26.10. (i). For fixed $x \leq 0$, $e^{-x} \geq 1$. Thus, $f_n(x) = ne^{-nx} \geq n$ and hence $\lim_{n\to\infty} f_n(x) = \infty$. Hence, $\sum_{n=1}^{\infty} f_n(x)$ diverges. Now for any $r > 0$, $\frac{f_{n+1}(r)}{f_n(r)} = \frac{n+1}{n} e^{-r} \to e^{-r} < 1$ as $n \to \infty$ and hence by the ratio test $\sum_{n=1}^{\infty} f_n(r)$ converges. Therefore, the series converges for $x \in (0, \infty)$.
(ii). For fixed $r > 0$ if $x \geq r$, then $e^{-nx} \leq e^{-nr}$ and $f_n(x) \leq f_n(r)$ for all n. Since $\sum_{n=1}^{\infty} f_n(r)$ converges, by the Weierstrass M-test, $\sum_{n=1}^{\infty} f_n(x)$ converges uniformly on $[r, \infty)$. Since for each fixed n, the function $f_n(x)$ is continuous and decreasing on $\mathcal{R}$, $\sup_{x \in (0,\infty)} f_n(x) = \sup_{x \in [0,\infty)} f_n(x) = f_n(0) = n \not\to 0$ as $n \to \infty$. Thus, $f_n(x) \not\to 0$ uniformly on $(0, \infty)$ as $n \to \infty$. Hence $\sum_{n=1}^{\infty} f_n(x)$ does not converge uniformly on $(0, \infty)$.
(iii). For every $r > 0$, since each $f_n(x)$ is continuous on $[r, \infty)$ and $\sum_{n=1}^{\infty} f_n(x)$ converges uniformly on $[r, \infty)$, the sum function f is continuous on $[r, \infty)$. Since this is true for each $r > 0$, f is continuous on $(0, \infty)$.
(iv). By the uniform convergence of $\sum_{n=1}^{\infty} f_n(x)$ on $[1, 2]$, $\int_1^2 f(x)dx = \sum_{n=1}^{\infty} \int_1^2 f_n(x)dx = \sum_{n=1}^{\infty} \int_1^2 ne^{-nx}dx = \sum_{n=1}^{\infty}(e^{-n} - e^{-2n}) = e/(e^2-1)$.

Chapter 27

Power and Taylor Series

In Chapters 9 and 10, we studied series of numbers, whereas Chapter 26 deals with the series of functions. In this and the next chapters, we will discuss a special type of series of functions known as power series centered at $\alpha \in \mathcal{R}$, which in the general form appears as

$$\sum_{n=0}^{\infty} a_n(x-\alpha)^n = a_0 + a_1(x-\alpha) + a_2(x-\alpha)^2 + \cdots. \tag{27.1}$$

Here, $a_n \in \mathcal{R}$, $n \geq 0$. Thus, when $\alpha = 0$ the power series (27.1) reduces to a polynomial of infinite degree, i.e., $\sum_{n=0}^{\infty} a_n x^n$. Because of their simplicity in differentiation and integration, power series play a dominate role in approximation theory, specially finding the solutions of ordinary differential equations. It is clear that for a given sequence $\{a_n\}$, the series (27.1) is a function of x whose domain consists of those values of x for which the series converges. In what follows, we are basically interested in addressing convergence results for (27.1) which do not depend on the point α, and hence for simplicity we shall assume that $\alpha = 0$. We begin with the following result.

Lemma 27.1. Suppose the power series $\sum_{n=0}^{\infty} a_n x^n$ converges for $x = x_1$ and diverges for $x = x_2$. Then it

(1). Converges absolutely and uniformly for each x such that $|x| < |x_1|$.

(2). Diverges for each x such that $|x| > |x_2|$.

Proof. (1). The case $x_1 = 0$ is obvious, so we assume that $x_1 \neq 0$. Since $\sum_{n=0}^{\infty} a_n x_1^n$ converges, $\{a_n x_1^n\}$ must be bounded, i.e., there exists a $M > 0$ such that $|a_n x_1^n| \leq M$, $n \geq 0$. Thus, for $|x| < |x_1|$ it follows that

$$|a_n x^n| = |a_n x_1^n| \left| \frac{x}{x_1} \right|^n \leq M \left| \frac{x}{x_1} \right|^n.$$

Since $|x/x_1| < 1$, the geometric series $\sum_{n=0}^{\infty} M|x/x_1|^n$ converges. Hence, by the Weierstrass test (Theorem 26.6) the power series $\sum_{n=0}^{\infty} a_n x^n$ converges absolutely and uniformly for $|x| < |x_1|$.

(2). Suppose x is such that $|x| > |x_2|$ and $\sum_{n=0}^{\infty} a_n x^n$ converges, then from part (1), $\sum_{n=0}^{\infty} a_n x_2^n$ also converges, which contradicts the assumption.

Remark 27.1. In Lemma 27.1 strict inequalities cannot be replaced by equalities. For this, we note that $\sum_{n=1}^{\infty} x^n/n$ converges for $x_1 = -1$ and diverges for $x_2 = 1$. Although, $|x_1| = |x_2| = 1$, the convergence of the series at x_1 does not imply the convergence at x_2, and the divergence at x_2 cannot be used to show its divergence at x_1.

Theorem 27.1. For the power series $\sum_{n=0}^{\infty} a_n x^n$ one of the following holds:

(1). It converges absolutely for every value of x.

(2). It converges only for $x = 0$.

(3). There exists a number $\rho > 0$ such that it converges absolutely for $|x| < \rho$ and diverges when $|x| > \rho$.

Proof. Let S be the set of all x for which the series $\sum_{n=0}^{\infty} a_n x^n$ converges. Since $0 \in S$, the set $S \neq \emptyset$. If S is unbounded, then Lemma 27.1 ensures the absolute convergence of the series for every x, i.e., (1) holds. If S is bounded, then it has a least upper bound, say, ρ. If $x > \rho$, then $x \notin S$ and thus the series diverges at x. If $x < -\rho$, we let $x < x_0 < -\rho$, i.e., $\rho < -x_0 < -x$. Thus, it follows from the definition of ρ that the series diverges at $-x_0$. Now since, $|x| = -x > -x_0$, from Lemma 27.1 the series diverges at x also. To complete the proof assume that the series diverges for some x_1 such that $|x_1| < \rho$. Then, Lemma 27.1 implies that the series diverges for all $x > |x_1|$, i.e., $|x_1|$ is an upper bound of S. But this contradicts the fact that ρ is the least upper bound of S. In conclusion, the series converges for all x such that $|x| < \rho$. Finally, the absolute convergence of the series follows from the fact that for any x such that $|x| < \rho$, there exists an x_2 such that $|x| < x_2 < \rho$, and Lemma 27.1 ensures absolute convergence at x_2.

In Theorem 27.1 the number $\rho = \sup\{r : \sum_{n=0}^{\infty} a_n r^n \text{ converges}\}$ is called the *radius of convergence*, whereas the largest interval $(-\rho, \rho), (-\rho, \rho], [-\rho, \rho)$, or $[-\rho, \rho]$ in which the series converges for all x, is called the *interval of convergence* of the series. It is clear that if, for example, $(-\rho, \rho)$ is the interval of convergence of $\sum_{n=0}^{\infty} a_n x^n$, then $(\alpha - \rho, \alpha + \rho)$ is the interval of convergence of (27.1). The following results show how the radius of convergence can be computed.

Theorem 27.2. Let ρ be the radius of convergence for the power series $\sum_{n=0}^{\infty} a_n x^n$ and suppose that $\lim_{n \to \infty} |a_{n+1}|/|a_n| = L$. Then, if $0 < L < \infty$, $\rho = 1/L$; if $L = 0$, $\rho = \infty$; if $L = \infty$, $\rho = 0$.

Proof. The proof follows from the ratio test, Theorem 9.8.

Theorem 27.3. Let ρ be the radius of convergence for the power series $\sum_{n=0}^{\infty} a_n x^n$ and suppose that $\limsup_{n \to \infty} \sqrt[n]{|a_n|} = L$. Then, if $0 < L < \infty$, $\rho = 1/L$; if $L = 0$, $\rho = \infty$; if $L = \infty$, $\rho = 0$.

Proof. The proof follows from the root test, Theorem 9.9.

Example 27.1. From Theorem 27.2 it follows that for the series $\sum_{n=0}^{\infty} x^n$, $\rho = 1$ and the interval of convergence is $(-1,1)$; for the series $\sum_{n=0}^{\infty} x^n/n^2$, $\rho = 1$ and the interval of convergence is $[-1,1]$; for the series $\sum_{n=0}^{\infty} x^n/n$, $\rho = 1$ and the interval of convergence is $[-1,1)$; for the series $\sum_{n=0}^{\infty} (1/n!)x^n$, $\rho = \infty$ and the interval of convergence is $\mathcal{R}$.

Example 27.2. Newton's binomial theorem states that for every rational number r,

$$(1+x)^r = \sum_{n=0}^{\infty} \binom{r}{n} x^n, \qquad (27.2)$$

where $\binom{r}{0} = 1$ and $\binom{r}{n+1} = \binom{r}{n}\frac{r-n}{n+1}$ for all $n = 0,1,\cdots$. Thus, if r is not a natural number, then since

$$\lim_{n\to\infty} \binom{r}{n+1} \bigg/ \binom{r}{n} = \lim_{n\to\infty} \frac{r-n}{n+1} = -1$$

the radius of convergence of the power series in (27.2) is $\rho = 1$. However, if r is a natural number, then since $\binom{r}{r+1} = \binom{r}{r} \times 0 = 0$, the binomial series contains only $r+1$ nonzero terms. Thus, in this case the radius of convergence is $\rho = \infty$.

Example 27.3. From Theorem 27.3 it follows that for the series $\sum_{n=0}^{\infty} n^n x^n$, $\rho = 0$; for the series $\sum_{n=0}^{\infty} x^n/3^k$, $\rho = 3$ and the interval of convergence is $(-3,3)$;

We shall now discuss calculus of power series.

Theorem 27.4. Let $\rho > 0$ be the radius of convergence of the power series $f(x) = \sum_{n=0}^{\infty} a_n x^n$. Then, $f(x)$ is continuous on $(-\rho, \rho)$.

Proof. Let $x \in (-\rho, \rho)$ and let $x_1, x_2 \in \mathcal{R}$ be such that $[x_1, x_2] \subset (-\rho, \rho)$. From Lemma 27.1 and Theorem 26.3 it follows that f is continuous on $[x_1, x_2]$, and hence at x.

The following result extends Theorem 27.4.

Theorem 27.5 (Abel's Theorem). Let $\rho > 0$ be the radius of convergence of the power series $f(x) = \sum_{n=0}^{\infty} a_n x^n$. If $f(x)$ converges at $\rho\,(-\rho)$, then $f(x)$ is continuous and converges uniformly on $[0,\rho]$ $([-\rho,0])$.

Proof. Let $f(x)$ converge at $x = \rho$ and assume that $x_0 \in (0,\rho]$. We set $b_n = a_n\rho^n$ and $c_n = x_0^n/\rho^n$, $n \in \mathcal{N}$. Since $\sum_{n=1}^{\infty} b_n$ converges, given $\epsilon > 0$ there exists an $N > 1$ such that $n > m \geq N$ implies that $|\sum_{j=m}^{n} b_j| < \epsilon/c_1$. Using the fact that $\{c_n\}$ is a decreasing sequence and Problem 10.3 (Abel's formula) it follows that for all $x_0 \in (0, \rho]$,

$$\left|\sum_{i=m}^{n} a_i x_0^i\right| = \left|\sum_{i=m}^{n} b_i c_i\right| = \left|c_n \sum_{i=m}^{n} b_i + \sum_{i=m}^{n-1}(c_i - c_{i+1}) \sum_{j=m}^{i} b_j\right|$$

$$< c_n \frac{\epsilon}{c_1} + (c_m - c_n)\frac{\epsilon}{c_1} = c_m \frac{\epsilon}{c_1} \leq \epsilon.$$

The above inequality obviously also holds for $x_0 = 0$. Thus, the power series $\sum_{n=0}^{\infty} a_n x^n$ converges uniformly on $[0, \rho]$. The case when $f(x)$ converges at $x = -\rho$ can be treated similarly.

Remark 27.2. A series $\sum_{n=0}^{\infty} a_n$ is said to be *Abel's summable*, to a sum s, if the associated power series $f(x) = \sum_{n=0}^{\infty} a_n x^n$ has the radius of convergence $\rho = 1$, and $\lim_{x \to 1-0} f(x) = s$. Theorem 27.5 ensures that if a series converges to a sum s then it is Abel's summable to s. However, there are nonconvergent series which are Abel's summable. For example, for the non-convergent series $1 - 2 + 3 - \cdots$ we associate the power series $\sum_{n=0}^{\infty}(-1)^n(n+1)x^n$. This series converges on $(-1, 1)$ to the function $f(x) = 1/(1+x)^2$ (see Example 27.2). Since $\lim_{x \to 1-0} 1/(1+x)^2 = 1/4$, the series $1 - 2 + 3 - \cdots$ is Abel's summable to $1/4$.

Theorem 27.6. Let $\rho > 0$ be the radius of convergence of the power series $f(x) = \sum_{n=0}^{\infty} a_n x^n$. Then

$$f'(x) = \sum_{n=1}^{\infty} n a_n x^{n-1}, \quad x \in (-\rho, \rho). \tag{27.3}$$

Proof. Let $x \in (-\rho, \rho)$ and the closed interval $[x_1, x_2] \subset (-\rho, \rho)$ be such that $x \in (x_1, x_2)$. Now from Problem 28.4, it follows that the radius of convergence of the power series $\sum_{n=1}^{\infty} n a_n x^{n-1}$ is also ρ. Also Lemma 27.1 ensures that the convergence of this series is uniform on $[x_1, x_2]$. Thus, Theorem 26.5 is applicable and the series $\sum_{n=0}^{\infty} a_n x^n$ can be differentiated term-by-term on (x_1, x_2), and hence at x. In conclusion, (27.3) follows.

Corollary 27.1. Let $\rho > 0$ be the radius of convergence of the power series $f(x) = \sum_{n=0}^{\infty} a_n x^n$. Then

$$f^{(k)}(x) = \sum_{n=k}^{\infty} \frac{n!}{(n-k)!} a_n x^{n-k}, \quad k \geq 0. \tag{27.4}$$

Proof. The proof is by induction whose initial step is Theorem 27.6.

Theorem 27.7. Let $\rho > 0$ be the radius of convergence of the power series $f(x) = \sum_{n=0}^{\infty} a_n x^n$. Then

$$\int_0^x f(t)dt = \sum_{n=0}^{\infty} \frac{a_n}{n+1} x^{n+1}, \quad x \in (-\rho, \rho). \tag{27.5}$$

Proof. Since the power series $f(x)$ converges on $(-\rho, \rho)$, Lemma 27.1 implies that it converges uniformly on $[-r, r]$, where $r < \rho$. Thus, from Theorem 26.4 this series can be integrated term by term. Since $r \in (0, \rho)$ is arbitrary, (27.5) follows.

Corollary 27.2. If the power series $f(x) = \sum_{n=0}^{\infty} a_n x^n$ converges at $x = -\rho$ and $x = \rho$, then

$$\int_{-\rho}^{\rho} f(t)dt = 2\sum_{n=0}^{\infty} \frac{a_{2n}}{2n+1}\rho^{2n+1}. \qquad (27.6)$$

Proof. The proof is clear from Theorems 27.5 and 27.7.

Corollary 27.3. If the power series $f(x) = \sum_{n=0}^{\infty} a_n x^n$ converges on $[-\rho, \rho)$ and if $\sum_{n=0}^{\infty} a_n \rho^{n+1}/(n+1)$ converges, then $f(x)$ is improperly integrable on $[-\rho, \rho)$ and (27.6) holds.

Proof. From the definition of improper integration and Theorem 27.5, we have

$$\int_{-\rho}^{\rho} f(t)dt = \lim_{x \to \rho^-} \int_{-\rho}^{x} f(t)dt$$

$$= \lim_{x \to \rho^-} \sum_{n=0}^{\infty} \frac{a_n}{n+1}x^{n+1} - \sum_{n=0}^{\infty} \frac{a_n}{n+1}(-\rho)^{n+1}$$

$$= \sum_{n=0}^{\infty} \frac{a_n}{n+1}\rho^{n+1} - \sum_{n=0}^{\infty} \frac{a_n}{n+1}(-\rho)^{n+1},$$

which is the same as (27.6).

Finally, in this chapter we state the following results.

Theorem 27.8. Consider two power series $f(x) = \sum_{n=0}^{\infty} a_n x^n$ and $g(x) = \sum_{n=0}^{\infty} b_n x^n$ which converge for $|x| < \rho_1$ and $|x| < \rho_2$, respectively. If $\rho = \min\{\rho_1, \rho_2\}$, then

$$f(x) \pm g(x) = \sum_{n=0}^{\infty} (a_n \pm b_n)x^n$$

and

$$f(x)g(x) = \sum_{n=0}^{\infty} c_n x^n,$$

where

$$c_n = \sum_{k=0}^{n} a_k b_{n-k} = \sum_{k=0}^{n} a_{n-k} b_k$$

converge for $|x| < \rho$.

Theorem 27.9 (Gauss Test). If at $x = -\rho$ (ρ) the successive terms of the series $\sum_{n=0}^{\infty} a_n x^n$ are of fixed sign, and if the ratio of the $(n+1)$th term to the nth term can be written in the form $1 - (c/n) + O(1/n^2)$, where c is independent of n, then the series converges at $x = -\rho$ (ρ) if $c > 1$ and diverges if $c \leq 1$.

Example 27.4. We shall find the sum of the power series

$$f(x) = \sum_{n=1}^{\infty}(-1)^{n+1}\frac{x^{n+1}}{n(n+1)}, \quad x \in (-1,1).$$

Since the radius of convergence of this power series is 1, from Theorem 27.6 it follows that

$$f'(x) = \sum_{n=1}^{\infty}(-1)^{n+1}\frac{x^n}{n}, \quad x \in (-1,1).$$

Since $(1+x)^{-1} = \sum_{n=0}^{\infty}(-1)^n x^n$, $x \in (-1,1)$ integrating both sides of this equation (Theorem 27.7), we get

$$\ln(1+x) = \sum_{n=0}^{\infty}(-1)^n\frac{x^{n+1}}{n+1} = f'(x), \quad x \in (-1,1).$$

Thus, it follows that

$$f(x) = f(x) - f(0) = \int_0^x f'(t)dt = \int_0^x \ln(1+t)dt = (1+x)\ln(1+x) - x.$$

We also note that $\sum_{n=1}^{\infty}(-1)^n 1/[n(n+1)]$ converges absolutely, thus by Theorem 27.5,

$$\lim_{x\to 1^-} f(x) = \lim_{x\to 1^-}\sum_{n=1}^{\infty}(-1)^{n+1}\frac{x^{n+1}}{n(n+1)} = \sum_{n=1}^{\infty}\frac{(-1)^{n+1}}{n(n+1)}.$$

Since the left-hand side is $\lim_{x\to 1^-}((1+x)\ln(1+x) - x) = 2\ln 2 - 1$, we find $\sum_{n=1}^{\infty}(-1)^{n+1}/[n(n+1)] = 2\ln 2 - 1$.

Example 27.5. We shall find the sum of the power series $f(x) = \sum_{n=1}^{\infty} n^2 x^{n-1}$, $x \in (-1,1)$. The radius of convergence of this power series is 1. For $x \in (-1,1)$, we have

$$\int_0^x f(t)dt = \sum_{n=1}^{\infty}\int_0^x n^2 t^{n-1}dt = \sum_{n=1}^{\infty} nx^n = \frac{x}{(1-x)^2},$$

and hence

$$f(x) = \left(\frac{x}{(1-x)^2}\right)' = \frac{1+x}{(1-x)^3}, \quad x \in (-1,1).$$

We also note that

$$\sum_{n=1}^{\infty}\frac{n^2}{2^n} = \frac{1}{2}\sum_{n=1}^{\infty}n^2\left(\frac{1}{2}\right)^{n-1} = \frac{1}{2}f\left(\frac{1}{2}\right) = \frac{1}{2}\frac{\left(1+\frac{1}{2}\right)}{\left(1-\frac{1}{2}\right)^3} = 6.$$

Chapter 28

Power and Taylor Series (Contd.)

In Chapter 27, we saw how power series (27.1) leads to an infinitely differential function in its interval of convergence. In this chapter, we will discuss the converse problem, i.e., given an arbitrary function, $f(x)$, we will find a power series whose sum is exactly the given function. Using the terminology of functions of complex variables, we begin with the following definition.

A function $f : (a, b) \to \mathcal{R}$ is said to be *analytic* on (a, b) if at each $\alpha \in (a, b)$, f can be represented as a power series (27.1), i.e.,

$$f(x) = \sum_{n=0}^{\infty} a_n (x - \alpha)^n \qquad (28.1)$$

that converges in some interval $I = (\alpha - \delta, \alpha + \delta)$, $\delta > 0$.

In view of Corollary 27.1 it is clear that an analytic function on (a, b) is infinitely differentiable, i.e., $f \in C^{\infty}(a, b)$. Further, the coefficients a_n in (28.1) can be calculated uniquely and appear as

$$a_n = \frac{f^{(n)}(\alpha)}{n!}, \quad n \geq 0. \qquad (28.2)$$

Thus, the power series (28.1) can be written as

$$f(x) = \sum_{n=0}^{\infty} \frac{f^{(n)}(\alpha)}{n!} (x - \alpha)^n, \qquad (28.3)$$

which in the literature is known as *Taylor's series* of f centered at α. The Taylor series (28.3) centered at $\alpha = 0$, i.e.,

$$f(x) = \sum_{n=0}^{\infty} \frac{f^{(n)}(0)}{n!} x^n, \qquad (28.4)$$

is usually known as *Maclaurin's series* of f.

In view of the results in Chapter 27, our above discussion can be summarized in the the following result.

Theorem 28.1. If f is analytic in (a, b), then $f \in C^\infty(a, b)$. Conversely, $f \in C^\infty(a, b)$ is analytic on (a, b) if and only if for each $\alpha \in (a, b)$ the Taylor series (28.3) converges to $f(x)$ for all x in some open interval $(\alpha - \delta, \alpha + \delta)$, $\delta > 0$.

The following example shows that every C^∞ function is not analytic.

Example 28.1. The *Cauchy function*

$$f(x) = \begin{cases} e^{-1/x^2}, & x \neq 0 \\ 0, & x = 0 \end{cases} \tag{28.5}$$

has derivatives of all orders for every $x \in \mathcal{R}$, but it is not analytic in any interval containing $x = 0$. For this, we note that $f^{(n)}(x) = e^{-1/x^2} P_{3n}(1/x)$, where $P_{3n}(1/x)$ is a polynomial of degree $3n$ in $1/x$. We also note that $\lim_{x \to 0} x^{-k} e^{-1/x^2}$ for all $k \in \mathcal{N}$. Now, we have

$$f'(0) = \lim_{x \to 0} \frac{f(x) - f(0)}{x} = \lim_{x \to 0} \frac{e^{-1/x^2}}{x} = 0,$$

and if we assume that $f^{(n)}(0) = 0$, then we find

$$f^{(n+1)}(0) = \lim_{x \to 0} \frac{f^{(n)}(x) - f^{(n)}(0)}{x} = \lim_{x \to 0} \frac{1}{x} P_{3n}(x) e^{-1/x^2} = 0.$$

Hence, induction shows that $f^{(n)}(0) = 0$, $n \in \mathcal{N}$. Thus, all terms of Maclaurin's series (28.4) are zero for any x, and thus it is convergent; its sum is equal to zero for any x, which is different from the Cauchy function (28.5).

Now we shall prove the following result which provides sufficient conditions for $f : C^\infty(a, b) \to \mathcal{R}$ to be analytic on (a, b).

Theorem 28.2. Let $f : C^\infty(a, b) \to \mathcal{R}$, and let there exist an $M > 0$ such that $|f^{(n)}(x)| \leq M^n$, $x \in (a, b)$ and $n \in \mathcal{N}$. Then, f is analytic on (a, b), i.e., for each $\alpha \in (a, b)$, (28.3) holds for all $x \in (a, b)$.

Proof. Comparing Taylor's series (28.3) with Taylor's formula (16.4), it suffices to show that the residue term R_{n+1}^p in (16.6) tends to zero as $n \to \infty$. For this, we have

$$|R_{n+1}^p| = \left| (x - \alpha)^{n+1} (1 - \theta)^{n+1-p} \frac{f^{(n+1)}(\alpha + \theta(x - \alpha))}{n! \, p} \right|$$

$$\leq |x - \alpha|^{n+1} \frac{M^{n+1}}{n! \, p}, \quad x \in (a, b), \quad 1 \leq p \leq n + 1.$$

Thus, if we set $K = \max\{M(\alpha - a), M(b - \alpha)\}$, then it follows that

$$|R_{n+1}^p| \leq \frac{K^{n+1}}{n! \, p}, \quad x \in (a, b), \quad 1 \leq p \leq n + 1.$$

Since $K^{n+1}/n!\,p \to 0$ as $n \to \infty$ for any K, the remainder term R_{n+1}^p in (16.6) tends to zero as $n \to \infty$.

Example 28.2. We shall compute the Taylor series for $f(x) = e^x$ at $x = a$. Since $f^{(n)}(x) = e^x$, $n \in \mathcal{N}$, and for any $r > 0$, $\max_{[-r,r]} |f^{(n)}(x)| = \max_{[-r,r]} |e^x| \le e^r = M \le M^n$, from Theorem 28.2 it follows that the function e^x is analytic on $\mathcal{R}$. Thus, the Taylor series (28.3) for e^x at a is

$$e^x = \left[1 + \frac{(x-a)}{1!} + \frac{(x-a)^2}{2!} + \frac{(x-a)^3}{3!} + \cdots \right] e^a = \sum_{n=0}^{\infty} \frac{(x-a)^n}{n!} e^a.$$

In particular, Maclaurin's series (28.4) of e^x is

$$e^x = 1 + \frac{x}{1!} + \frac{x^2}{2!} + \frac{x^3}{3!} + \cdots = \sum_{n=0}^{\infty} \frac{x^n}{n!}.$$

Thus, for $x = 1$, we have

$$e = 1 + \frac{1}{1!} + \frac{1}{2!} + \frac{1}{3!} + \cdots = \sum_{n=0}^{\infty} \frac{1}{n!}. \qquad (28.6)$$

This series can be used only to compute a crude value of e. As in Chapter 16, here we shall use this expansion to show the irrationality of e. Suppose for contrary $e = p/q$, where p and q are integers and $q > 1$. Thus, we have $e = p/q = \sum_{n=0}^{\infty} 1/n!$, which is the same as

$$p(q-1)! = q! \sum_{k=0}^{q} \frac{1}{k!} + q! \sum_{k=q+1}^{\infty} \frac{1}{k!}.$$

Now, we observe that

$$0 < p(q-1)! - q! \sum_{k=0}^{q} \tfrac{1}{k!} = \tfrac{1}{q+1} + \tfrac{1}{(q+1)(q+2)} + \cdots$$
$$< \tfrac{1}{q+1} + \tfrac{1}{(q+1)^2} + \cdots = \tfrac{1}{q} < 1.$$

But, $p(q-1)! - q! \sum_{k=0}^{q} 1/k!$ is a positive integer.

Example 28.3. For the function $f(x) = \sin x$, we have

$$f^{(n)}(x) = \begin{cases} (-1)^{n/2} \sin x, & n \text{ even} \\ (-1)^{(n-1)/2} \cos x, & n \text{ odd}. \end{cases}$$

Thus, $|f^{(n)}(x)| \le 1$, $n \in \mathcal{N}$, and hence $\sin x$ is analytic on $\mathcal{R}$. Now, since

$$f^{(n)}(0) = \begin{cases} 0, & n \text{ even} \\ (-1)^{(n-1)/2}, & n \text{ odd} \end{cases}$$

the Maclaurin series of $\sin x$ can be written as

$$\sin x = x - \frac{x^3}{3!} + \frac{x^5}{5!} - \cdots = \sum_{n=0}^{\infty} \frac{(-1)^n}{(2n+1)!} x^{2n+1}. \qquad (28.7)$$

Similarly the function $f(x) = \cos x$ is analytic on $\mathcal{R}$, and its Maclaurin series can be obtained directly, or in view of the uniqueness, by term-by-term differentiation of (28.7)

$$\cos x = 1 - \frac{x^2}{2!} + \frac{x^4}{4!} - \cdots = \sum_{n=0}^{\infty} \frac{(-1)^n}{(2n)!} x^{2n}. \qquad (28.8)$$

Next, we shall prove *Bernstein criterion* for $f : C^{\infty}(a,b) \to \mathcal{R}$ to be analytic on (a,b).

Theorem 28.3. Let $f : C^{\infty}(a,b) \to \mathcal{R}$ and $f^{(n)}(x) \geq 0$, $x \in (a,b)$, $n \in \mathcal{N}$. Then, f is analytic on (a,b). In fact, if $\alpha \in (a,b)$ and $f^{(n)}(x) \geq 0$, $x \in [\alpha, b)$, $n \in \mathcal{N}$, then (28.3) holds for all $x \in [\alpha, b)$.

Proof. We assume that $\alpha \geq 0$, and the case $\alpha \leq 0$ can be treated similarly. From Problem 20.7 and the fact that $f^{(n)}(x) \geq 0$, $x \in [\alpha, b)$ it follows that

$$0 \leq R_n(x) = \frac{1}{(n-1)!} \int_{\alpha}^{x} (x-t)^{n-1} f^{(n)}(t) dt$$
$$= f(x) - \sum_{k=0}^{n-1} \frac{f^{(k)}(\alpha)}{k!} (x-\alpha)^k \leq f(x),$$

i.e.,

$$0 \leq R_n(x) \leq f(x), \quad x \in [\alpha, b). \qquad (28.9)$$

Now in $R_n(x)$ we use the substitution $t = xu$, to find

$$0 \leq R_n(x) = \frac{x^n}{(n-1)!} \int_{\alpha/x}^{1} (1-u)^{n-1} f^{(n)}(xu) du. \qquad (28.10)$$

We need to show that $R_n(x) \to 0$ for each $x \in [\alpha, b)$. Note that $R_n(\alpha) = 0$. We fix $x \in (\alpha, b)$. Then, there exists a β with $\alpha < x < \beta < b$, and thus in view of (28.9) and (28.10), it follows that

$$0 \leq R_n(x) \leq \frac{x^n}{(n-1)!} \int_{\alpha/\beta}^{1} (1-u)^{n-1} f^{(n)}(\beta u) du$$
$$\leq \left(\frac{x}{\beta}\right)^n R_n(\beta) \leq \left(\frac{x}{\beta}\right)^n f(\beta).$$

Now since $x/\beta < 1$, $(x/\beta)^n \to 0$ as $n \to \infty$.

Example 28.4. Since $d^n e^x / dx^n = e^x > 0$, $x \in \mathcal{R}$, the function e^x is analytic on $\mathcal{R}$. Similarly, since $d^n (1-x)^{-1}/dx^n = n!(1-x)^{-n-1} > 0$, $x \in (-1,1)$, the function $(1-x)^{-1}$ is analytic on $(-1,1)$.

Finally, for analytic functions we state an analog of Theorem 27.8.

Theorem 28.4. If $f : C^\infty(a, b) \to \mathcal{R}$ and $g : C^\infty(c, d) \to \mathcal{R}$ are analytic on (a, b) and (c, d), respectively, then $f(x) + g(x)$ and $f(x)g(x)$ are analytic on $(a, b) \cap (c, d)$.

Example 28.5. Since $\sinh x = (e^x - e^{-x})/2$ and $\cosh x = (e^x + e^{-x})/2$, from Example 28.4 and Theorem 28.4 and both of these functions are analytic on $\mathcal{R}$, we also have

$$\sinh x = \sum_{n=0}^{\infty} \frac{x^{2n+1}}{(2n+1)!} \quad \text{and} \quad \cosh x = \sum_{n=0}^{\infty} \frac{x^{2n}}{(2n)!}, \quad x \in \mathcal{R}.$$

Example 28.6. The function $f(x) = x^2$ is analytic on $\mathcal{R}$, and the function $2/(1-x)^2$ is analytic on $(-1, 1)$. From Theorem 28.4, it follows that both the functions $f(x) + g(x) = x^2 + 2/(1-x)^2$ and $f(x)g(x) = 2x^2/(1-x)^2$ are analytic on $(-1, 1)$.

Problems

28.1. Find the radius of convergence of each of the following power series.

(i). $\sum_{n=1}^{\infty} 2^{-n}(3 + (-1)^n)^n x^n$.

(ii). $\sum_{n=1}^{\infty} (-1)^n 3^{-n} x^{2n}$.

(iii). $\sum_{n=1}^{\infty} n^{-1} 4^{-n} x^{2n}$.

(iv). $\sum_{n=1}^{\infty} n! n^{-n} x^n$.

(v). $\sum_{n=1}^{\infty} \frac{1 \cdot 3 \cdots (2n-1)}{(n+1)!} x^{2n}$.

28.2. Find the interval of convergence of each power series in Problem 28.1.

28.3. Assume that for the power series $\sum_{n=0}^{\infty} a_n x^n$ the radius of convergence is ρ. Find the radius of convergence of each of the following power series, where k is a fixed positive integer.

(i). $\sum_{n=0}^{\infty} a_n^k x^n$.

(ii). $\sum_{n=0}^{\infty} a_n x^{kn}$.

28.4. Let $a_n \in \mathcal{R}$ for $n \in \mathcal{N}$. Show that $\limsup_{n \to \infty} \sqrt[n]{n|a_n|} = \limsup_{n \to \infty} \sqrt[n]{|a_n|}$.

28.5. Find the sum of the power series

(i). $\sum_{n=0}^{\infty} (n+1)x^{n+2}$.

(ii). $\sum_{n=1}^{\infty} (-1)^{n-1} x^n/n$.

(iii). $\sum_{n=0}^{\infty} x^n/(n+1)$.

(iv). $\sum_{n=1}^{\infty} 2n(1-x)^n/(n+1)$.

28.6.   Show that the following series are Abel summable, and find their Abel sums:

(i). $1 - 1 + 1 - 1 + \cdots$.

(ii). $1 - 2 + 3 - 4 + \cdots$.

(iii). $-\frac{1}{1} + \frac{1}{2} - \frac{1}{3} + \frac{1}{4} - \cdots$.

(iv). $1 - \frac{1}{3} + \frac{1}{5} - \frac{1}{7} + \cdots$.

28.7.   Suppose $|a_n| \le |b_n|$ for large $n \in \mathcal{N}$. Show that if $\sum_{n=1}^{\infty} b_n x^n$ converges on an open interval I, then $\sum_{n=1}^{\infty} a_n x^n$ also converges on I. Is the result true on the closed interval?

28.8.   Show that $\sum_{n=0}^{\infty} a_n x^n$ with $a_n = \begin{cases} 1, & n = \text{odd} \\ 1/n, & n = \text{even} \end{cases}$ has the radius of convergence $\rho = 1$, and that $\lim_{n \to \infty} |a_{n+1}/a_n|$ does not exist.

28.9   If f is represented by two power series $\sum_{n=0}^{\infty} a_n x^n$ and $\sum_{n=0}^{\infty} b_n x^n$, which have radius of convergence $\rho > 0$, then show that $a_n = b_n$, $n \in \mathcal{N}$. Thus, if $\sum_{n=0}^{\infty} a_n x^n = 0$ for $|x| < \rho$, then $a_n = 0$, $n \in \mathcal{N}$.

28.10.   Find the radius of convergence and the interval of convergence of the *hypergeometric series*

$$f(x) = 1 + \frac{ab}{c\,1!}x + \frac{a(a+1)b(b+1)}{c(c+1)\,2!}x^2 + \frac{a(a+1)(a+2)b(b+1)(b+2)}{c(c+1)(c+2)\,3!}x^3 + \cdots .$$

28.11.   Prove that, for any real r, the binomial expansion

$$(1+x)^r = 1 + rx + \frac{r(r-1)}{2!}x^2 + \cdots + \frac{r(r-1)\cdots(r-n+1)}{n!}x^n + \cdots$$

converges in $|x| < 1$. (This is an extension of Example 27.2.) Also, observe that this expansion holds when $x = -1$ if $r > 0$, and $x = 1$ if $r > -1$.

28.12. (i). Use the binomial expansion $(1+x)^{-1} = 1 - x + x^2 - \cdots$, $-1 < x < 1$ to obtain the Taylor series

$$\ln(1+x) = x - \frac{x^2}{2} + \frac{x^3}{3} - \cdots = \sum_{n=1}^{\infty} \frac{(-1)^{n-1}}{n} x^n, \quad -1 < x \le 1.$$

(ii). Use the binomial expansion $(1+x^2)^{-1} = 1 - x^2 + x^4 - \cdots$, $-1 < x < 1$ to obtain the Taylor series

$$\tan^{-1} x = x - \frac{x^3}{3} + \frac{x^5}{5} - \cdots = \sum_{n=0}^{\infty} \frac{(-1)^n}{2n+1} x^{2n+1}, \quad -1 \le x \le 1.$$

(iii). Use binomial expansion

$$\frac{1}{(1-x^2)^{1/2}} = 1 + \frac{x^2}{2} + \frac{1 \cdot 3}{2!}\frac{x^4}{2^2} + \frac{1 \cdot 3 \cdot 5}{3!}\frac{x^6}{2^3} + \cdots, \quad -1 < x < 1$$

to obtain the Taylor series

$$\sin^{-1} x = x + \frac{x^3}{2 \cdot 3} + \frac{1 \cdot 3}{2!}\frac{x^5}{2^2 5} + \frac{1 \cdot 3 \cdot 5}{3!}\frac{x^7}{2^3 7} + \cdots, \quad -1 \leq x \leq 1.$$

Answers or Hints

28.1. (i). Let $a_n = \frac{1}{2^n}(3 + (-1)^n)^n$. Then, $\overline{\lim}_{n \to \infty} |a_n|^{1/n} = \overline{\lim}_{n \to \infty} \frac{1}{2}(3 + (-1)^n) = \frac{1}{2}(3 + 1) = 2$, and hence $\rho = 1/2$.

(ii). Let $a_{2n} = (-1)^n/3^n$ and $a_{2n-1} = 0$, $n \geq 1$. Then, $\overline{\lim}_{n \to \infty} |a_n|^{1/n} = \overline{\lim}_{n \to \infty} |a_{2n}|^{1/2n} = \overline{\lim}_{n \to \infty} \left(\frac{1}{3^n}\right)^{1/2n} = \frac{1}{\sqrt{3}}$, and hence $\rho = \sqrt{3}$.

(iii). $\rho = 2$.

(iv). $\rho = e$.

(v). $\rho = 1/\sqrt{2}$.

28.2. (i). $(-1/2, 1/2)$.

(ii). $(-\sqrt{3}, \sqrt{3})$.

(iii). $(-2, 2)$.

(iv). $(-e, e)$.

(v). $[-1/\sqrt{2}, 1/\sqrt{2}]$. (Use Problem 9.9.)

28.3. (i). ρ^k.

(ii). $\rho^{1/k}$.

28.4. Since $\lim_{n \to \infty} \sqrt[n]{n} = 1$, for a given $\epsilon > 0$ there exists $N \in \mathcal{N}$ such that for all $n \geq N$, $1 - \epsilon < \sqrt[n]{n} < 1 + \epsilon$. Thus, it follows that $(1 - \epsilon) \limsup_{n \to \infty} \sqrt[n]{|a_n|} \leq \limsup_{n \to \infty} \sqrt[n]{n|a_n|} \leq (1+\epsilon) \limsup_{n \to \infty} \sqrt[n]{|a_n|}$. The result now follows as $\epsilon \to 0$.

28.5. (i). $x^2/(1-x)^2$.

(ii). $\ln(1 + x)$.

(iii). $-\ln(1 - x)/x$.

(iv). $2(\ln x + 1/x - 1)/(1 - x)$.

28.6. (i). $1/2$.

(ii). $1/4$.

(iii). $-\ln 2$.

(iv). $\pi/4$.

28.7. The answer is no. For this, notice that $\frac{1}{2} \ln \frac{1+x}{1-x} = \sum_{n=0}^{\infty} \frac{1}{2n+1} x^{2n+1}$

converges on $(-1, 1)$, whereas $\arctan x = \sum_{n=0}^{\infty} (-1)^n \frac{1}{2n+1} x^{2n+1}$ converges on $[-1, 1]$. If we take $a_n = \frac{1}{2n+1}$ and $b_n = (-1)^n \frac{1}{2n+1}$, then we have $|a_n| = |b_n|$, $n \geq 0$.

28.8. Since $\limsup_{n \to \infty} \sqrt[n]{|a_n|} = 1$, the radius of convergence is $\rho = 1$. Notice that $\lim_{n \to \infty} |a_{n+1}|/|a_n|$ is 0 if n is odd, and ∞ if n is even.

28.9. If $f(x) = \sum_{n=0}^{\infty} a_n x^n$ and $g(x) = \sum_{n=0}^{\infty} b_n x^n$, then since $f(x) = g(x)$, $x \in (-\rho, \rho)$ from Corollary 27.1 it follows that $a_k = f^{(k)}(0) = g^{(k)}(0) = b_k$, $k \in \mathcal{N}$.

28.10. The ratio of the coefficients of x^{n+1} and x^n in the series is $[(a + n)(b + n)]/[(c + n)(n + 1)]$, which tends to 1 uniformly as $n \to \infty$, regardless of the values of a, b and c. Hence, by Theorem 27.2 the series has radius of convergence $\rho = 1$. Also, since this ratio can be written as $1 - \frac{1+c-a-b}{n} + O\left(\frac{1}{n^2}\right)$ the series converges absolutely at $x = \pm 1$ by Theorem 27.9 provided $c > a + b$. Thus, if $c > a + b$ the interval of convergence is $[-1, 1]$.

28.11. From (16.6) with $p = 1$ and $\alpha = 0$, we have
$$R_{n+1}^1 = \frac{r(r-1)\cdots(r-n)}{n!} x^{n+1} \left(\frac{1-\theta}{1+\theta x}\right)^n (1 + \theta x)^{r-1}, \quad 0 < \theta < 1.$$
Now, note that $\lim_{n \to \infty} [r(r-1) \cdots (r-n)x^{n+1}]/n! = 0$, $0 < (1-\theta)/(1+\theta x) < 1$, $(1+\theta x)^{r-1} < (1+|x|)^{r-1}$ if $r > 1$, and $(1+\theta x)^{r-1} < (1-|x|)^{r-1}$ if $r < 1$. Hence, $\lim_{n \to \infty} R_{n+1}^1 = 0$.

28.12. In all the parts use term-by-term integration.

Chapter 29

Metric Spaces

Chapter 3 describes that the real number system $\mathcal{R}$ is inherited with two important properties, namely order and completeness. The order relation leads to the notion of distance between any two real numbers, which foster a topological structure in $\mathcal{R}$. The main purpose of this and the next chapter is to develop some of these topological properties in a general setting. We begin with the following definitions.

Let S be any nonempty set. A function $\rho : S \times S \to \mathcal{R}$ is called a *metric* on S if it satisfies the following properties for all $x, y, z \in S$:

Positive Definite	$\rho(x, y) \geq 0$ with $\rho(x, y) = 0$ if and only if $x = y$,
Symmetric	$\rho(x, y) = \rho(y, x)$,
Triangle Inequality	$\rho(x, y) \leq \rho(x, z) + \rho(z, y)$.

If ρ is a metric on S, then the ordered pair (S, ρ) is called a *metric space*.

Example 29.1. The function $\rho(x, y) = |x - y|$ is obviously a metric (absolute value metric) for the set $S = \mathcal{R}$. We denote the metric space $(\mathcal{R}, \rho)$ by R^1.

Example 29.2. The function $\rho(x, y) = \begin{cases} 0 & \text{if } x = y \\ 1 & \text{if } x \neq y \end{cases}$ is also a metric (discrete metric) for the set $S = \mathcal{R}$. We denote the discrete metric space $(\mathcal{R}, \rho)$ by R_d. From Examples 29.1 and 29.2 it is clear that a given set may have more than one metric.

Example 29.3. For ordered n-tuples $x = (a_1, \cdots, a_n)$ and $y = (b_1, \cdots, b_n)$, we define $\rho(x, y) = \left(\sum_{k=1}^{n} (a_k - b_k)^2 \right)^{1/2}$. (Note that for $n = 2$, $\rho(x, y)$ is the usual distance formula for points in the Cartesian plane.) We shall show that this ρ, known as *Euclidean metric*, satisfies the triangle inequality, i.e., if $z = (c_1, \cdots, c_n)$ then $\rho(x, y) \leq \rho(x, z) + \rho(z, y)$. For this, we let $x_k = a_k - c_k$, $y_k = c_k - b_k$, $k = 1, \cdots, n$. Then, $\rho(x, z) = \left(\sum_{k=1}^{n} x_k^2 \right)^{1/2}$, $\rho(z, y) = \left(\sum_{k=1}^{n} y_k^2 \right)^{1/2}$, and $\rho(x, y) = \left(\sum_{k=1}^{n} (x_k + y_k)^2 \right)^{1/2}$. The triangle inequality now immediately follows from the Minkowski inequality (Problem 6.6(ii)). The space of all ordered n-tuples with this metric is called the

Euclidean n-space and denoted as R^n. Note that for $n = 1$, R^n is the same as R^1, in fact, we have $\left(\sum_{k=1}^{1}(a_k - b_k)^2\right)^{1/2} = |a_1 - b_1|$.

Example 29.4. Let $C[a, b]$ be the set of all real-valued continuous functions defined on $[a, b]$. For all $f, g \in C[a, b]$ it can be easily seen that $\rho(f, g) = \sup_{x \in [a,b]} |f(x) - g(x)|$ and $\rho_1(f, g) = \int_a^b |f(x) - g(x)|dx$ are metrics. ($\int_a^b |f(x) - g(x)|dx = 0$ implies that $|f(x) - g(x)| = 0$ for every $x \in [a, b]$, and since $|f(x) - g(x)|$ is a nonnegative continuous function, we have $f = g$.)

If S is a metric space with metric ρ and $T \subset S$, then it is clear that T is a metric space (known as *subspace* of S) with metric ρ.

Now let (S, ρ) be a metric space, let $x_0 \in S$, and let $\epsilon > 0$. The *neighborhood* of x_0 of radius ϵ is given by $N_\epsilon(x_0) = N(x_0, \epsilon) = \{x \in S : \rho(x, x_0) < \epsilon\}$. $N(x_0, \epsilon)$ is also called a *ball* centered at x_0 with radius ϵ. Clearly, $N(x_0, \epsilon) \subseteq N(x_0, \tau)$ whenever $0 < \epsilon \leq \tau$. Exactly as in Chapter 4, this definition can be used to characterize terms such as interior points, boundary points, open sets, and closed sets. These concepts can then be employed to carry over several results of Chapter 4 to this general setting with only minor changes. In what follows we shall state several results which are analogous to those established in earlier chapters and prove a few of them.

Theorem 29.1. Let (S, ρ) be a metric space. Then, the following hold:

(1). The empty set $\emptyset$ and the whole space S are open as well as closed.

(2). Every finite subset of S is closed.

(3). Any neighborhood of a point in S is an open set.

Proof. (1) and (2) follow from the definitions of open and closed sets. To show (3), let $x \in S$ and $\epsilon > 0$. We need to show that $N(x, \epsilon)$ is an open set. For this, we shall prove that any point $y \in N(x, \epsilon)$ is an interior point of $N(x, \epsilon)$. If $y \in N(x, \epsilon)$, then $d = \epsilon - \rho(x, y) > 0$. We claim that $N(y, d) \subseteq N(x, \epsilon)$. If $z \in N(y, d)$, then $\rho(z, y) < d$. Thus, we have

$$\rho(z, x) \leq \rho(z, y) + \rho(y, x) < d + \rho(y, x) = [\epsilon - \rho(x, y)] + \rho(y, x) = \epsilon$$

and hence, $z \in N(x, \epsilon)$. This shows that $N(y, d) \subseteq N(x, \epsilon)$, and therefore y is an interior point of $N(x, \epsilon)$.

Theorem 29.2. Let (S, ρ) be a metric space, and I be an index set. Then, the following hold:

(1). If $\{T_i\}_{i \in I}$ is any collection of open sets in S, then $\cup_{i \in I} T_i$ is open.

(2). If $\{T_i\}_{i=1}^{n}$ is a finite collection of open sets in S, then $\cap_{i=1}^{n} T_i$ is open.

(3). If $\{T_i\}_{i \in I}$ is any collection of closed sets in S, then $\cap_{i \in I} T_i$ is closed.

(4). If $\{T_i\}_{i=1}^{n}$ is a finite collection of closed sets in S, then $\cup_{i=1}^{n} T_i$ is closed.

(5). If T is open in S and U is closed in S, then $T\backslash U$ is open and $U\backslash T$ is closed.

(6). If $T \subseteq S$, then the *interior* of T denoted as T^i is the largest open set contained in T.

(7). If $T \subseteq S$, then the *closure* of T represented by $\overline{T}$ is closed, where $\overline{T} = T \cup T'$ and T' is the *derived* set of T. In particular, $T^i \subseteq T \subseteq \overline{T}$.

(8). T is closed if and only if $T = \overline{T}$.

(9). If $\{T_i\}_{i \in I}$ is any collection of compact sets in S, then $\cap_{i \in I} T_i$ is compact.

(10). If $\{T_i\}_{i=1}^n$ is a finite collection of compact sets in S, then $\cup_{i=1}^n T_i$ is compact.

Remark 29.1. The preceding results are perceived as though every result known for $\mathcal{R}$ continue to apply in a metric space; however, when it comes to compact sets we need to be careful. In fact, the Heine-Borel theorem no longer holds for arbitrary metric spaces. For this, as in Example 29.2 consider the discrete metric space R_d. Since $N_{1/2}(x) = \{x\}$ for each $x \in S = \mathcal{R}$, every point in S is open. Thus, $\{x\}_{x \in S}$ is an uncountable open cover of S, which does not have a finite subcover.

A set T in the metric space (S, ρ) is said to be *dense* in S if T has at least one point in common with every open ball. In Chapter 3 we have seen that the set of rationals $\mathcal{Q}$ as well as of irrationals $\mathcal{R}\backslash\mathcal{Q}$ are dense in R^1. We note that the only dense set of a discrete metric space (S, ρ) is S itself.

Theorem 29.3. The set T is dense in the metric space (S, ρ) if and only if $\overline{T} = S$.

Proof. Assume that T is dense in S and $y \in S\backslash T$. Then, every open ball centered at y contains a point of T different from y. But, then from the definition of a limit point $y \in T'$, this implies that $T \cup T' = S$, i.e., $\overline{T} = S$. Conversely, let T be not dense in S. Then, there exists an open ball $S_\rho(x)$ such that $S_\rho(x) \cap T = \emptyset$. But, then $T \subseteq S\backslash S_\rho(x)$. Now since $S\backslash S_\rho(x)$ is closed, it follows that $\overline{T} \subseteq S\backslash S_\rho(x)$, which implies that $\overline{T} \subset S$.

Remark 29.2. A metric space (S, ρ) is said to be *separable* if there is a countable set which is dense in S. The set $\mathcal{Q}$ is countable which is dense in R^1, and hence R^1 is separable. However, from Remark 29.1 it is clear that the discrete metric space R_d is not separable.

Now we shall extend the theory of real sequences discussed in Chapters 7 and 8 in a general metric space. We begin with the following definitions.

Let (S, ρ) be a metric space and let $\{u_n\}_{n=1}^\infty$ be a sequence of points in S. The sequence $\{u_n\}$ is said to be *bounded* if given $x_0 \in S$ there exists an $M > 0$ such that $\rho(u_n, x_0) < M$, $n \in \mathcal{N}$. The sequence $\{u_n\}$ is said to *converge* to

$u \in S$ (known as the *limit*) if given $\epsilon > 0$ there exists an $N \in \mathcal{N}$ such that $\rho(u_n, u) < \epsilon$, $n \geq N$. In this case we write $\lim_{n\to\infty} u_n = u$, or $u_n \to u$ as $n \to \infty$. We say $\{u_n\}$ is a *Cauchy sequence* if given $\epsilon > 0$ there exists an $N \in \mathcal{N}$ such that $\rho(u_m, u_n) < \epsilon$, $m, n \geq N$. A metric space (S, ρ) is said to be *complete* if every Cauchy sequence in S converges to some point in S.

Theorem 29.4. Let (S, ρ) be a metric space and $\{u_n\}_{n=1}^{\infty}$ be a sequence of points in S. Then, the following hold:

(1). If $\{u_n\}_{n=1}^{\infty}$ converges, then it is bounded.

(2). $\{u_n\}_{n=1}^{\infty}$ converges to a unique limit.

(3). $\{u_n\}_{n=1}^{\infty}$ converges to u if and only if every subsequence $\{u_{n_j}\}$ converges to the same u.

(4). If $\{u_n\}_{n=1}^{\infty}$ converges then it is a Cauchy sequence.

Remark 29.3. In arbitrary metric spaces the Bolzano-Weierstrass theorem does not hold, i.e., a metric space may contain bounded sequences which have no convergent subsequences. For this, we consider the discrete metric space R_d. Since $\rho(0, n) = 1$ for all $n \in \mathcal{N}$, $\{n\}$ is a bounded sequence in R_d. Assume that there exists a subsequence $n_1 < n_2 < \cdots$ and an $u \in R_d$ such that $n_k \to u$ as $k \to \infty$. Then there is an $N \in \mathcal{N}$ such that $\rho(n_k, u) < 1/2$ for all $k \geq N$, but this means $n_k = u$ for all $k \geq N$, which is a contradiction.

Remark 29.4. The converse of Theorem 29.4(4) does not hold, i.e., in a metric space a Cauchy sequence may not converge. To show this, we let S be the set of all points (x, y) in the Euclidean plane R^2 such that $x^2 + y^2 < 1$. The sequence $\{u_n\}$ where $u_n = (0, n/(n+1))$ is a Cauchy sequence of points in S but there is no $u \in S$ such that $\{u_n\}$ converges to u. Hence, the sequence $\{u_n\}$ of points in S does not converge in S. As another example, let $u_n \in Q = S$ such that $u_n \to \sqrt{3}$. Then, $\{u_n\}$ is a Cauchy sequence, but it does not converge in S.

Remark 29.5. From Remark 29.4 it is clear that a metric space may not be complete.

Example 29.5. The space $(B[0, 1], \rho)$, where $B[0, 1]$ is the set of all bounded functions on $[0, 1]$, and for $f, g \in B[0, 1]$, $\rho(f, g) = \sup_{x \in [0,1]} |f(x) - g(x)|$ is a complete metric space. The verification of $(B[0, 1], \rho)$ is a metric space is straightforward. To show its completeness let $\{f_n\}$ be a Cauchy sequence in $(B[0, 1], \rho)$, then for each $x \in [0, 1]$, we have $|f_n(x) - f_m(x)| \leq \rho(f_n, f_m)$. Thus, $\{f_n(x)\}$ is a Cauchy sequence of real numbers, and hence in view of the completeness of $\mathcal{R}$, it converges. For each $x \in [0, 1]$ let $f(x) = \lim_{n\to\infty} f_n(x)$. We need to show that $f \in B[0, 1]$. For this, there exists an $N \in \mathcal{N}$ such that $\rho(f_n, f_m) < 1$ for all $n, m > N$. Therefore, if $n > N$, $|f_n(x) - f_{N+1}(x)| < 1$ for all $x \in [0, 1]$. Now, there exists an $M > 0$ such that $|f_{N+1}(x)| \leq M$ for all $x \in [0, 1]$. Thus, for $n > N$ it follows that

$|f_n(x)| < 1 + |f_{N+1}(x)| \le 1 + M$ for all $x \in [0,1]$. Since $f(x) = \lim_{n \to \infty} f_n(x)$, for each $x \in [0,1]$ there exists an $n > N$ with $|f_n(x) - f(x)| < 1$. Thus, we have $|f(x)| \le |f(x) - f_n(x)| + |f_n(x)| < 1 + (1 + M) = 2 + M$. Hence, $f(x)$ is bounded on $[0,1]$. Finally, we need to show that $\{f_n\}$ converges to f in $(B[0,1], \rho)$. For this, let $\epsilon > 0$. Since $\{f_n\}$ is Cauchy, there exists an $N \in \mathcal{N}$ such that $\rho(f_n, f_m) < \epsilon/2$ whenever $n, m > N$. Now for each $x \in [0,1]$ choose $k > N$ such that $|f(x) - f_k(x)| < \epsilon/2$. Then, for $n > N$, it follows that $|f(x) - f_n(x)| \le |f(x) - f_k(x)| + |f_k(x) - f_n(x)| < \epsilon/2 + \epsilon/2 = \epsilon$. But, this implies that $\rho(f, f_n) < \epsilon$ for all $n > N$. Hence $(B[0,1], \rho)$ is complete.

Example 29.6. The space $(C[0,1], \rho)$ considered in Example 29.4 (with $a = 0$ and $b = 1$) is a complete metric space. Clearly, $(C[0,1], \rho) \subseteq (B[0,1], \rho)$. Thus, in view of Problem 30.8 it suffices to show that $C[0,1]$ is a closed subset of $B[0,1]$. Let f be a limit point of $C[0,1]$. This means that for each $n \in \mathcal{N}$ there exists $f_n \in C[0,1] \cap N_{1/n}(f)$. Obviously, $\{f_n\}$ converges to f. We need to show that $f \in C[0,1]$. Let $x_0 \in [0,1]$ be arbitrary and $\epsilon > 0$ be given. Let $n \in \mathcal{N}$ be large so that $1/n < \epsilon/3$. Then, $\rho(f_n, f) < \epsilon/3$, i.e., $|f_n(x) - f(x)| < \epsilon/3$ for every $x \in [0,1]$. Since $f_n \in C[0,1]$, there exists a $\delta > 0$ such that $|f_n(x) - f_n(x_0)| < \epsilon/3$ provided $x \in [0,1]$ and $|x - x_0| < \delta$. Thus, if $x \in [0,1]$ and $|x - x_0| < \delta$, it follows that

$$
\begin{aligned}
|f(x) - f(x_0)| &\le |f(x) - f_n(x)| + |f_n(x) - f_n(x_0)| + |f_n(x_0) - f(x_0)| \\
&< \tfrac{\epsilon}{3} + \tfrac{\epsilon}{3} + \tfrac{\epsilon}{3} = \epsilon,
\end{aligned}
$$

and hence g is continuous at x_0. Since $x_0 \in [0,1]$ is arbitrary, $f \in C[0,1]$. Finally, since f is an arbitrary limit point of $C[0,1]$, it is clear that $C[0,1]$ is closed.

Example 29.7. The set of all polynomials P with real coefficients is dense in the metric space $(C[0,1], \rho)$. Thus, for any given function $f \in C[0,1]$ and $\epsilon > 0$ there exists a polynomial $P_n(x)$ such that $|f(x) - P_n(x)| < \epsilon$ for all $x \in [0,1]$. This well-known result is called the *Weierstrass Approximation Theorem*.

Chapter 30

Metric Spaces (Contd.)

Our discussion in Chapter 12 and later chapters on the continuity of a real-valued function at a point in R^1 was based on the metric for R^1. In this chapter, we will extend the concept of continuity to functions from one metric space into another, and discuss some interesting results. We begin with the following definitions.

Let (S, ρ_1) and (T, ρ_2) be metric spaces. We say $f : S \to T$ approaches $L \in T$ as x approaches $x_0 \in S$ if given $\epsilon > 0$ there exists a $\delta > 0$ such that $\rho_2(f(x), L) < \epsilon$ provided $0 < \rho_1(x, x_0) < \delta$. In this case, we write $\lim_{x \to x_0} f(x) = L$, or $f(x) \to L$ as $x \to x_0$. The function $f : S \to T$ is called *continuous at a point* $x_0 \in S$ if for every $\epsilon > 0$ there exists a $\delta > 0$ such that $\rho_2(f(x), f(x_0)) < \epsilon$ whenever $\rho_1(x, x_0) < \delta$. If f is continuous at each point of a set $D \subseteq S$, then f is said to be *continuous on D*.

Example 30.1. Let (S, ρ) be a metric space and let $x_0 \in S$. Then, the function $f : S \to \mathcal{R}$ defined by $f(x) = \rho(x, x_0)$ is continuous on S. For this, let $x, y \in S$. From the triangle inequality, we have $\rho(y, x_0) \leq \rho(y, x) + \rho(x, x_0)$, and hence $\rho(y, x_0) - \rho(x, x_0) \leq \rho(x, y)$. Similarly, we have $\rho(x, x_0) - \rho(y, x_0) \leq \rho(x, y)$. Thus, it follows that $|\rho(x, x_0) - \rho(y, x_0)| \leq \rho(x, y)$. Hence, given any $\epsilon > 0$, if $\rho(x, y) < \epsilon$, we have $|f(x) - f(y)| = |\rho(x, x_0) - \rho(y, x_0)| \leq \rho(x, y) < \epsilon$. Thus, f is continuous at the arbitrary point y and hence continuous on S.

Theorem 30.1. Let (S, ρ_1) and (T, ρ_2) be metric spaces and $f : S \to T$, and let $x_0 \in S$. Then, the following statements are equivalent:

(1). f is continuous at x_0.

(2). If $\{x_n\}$ is any sequence in S which converges to x_0, then $\{f(x_n)\}$ converges to $f(x_0)$ in T.

(3). For every neighborhood V of $f(x_0)$ in T, there exists a neighborhood U of x_0 in S such that $f(U) \subseteq V$.

Proof. $(1) \Rightarrow (2)$. Since f is continuous at x_0, given $\epsilon > 0$ there exists a $\delta > 0$ such that $\rho_2(f(x), f(x_0)) < \epsilon$ whenever $\rho_1(x, x_0) < \delta$. Since $x_n \to x_0$, there exists an $N \in \mathcal{N}$ such that $\rho_1(x_n, x_0) < \delta$. Thus, for $n > N$ we have $\rho_2(f(x_n), f(x_0)) < \epsilon$, and hence $f(x_n) \to f(x_0)$.

(2)⇒(3). We shall prove the contrapositive form, i.e., we assume that there exists a neighborhood $V = N(f(x_0), \epsilon)$ of $f(x_0)$ such that, for all neighborhoods U of x_0, $f(U) \not\subseteq V$. For this, we need to find a sequence $\{x_n\}$ in S such that $x_n \to x_0$ but $\{f(x_n)\}$ does not converge to $f(x_0)$. Let for each $n \in \mathcal{N}$, $U_n = N(x_0, 1/n)$. Since $f(U_n) \not\subseteq V$, there exists a point x_n in U_n such that $f(x_n) \notin V$. Thus, although $x_n \to x_0$, $\{f(x_n)\}$ does not converge to $f(x_0)$ because none of the $f(x_n)$ are in V. Thus, we must have $\rho_2(f(x_n), f(x_0)) \geq \epsilon > 0$ for all n.

(3)⇒(1). Let $\epsilon > 0$ be given and $V = N(f(x_0), \epsilon)$. In view of (1) there exists a neighborhood $U = N(x_0, \delta)$ such that $f(U) \subseteq V$. Thus, whenever $\rho_1(x, x_0) < \delta$ it follows that $x \in U$, and hence $f(x) \in V$ and $\rho_2(f(x), f(x_0)) < \epsilon$. Therefore, f is continuous at x_0.

The following result is an analog of Theorem 12.1.

Theorem 30.2. Let (S, ρ_1) and (T, ρ_2) be metric spaces and $f : S \to T$. Then, f is continuous on S iff $f^{-1}(G)$ is an open set in S whenever G is an open set in T.

Example 30.2. In the discrete metric space R_d considered in Example 29.2 every function $f : S \to T$ is continuous on S. In fact, for any $x_0 \in S$ if we choose $\delta < 1$, then $N_\delta(x_0) = \{x_0\}$, and thus the condition $f(N_\delta(x_0)) \subseteq N_\epsilon(f(x_0))$ holds for each $\epsilon > 0$.

The following result is an analog of Theorem 12.2.

Theorem 30.3. Let (S, ρ_1) and (T, ρ_2) be metric spaces and $f, g : S \to T$. If f and g are continuous at $x_0 \in S$, then $f \pm g$, $f \cdot g$ are continuous at x_0, and f/g is continuous at x_0 provided $g(x_0) \neq 0$.

The following result is an analog of Theorem 12.3.

Theorem 30.4. Let (S, ρ_1), (T, ρ_2), and (U, ρ_3) be metric spaces and $f : S \to T$, $g : T \to U$. If f is continuous at $x_0 \in S$ and g is continuous at $f(x_0) \in T$, then $g \circ f(x) = g(f(x))$ is continuous at x_0.

We shall now prove the following result.

Theorem 30.5. Let (S, ρ_1) and (T, ρ_2) be metric spaces, let $f : S \to T$ be continuous on S, and let C be a compact subset of S. Then, $f(C)$ is a compact subset of T.

Proof. Let $\mathcal{O} = \{O_i\}$ be an open cover of $f(C)$. Since f is continuous on S, Theorem 30.2 implies that $f^{-1}(O_i)$ is open in S. Further, since $f(C) \subseteq \cup_i O_i$, it follows that $C \subseteq \cup_i \{f^{-1}(O_i)\}$. But, this means that $\cup_i \{f^{-1}(O_i)\}$ is an open cover for C. Since C is compact, there exists a finite number of sets, say, $O_{k_1}, \cdots, O_{k_n}$ in $\mathcal{O}$ such that $C \subseteq f^{-1}(O_{k_1}) \cup \cdots \cup f^{-1}(O_{k_n})$. Now, since for every $D \subseteq T$, $f(f^{-1}(D)) \subseteq D$, it follows that $f(C) \subseteq O_{k_1} \cup \cdots \cup O_{k_n}$.

Thus, $\{O_{k_1}, \cdots, O_{k_n}\}$ is a finite subcover of $\mathcal{O}$ for $f(C)$, and hence $f(C)$ is compact.

The following corollary of Theorem 30.5 extends Theorem 12.4.

Corollary 30.1. Let f be a continuous real-valued function defined on a metric space (S, ρ), and let C be a compact subset of S. Then, f attains a maximum and a minimum value on C.

Let (S, ρ_1) and (T, ρ_2) be metric spaces. A function $f : S \to T$ is called *uniformly continuous* if given any $\epsilon > 0$ there exists a $\delta > 0$ such that $\rho_2(f(x_1), f(x_2)) < \epsilon$ whenever $\rho_1(x_1, x_2) < \delta$.

Example 30.3. The function defined in Example 30.1 is uniformly continuous on S, because $|f(x) - f(y)| \leq \rho(x, y)$ for all $x, y \in S$.

The following result generalizes Theorem 14.1.

Theorem 30.6. Let (S, ρ_1) and (T, ρ_2) be metric spaces, let $f : S \to T$ be continuous on S, and let C be a compact subset of S. Then, f is uniformly continuous on C.

The following result extends Problem 14.8(i).

Theorem 30.7. Let (S, ρ_1) and (T, ρ_2) be metric spaces. If $f : S \to T$ is uniformly continuous and $\{x_n\}$ is a Cauchy sequence in S, then $\{f(x_n)\}$ is a Cauchy sequence in T.

Proof. Let $\epsilon > 0$ be given. Then, there exists a $\delta > 0$ such that $\rho_2(f(u), f(v)) < \epsilon$ whenever $\rho_1(u, v) < \delta$. Since $\{x_n\}$ is a Cauchy sequence in S, there exists an $N \in \mathcal{N}$ such that $\rho_1(x_m, x_n) < \delta$ for all $m, n > N$. Hence, $\rho_2(f(x_m), f(x_n)) < \epsilon$ for all $m, n > N$. This implies that $\{f(x_n)\}$ is a Cauchy sequence in T.

Our last result in this chapter is a well-known classical result about the existence and uniqueness of a fixed point (see Theorem 12.7) of a *contraction*, that is, a function f from a metric space (S, ρ) to itself satisfying

$$\rho(f(x), f(y)) \leq \theta\rho(x, y) \tag{30.1}$$

for all $x, y \in S$ and some θ with $0 \leq \theta < 1$. We shall also give an application of this result to the solution of an initial value problem.

Theorem 30.8 (Banach's Contraction Mapping Principle). Let (S, ρ) be a complete metric space and $f : S \to S$ be such that for some $k \in \mathcal{N}$, the function $f^k = f(f^{k-1})$ is a contraction. Then, f has a unique fixed point in S.

Proof. Consider any $x_0 \in S$ and define

$$x_n = f(x_{n-1}), \quad n = 1, 2, \cdots . \tag{30.2}$$

First we assume that f is itself a contraction. We claim that the sequence $\{x_n\}$ generated by (30.2) is a Cauchy sequence in S. Note that $x_n = f^n(x_0)$ for $n = 1, 2, \cdots$. If $1 \leq m < n$, then we have

$$\begin{aligned}
\rho(x_m, x_n) &= \rho(f^m(x_0), f^n(x_0)) \\
&\leq \theta^m \rho(x_0, f^{n-m}(x_0)) \\
&\leq \theta^m [\rho(x_0, f(x_0)) + \rho(f(x_0), f^2(x_0)) \\
&\quad + \cdots + \rho(f^{n-m-1}(x_0), f^{n-m}(x_0))] \\
&\leq \theta^m \rho(x_0, f(x_0))[1 + \theta + \cdots + \theta^{n-m-1}] \\
&\leq \frac{\theta^m}{1-\theta} \rho(x_0, f(x_0)).
\end{aligned}$$

Since $\theta < 1$, $\theta^m \to 0$ as $m \to \infty$. Hence, $\{x_m\}$ is a Cauchy sequence. As S is a complete metric space, let $x_n \to x^*$ in S. Then, because of the continuity of f, it follows that

$$f(x^*) = \lim_{n \to \infty} f(x_n) = \lim_{n \to \infty} x_{n+1} = x^*,$$

i.e., x^* is a fixed point of f.

To show the uniqueness of a fixed point of f, let $f(\hat{x}) = \hat{x}$ for some $\hat{x} \in S$. Then, we have

$$\rho(x^*, \hat{x}) = \rho(f(x^*), f(\hat{x})) \leq \theta \rho(x^*, \hat{x}).$$

Again, since $\theta < 1$, we find $\rho(x^*, \hat{x}) = 0$, i.e., $x^* = \hat{x}$.

Now assume that f^k is a contraction for some integer $k \geq 2$. By what we have just proved, the function $g = f^k$ has a unique fixed point x^* in S. Since $g(f(x^*)) = f(g(x^*)) = f(x^*)$, it follows that $f(x^*)$ is also a fixed point of g. The uniqueness of the fixed point of g shows that $f(x^*) = x^*$, i.e., x^* is a fixed point of f itself. Also, if $\hat{x}$ is a fixed point of f, then clearly it is a fixed point of g. Again, the uniqueness of the fixed point of g shows that $\hat{x} = x^*$. Thus, f has a unique fixed point in S.

Remark 30.1. An important aspect of the contraction mapping principle is that the desired unique fixed point is obtained by starting with an arbitrary point x_0 and employing a very simple iterative process. We also note that if x^* is the unique fixed point, then the successive approximations $x_n = f^n(x_0)$ satisfy

$$\rho(x^*, x_n) = \rho(f(x^*), f(x_{n-1})) \leq \theta \rho(x^*, x_{n-1}) \leq \cdots \leq \theta^n \rho(x^*, x_0).$$

Further, since

$$\begin{aligned} \rho(x^*, x_n) &\leq \rho(x^*, x_{n+1}) + \rho(x_{n+1}, x_n) \\ &= \rho(f(x^*), f(x_n)) + \rho(x_{n+1}, x_n) \\ &\leq \theta \rho(x^*, x_n) + \rho(x_{n+1}, x_n), \end{aligned}$$

it follows that $\rho(x^*, x_n) \leq \rho(x_{n+1}, x_n)/(1-\theta)$. Thus, if we require $\rho(x^*, x_n) < \epsilon$, then we only need to continue until $\rho(x_{n+1}, x_n) < \epsilon(1-\theta)$.

Example 30.4. We shall apply Theorem 30.8 to demonstrate the following result: Let $a > 0, b > 0, (t_0, y_0) \in \mathcal{R}^2$ and $E = [t_0 - a, t_0 + a] \times [y_0 - b, y_0 + b]$. Let $h : E \to \mathcal{R}$ be continuous and $|h(t, y)| \leq \alpha$ for all $(t, y) \in E$. Let $\delta = \min\{a, b/\alpha\}$. Then, there exists a unique continuously differentiable function y on $[t_0 - \delta, t_0 + \delta]$ such that for all $t \in [t_0 - \delta, t_0 + \delta]$,

$$y'(t) = h(t, y(t)) \quad \text{and} \quad y(t_0) = y_0 \tag{30.3}$$

provided h satisfies the Lipschitz condition in the second variable, i.e., $|h(t, y_1) - h(t, y_2)| \leq L|y_1 - y_2|$ for all (t, y_1) and (t, y_2) in E and some $L \geq 0$. For this, note that for $t \in [t_0 - \delta, t_0 + \delta]$, (30.3) in view of Theorem 20.6 is equivalent to the integral equation

$$y(t) = y_0 + \int_{t_0}^t h(s, y(s)) ds. \tag{30.4}$$

Consider $S = \{y \in C([t_0 - \delta, t_0 + \delta]) : |y(t) - y_0| \leq b \text{ for all } t \in [t_0 - \delta, t_0 + \delta]\}$ and define

$$f(y)(t) = y_0 + \int_{t_0}^t h(s, y(s)) ds, \quad y \in S, \quad t \in [t_0 - \delta, t_0 + \delta].$$

Since for all $t \in [t_0 - \delta, t_0 + \delta]$ and $y \in S$,

$$|f(y)(t) - y_0| \leq \int_{t_0}^t |h(s, y(s))| ds \leq \alpha|t - t_0| \leq \alpha\delta \leq b,$$

it follows that f maps S into S. We also note that S is a closed subset of $C([t_0 - \delta, t_0 + \delta])$ and hence it is complete in the metric $\rho(y, z) = \sup_{[t_0 - \delta, t_0 + \delta]} |y(t) - z(t)| = \|y - z\|_\infty$. Also, for $y, z \in S$ and $t \in [t_0 - \delta, t_0 + \delta]$, we have

$$\begin{aligned} |f(y)(t) - f(z)(t)| &= \left| \int_{t_0}^t [h(s, y(s)) - h(s, z(s))] ds \right| \\ &\leq \left| \int_{t_0}^t L|y(s) - z(s)| ds \right| \\ &\leq L|t - t_0| \|y - z\|_\infty, \end{aligned}$$

so that $\|f(y) - f(z)\|_\infty \leq L\delta \|y - z\|_\infty$. If $L\delta < 1$, then f is a contraction.

Otherwise, consider the map f^2. Now for all $y, z \in S$ and $t \in [t_0 - \delta, t_0 + \delta]$, we have

$$
\begin{aligned}
|f^2(y)(t) - f^2(z)(t)| &= \left| \int_{t_0}^t [h(s, f(y)(s)) - h(s, f(z)(s))]ds \right| \\
&\leq \left| \int_{t_0}^t L|f(y)(s) - f(z)(s)|ds \right| \\
&\leq L \left| \int_{t_0}^t |s - t_0| \|y - z\|_\infty ds \right| \\
&= L \frac{|t - t_0|^2}{2} \|y - z\|_\infty,
\end{aligned}
$$

so that $\|f^2(y) - f^2(z)\|_\infty \leq (L\delta)^2/2 \|y - z\|_\infty$. Now, by induction it follows that

$$
\|f^n(y) - f^n(z)\|_\infty \leq \frac{(L\delta)^n}{n!} \|y - z\|_\infty, \quad n = 1, 2, \cdots
$$

for all $y, z \in S$. Since $(L\delta)^n/n! \to 0$ as $n \to \infty$, we can choose k sufficiently large such that $(L\delta)^k/k! < 1$, so that f^k is a contraction. Thus, in view of Theorem 30.8, f has a unique fixed point y is S, i.e., there exists a unique $y \in C([t_0 - \delta, t_0 + \delta])$ such that

$$
y(t) = f(y)(t) = y_0 + \int_{t_0}^t h(s, y(s))ds
$$

for all $t \in [t_0 - \delta, t_0 + \delta]$. This in turn implies that there is a unique continuously differentiable function y on $[t_0 - \delta, t_0 + \delta]$ such that $y'(t) = h(t, y(t))$ for all $t \in [t_0 - \delta, t_0 + \delta]$ and $y(t_0) = y_0$.

Problems

30.1. For the set of all ordered n-tuples $x = (a_1, \cdots, a_n)$ and $y = (b_1, \cdots, b_n)$, show that

(i). $\rho_1(x, y) = \sum_{k=1}^n |a_k - b_k|$ is a metric.

(ii). $\rho_2(x, y) = \max_{1 \leq k \leq n} |a_k - b_k|$ is a metric.

(iii). $\rho_2(x, y) \leq \rho(x, y) \leq \sqrt{n}\rho_2(x, y)$, where ρ is the Euclidean metric (see Example 29.3).

30.2. ℓ^∞ denotes the set of bounded sequences of real numbers. If $x = \{a_n\}_{n=1}^\infty$ and $y = \{b_n\}_{n=1}^\infty$ are points in ℓ^∞, then show that $\rho(x, y) = \sup_n |a_n - b_n|$ is a metric on ℓ^∞.

30.3. Let (S, ρ) be a metric space and let $\rho_1 : S \times S \to \mathcal{R}$ be defined by $\rho_1(x, y) = \rho(x, y)/[1 + \rho(x, y)]$. Show that (S, ρ_1) is also a metric space.

30.4. Let S be the set of all Riemann-integrable functions defined on $[0, 1]$ and let for all $f, g \in S$, ρ be defined by $\rho(x, y) = \int_0^1 |f(x) - g(x)|dx$. Show that (S, ρ) is not a metric space.

30.5. Let (S, ρ) be a metric space and let $T, U \subseteq S$. Show that

(i). $(T \cup U)^i \supseteq T^i \cup U^i$ and $(T \cap U)^i = T^i \cap U^i$.

(ii). $(T \cup U)' \subseteq T' \cup U'$.

(iii). $\overline{T \cup U} = \overline{T} \cup \overline{U}$ and $\overline{T \cap U} \subseteq \overline{T} \cap \overline{U}$.

30.6. Let T be a nonempty set in a metric space (S, ρ). The *diameter* of T is defined by $d(T) = \sup_{x,y \in T} \rho(x, y)$. Clearly, a nonempty set T is bounded iff $d(T) < \infty$, and a set has the diameter 0 iff the set is a singleton. Show that $d(\overline{T}) = d(T)$.

30.7. Let T and U be nonempty sets in a metric space (S, ρ). The *distance* between these sets is defined by $d(T, U) = \inf_{x \in T, y \in U} \rho(x, y)$. If $T = \{a\}$, then $d(a, U) = \inf_{y \in U} \rho(a, y)$ is the distance from the point $a \in S$ to the set U. Show that

(i). y_0 is a limit point of U iff $d(y_0, U - \{y_0\}) = 0$.

(ii). $\overline{U} = \{x \in S : d(x, U) = 0\}$.

30.8. Let (S, ρ) be a complete metric space and $\emptyset \neq T \subseteq S$. Show that the subspace (T, ρ) is complete iff T is closed in the space (S, ρ).

30.9. Show that the space $(C[0,1], \rho_1)$ considered in Example 29.4 (with $a = 0$ and $b = 1$) is not complete.

30.10 (Cantor Intersection Theorem). Let (S, ρ) be a complete metric space. Let $\{H_n\}$ be a decreasing sequence of nonempty closed subsets of S such that diameter of H_n tends to 0 as $n \to \infty$. Show that $\cap_n H_n$ contains exactly one point.

30.11. Prove Theorem 30.2.

30.12. Let (S, ρ_1) and (T, ρ_2) be metric spaces and $f : S \to T$. Then, f is continuous on S iff $f^{-1}(G)$ is a closed set in S whenever G is a closed set in T.

30.13. Let (S, ρ_1) and (T, ρ_2) be metric spaces and let $f, g : S \to T$ both be continuous on S. Show that if $C \subseteq S$ and $f(x) = g(x)$ for all $x \in C$, then $f(x) = g(x)$ for all $x \in \overline{C}$.

30.14. Let (S, ρ_1), (T, ρ_2), and (U, ρ_3) be metric spaces and $f : S \to T$, $g : T \to U$. Show that if f is uniformly continuous on S and g is uniformly continuous on T, then $g \circ f$ is uniformly continuous on S.

30.15. The family $\mathcal{F}$ of functions from the metric space (S, ρ_1) to the metric space (T, ρ_2) is called *equicontinuous* on S if given $\epsilon > 0$ there is a $\delta > 0$ such that for every $f \in \mathcal{F}$, $\rho_2(f(x_1), f(x_2)) < \epsilon$ whenever $\rho_1(x_1, x_2) < \delta$. Prove that if (S, ρ_1) is a compact metric space and the sequence $f_n : S \to T$

is equicontinuous on S and $f(x) = \lim_{n\to\infty} f_n(x)$ for each $x \in S$ then the sequence $\{f_n\}$ converges uniformly to f on S.

30.16 (Ascoli's Theorem). Show that a closed subspace S of $C[0,1]$ is compact iff S is equicontinuous and uniformly bounded.

Answers or Hints

30.1. (i). Verify the definition of metric directly.
(ii). Verify the definition of metric directly.
(iii). Note that $\max_{1\le k\le n} |a_k - b_k| \le \left(\sum_{k=1}^{n}(a_k - b_k)^2\right)^{1/2}$, and for all $1 \le k \le n$, $|a_k - b_k| \le \max_{1\le k\le n} |a_k - b_k|$.

30.2. Clearly, $\rho(x,y) \ge 0$ for all $x, y \in \ell^\infty$. If $\rho(x,y) = 0$ then $\sup_n |a_n - b_n| = 0$ and this implies that for every $n \in \mathcal{N}$, $|a_n - b_n| = 0$, and thus $x = y$. Conversely, if $x = y$ then $|a_n - b_n| = 0$ for all n, and this implies that $\rho(x,y) = 0$. Clearly, $\rho(x,y) = \sup_n |a_n - b_n| = \sup_n |b_n - a_n| = \rho(y,x)$. Now let $z = \{c_n\}_{n=1}^\infty \in \ell^\infty$, then we have $\rho(x,y) = \sup_n |a_n - b_n| \le \sup_n(|a_n - c_n| + |c_n - b_n|) \le \sup_n |a_n - c_n| + \sup_n |c_n - b_n| = \rho(x,z) + \rho(z,y)$.

30.3. Note that $f(x) = x/(1+x)$ is an increasing function.

30.4. From Problem 22.4 it follows that there exists functions f, g which are Riemann integrables on $[0,1]$ with $f \ne g$ and $\rho(f,g) = 0$.

30.5. (i). Let $x \in T^i \cup U^i$. If $x \in T^i$, then there is an open ball $I(x)$ such that $x \in I(x) \subset T \subset T \cup U$, thus $x \in (T \cup U)^i$. Conversely, it is not true. For this, let $T = [2,3], U = (3,4)$. Then, we have $T \cup U = [2,4)$ and $(T \cup U)^i = (2,4)$, and $T^i \cup U^i = (2,3) \cup (3,4) \subset (2,4)$.
If $x \in T^i \cap U^i$, then $x \in T^i$ and $x \in U^i$. Thus, there exists open balls $I(x)$ and $I^*(x)$ such that $x \in I(x) \subset T$ and $x \in I^*(x) \subset U$. Clearly, $I(x) \cap I^*(x) = J(x)$ is an open ball such that $J(x) \subset T$ and $J(x) \subset U$, thus $J(x) \subset T \cap U$ and hence $x \in (T \cap U)'$. Conversely, if $x \in (T \cap U)'$, there is an open ball $I(x)$ such that $x \in I(x) \subset T \cap U$. This means that $x \in I(x) \subset T$ and $x \in I(x) \subset U$, and hence $x \in T^i$ and $x \in U^i$.
(ii). Let $x \in (T \cup U)'$. Then, for every nbd N of x, $N \cap (T \cup U) = (N \cap T) \cup (N \cap U)$ is an infinite set. Thus, at least one of the sets $N \cap T$ or $N \cap U$ is infinite. Assuming that $N \cap T$ is infinite, then we have $x \in T'$.
(iii). Let $x \in \overline{T \cup U}$. Then, $x \in (T \cup U) \cup (T \cup U)' \subseteq T \cup U \cup T' \cup U'$, thus x is at least in one of the sets T, U, T', U'. Therefore, $x \in \overline{T}$ or $x \in \overline{U}$. Conversely, if $x \in \overline{U} = U \cup U'$, we have two cases: (a) $x \in U$, then $x \in T \cup U \subseteq \overline{T \cup U}$. (b) $x \in U'$, then for for every nbd N of x, $N \cap U$ is an infinite set and from this it follows that $N \cap (U \cup T)$ is an infinite set. Hence, $x \in \overline{T \cup U}$.
Let $x \in \overline{T \cap U} = (T \cap U) \cup (T \cap U)'$. If $x \in T$ or $x \in U$, the result is clear. If $x \in (T \cap U)'$, then for every nbd N of x, $N \cap (T \cap U)$ is an infinite set, thus

$N \cap T$ and $N \cap U$ are infinite sets. Hence, it follows that $x \in \overline{T} \cap \overline{U}$. For the converse, let $T = (0, 1), U = (1, 2)$. We have $T \cap U = \emptyset$ and $\overline{T} \cap \overline{U} = \{1\}$.

30.6. Since $T \subseteq \overline{T}$, it is clear that $d(T) \leq d(\overline{T})$. For the converse, let $\epsilon > 0$ and $x, y \in \overline{T}$. From the definition there are $x', y' \in T$ such that $d(x, x') < \epsilon$ and $d(y, y') < \epsilon$. Thus, it follows that $d(x, y) \leq d(x, x') + d(x', y') + d(y', y) < 2\epsilon + d(T)$. Therefore, from the definition of the supremum, we have $d(\overline{T}) \leq 2\epsilon + d(T)$, which implies that $d(\overline{T}) \leq d(T)$.

30.7. (i). If y_0 is a limit point of U, then given any $\epsilon > 0$ there is a point y of $U - \{y_0\}$ in the open ball of radius ϵ centered at y_0. Since $d(y_0, y) < \epsilon$, $d(y_0, U - \{y_0\}) < \epsilon$. But, since $\epsilon > 0$ is arbitrary, $d(y_0, U - \{y_0\}) = 0$. Conversely, if $d(y_0, U - \{y_0\}) = 0$ then every open ball $N_r(y_0)$ contains a point of $U - \{y_0\}$, for otherwise $d(y_0, y) \geq r$ for every $y \in U - \{y_0\}$ and so $d(y_0, U - \{y_0\}) \geq r$.

(ii). Let $y_0 \in \overline{U}$. If $y_0 \in U$, then clearly $d(y_0, U) = 0$. Let $y_0 \in U' \backslash U$ and let $\epsilon > 0$ be given. Then there is a point $y \in U \cap N_\epsilon(y_0)$. Since $d(y_0, y) < \epsilon$, it follows that $d(y_0, U) < \epsilon$. Since $\epsilon > 0$ is arbitrary $d(y_0, U) = 0$. Conversely, let $d(y_0, U) = 0$. If $y_0 \in U$ then $y_0 \in \overline{U}$ and we are done. Suppose $y_0 \notin U$ and consider any open ball $N_r(y_0)$. Since $d(y_0, U) = 0$, there is a point $y \in U$ with $d(y_0, y) < r$. Thus, $y \in N_r(y_0)$ and $y \neq y_0$ since $y_0 \notin U$. Thus, y_0 is a limit point of U and therefore $y_0 \in \overline{U}$.

30.8. Let (T, ρ) be complete and x_0 be a limit point of T. Then, for each $n \in \mathcal{N}$ there exists a point $x_n \in T \cap N_{1/n}(x_0)$. From Theorem 29.4(4) it is clear that $\{x_n\}$ is a Cauchy sequence and $\lim_{n \to \infty} x_n = x_0$. Now, the completeness of (T, ρ) implies that $x_0 \in T$, and hence T is closed. Conversely, assume that T is closed and $\{x_n\}$ is a Cauchy sequence in (T, ρ). Then, $\{x_n\}$ is also a Cauchy sequence in (S, ρ). Thus, there exists an $x_0 \in S$ such that $\lim_{n \to \infty} x_n = x_0$. Now since each $x_n \in T$, $x_0 \in \overline{T}$. But, T is closed and hence $x_0 \in T$. This implies that the Cauchy sequence $\{x_n\}$ converges in (T, ρ).

30.9. We claim that the sequence of continuous functions defined by

$$f_1(x) = 1 \text{ and for } n \geq 2, \quad f_n(x) = \begin{cases} 0, & 0 \leq x \leq 1/2 - 1/n \\ nx + 1 - n/2, & 1/2 - 1/n < x < 1/2 \\ 1, & 1/2 \leq x \leq 1 \end{cases}$$

is a Cauchy sequence. For this, let $\epsilon > 0$ be given. Then, we have $\rho_1(f_n, f_m) \leq \int_0^1 (|f_n(x)| + |f_m(x)|) dx = 1/(2n) + 1/(2m) < \epsilon$ for all $n > 1/\epsilon$ and $m > 1/\epsilon$. Now assume that this sequence converges to a continuous function f, then $\int_0^{1/2 - 1/n} |0 - f(x)| dx + \int_{1/2 - 1/n}^{1/2} |f(x) - nx - 1 + n/2| dx + \int_{1/2}^1 |1 - x| dx$ must go to zero as $n \to \infty$. Since each integrand is positive, this implies that

$$f(x) = \begin{cases} 0, & 0 \leq x < 1/2 \\ 1, & 1/2 \leq x \leq 1, \end{cases} \text{ which is not continuous.}$$

30.10. Let $H = \cap_n H_n$. If H contains two points x and y then we have a contradiction, namely, $d(H) = \sup_{x,y \in H} \rho(x, y) > 0$. Now for every n let

$x_n \in H_n$. Since $d(H_n) \to 0$ as $n \to \infty$, $\{x_n\}$ is a Cauchy sequence. Since (S, ρ) is complete, there exists $x_0 \in S$ such that $\{x_n\}$ converges to x_0. To complete the proof we will show that $x_0 \in H$. Let $\hat{n} \in \mathcal{N}$ be arbitrary. Then, $x_k \in H_{\hat{n}}$ for all $k \geq \hat{n}$. Thus, it follows that $x_0 \in \overline{H}_{\hat{n}}$. Now since $H_{\hat{n}}$ is closed, $x_0 \in H_{\hat{n}}$. Since $\hat{n}$ is arbitrary it is clear that $x_0 \in H$.

30.11. Suppose that $f : S \to T$ is continuous on S and let G be an open set in T. If $f^{-1}(G) = \emptyset$, it is open. We assume that $x_0 \in f^{-1}(G)$, then $f(x_0) \in G$, and so there exists an $\epsilon > 0$ such that $N_\epsilon(f(x_0)) \subseteq G$. Since f is continuous at x_0, there exists a $\delta > 0$ such that $f(N_\delta(x_0)) \subseteq N_\epsilon(f(x_0))$. Thus, $f(N_\delta(x_0)) \subseteq G$, and hence $N_\delta(x_0) \subseteq f^{-1}(G)$. Therefore, $f^{-1}(G)$ is open. Conversely, suppose that $f^{-1}(G)$ is open in S for every open set G in T. Let $x_0 \in S$ and consider $N_\epsilon(f(x_0))$. Then, $f^{-1}(N_\epsilon(f(x_0))$ is open and contains x_0. Hence, there exists a $\delta > 0$ such that $N_\delta(x_0) \subseteq f^{-1}(N_\epsilon(f(x_0))$. This implies that $f(N_\delta(x_0)) \subseteq N_\epsilon(f(x_0))$, and hence f is continuous at x_0. Since $x_0 \in S$ is arbitrary, f is continuous on S.

30.12. From Theorem 2.2(5), we have $f^{-1}(T \backslash G) = S \backslash f^{-1}(G)$. We also know that a set $G \subset T$ is closed iff $T \backslash G$ is an open set. The result now follows from Theorem 30.2.

30.13. Recall that x_0 is a point of closure of C if for every $r > 0$, there is a y in C such that $\rho_1(x_0, y) < r$. Let $h(x_0) = f(x_0) - g(x_0)$, which is a continuous function. If $h(x_0) \neq 0$, we have $\rho_2(h(x_0), h(y)) = \rho_2(h(x_0), 0) > 0$. Let $\epsilon < \rho_2(h(x_0), 0)$, then if $\rho_1(x_0, y) < r$, we have $\rho_2(h(x_0), h(y)) > \epsilon$, but this contradicts the continuity of h, and hence $h(x_0) = 0$.

30.14. See Problem 14.4.

30.15. See Problem 25.8.
30.16. See Problem 25.8.

Bibliography

[1]. T.M. Apostol, *Mathematical Analysis*, Addison Wesley, Massachusetts, 1974.

[2]. R.G. Bartle and D.R. Sherbert, *Introduction to Real Analysis*, second edition, John Wiley & Sons, New York, 1992.

[3]. R.P. Boas, Jr., *Primer of Real Functions*, John Wiley & Sons, New York, 1960.

[4]. A.M. Bruckner, J.B. Bruckner, and B.S. Thomson, *Real Analysis*, Prentice Hall, New Jersey, 1997.

[5]. E. Hewitt and K. Stromberg, *Real and Abstract Analysis*, Springer-Verlag, Berlin, Heidelberg, 1955.

[6]. A.N. Kolmogorov and S.V. Fomin, *Introductory Real Analysis*, Dover, New York, 1970.

[7]. H.L. Royden, *Real Analysis*, Macmillan, New York, 1968.

[8]. W. Rudin, *Real and Complex Analysis*, McGraw-Hill, New York, 1966.

[9]. W. Rudin, *Principles of Mathematical Analysis*, third edition, McGraw-Hill, New York, 1990.

[10]. I. Stewart and D. Tall, *The Foundations of Mathematics*, Oxford University Press, New York, 1977.

Index

A

Abel's formula   87
Abel's test   231
Abel's theorem   239
absolutely convergence for a series   84
absolutely integrable function   190
absolute-value function   46
accumulation point   30
additive function   106
algebraic number   39
almost everywhere continuous   198
Alternative Series Test   83
analytic function   243
antiderivative   181
Archimedean property   23
arithmetic-geometric mean inequality   7, 156
Ascoli's theorem   264
axiom   3

B

Banach's contraction mapping principle   259
Bernoulli's inequality   7
Bernstein's criterion   246
bijective function   14
Bolzano-Weierstrass theorem   31
Bonnet's mean-value theorem   185
boundary of a set   29
boundary point   29
bounded function   47
bounded set   21
bounded variation   121

C

Cantor intersection theorem	263
Cantor set	42
cardinal of a set	37
Cartesian product	13
Cauchy function	244
Cauchy sequence	57
Cauchy's condensation criterion	75
Cauchy's convergence criterion	57, 74
Cauchy's criterion for uniform convergence	214, 230
Cauchy's form of the remainders	142
Cauchy's inequality	147
Cauchy's mean-value theorem	133
Cauchy's Principal Value (C.P.V.)	193
Cauchy's rule	163
Cesaro-Stolz's theorem	66
Chain rule	131
chain rule for integrals	182
change of variables	183
characteristic function	46
closed set	30
closure of a set	30, 253
compact set	32
comparison test	76
complement of a set	12
complete metric space	254
complete ordered field	21
composite function	14
conditionally integrable	190
conjecture	6
constant function	45
continuous function	99, 101, 257
continuum hypothesis	41
contractive sequence	60
convergent sequence	53
convex downward	151
convex upward	151
corollary	3
countable set	37
counterexample	6
cross product	13

D

Darboux integrals 169
Darboux sums 169
Darboux theorem 133
Dedekind's property 23
deleted nbd 29
dense set 24, 253
derivative of a function 129
derived set 30, 253
differentiable function 129
Dini's theorem 215, 230
directed straight line 24
Dirichlet's function 95
Dirichlet's test 232
discontinuity of the first kind 111
discontinuity of the second kind 112
discontinuous function 99, 111
distance between any two real numbers 46
divergent sequence 53
domain 14

E

empty set 12
equicontinuous sequence 223
equivalence relation 14
equivalent sets 37
Euclidean metric 251
even function 48
extension of a function 14
exterior of a set 29
exterior point 29

F

family of sets 13
Fibonacci sequence 64

field   20
Fixed Point Property   103
Froda's theorem   113
function between A and B   14
function piecewise smooth   136
function Riemann-Stieltjes integrable   203
function bounded above   47
function bounded below   47
fundamental theorem of calculus   180

     G

Gauss test   241
geometric series   73
golden ratio   64
graph of a function   14
greatest-integer function   47

     H

harmonic series   74
Heine-Borel theorem   32, 69
Hölder's inequality   147
hypergeometric series   248

     I

identity function   45
image of a set   15
improper integral   189
increasing sequence   63
infima   32
infimum   21
infinite discontinuity   112
infinite series   73
injective function   14

integration by parts 183
interior of a set 29, 253
interior point 29
intermediate-value property 116
intermediate-value theorem 101
intersection of two sets 12
interval of convergence 238
intervals 12
inverse image of a set 15

J

Jensen's inequality 156
Jordan decomposition theorem 123

K

Kummer's test 78

L

L'Hôpital's rule 159
Lagrange's form of remainders 142
Laws of reflection and refraction of light 145
Lebesgue measure 197
left-hand limit 92
left-continuous 99
Leibniz's rule 146
limit comparison test 76
limit of a sequence 53
limit point 30
Lipschitz continuous 105
logarithmic ratio test 79
lower bound 21, 54
lower limit 32
lower Riemann sum 169

lower Riemann-Stieltjes sum   203
lower semicontinuous   106

M

Maclaurin's series of a function   243
Maclaurin's theorem   142
mathematical induction   5
maximal element of a set   21
mean-value theorem   132
mean-value theorem of integral calculus   182
measure zero   197
metric on a nonempty set   251
metric space   251
minimal element of a set   21
Minkowski's inequality   147
mixed discontinuity   112
monotone decreasing function   48
monotone function   48
monotone increasing function   48
monotone sequence   63
multiplicative function   106

N

negative part of a number   46
neighborhood   29
nested interval property   65
Newton-Leibniz theorem   181
nowhere dense set   24

O

odd function   48
open cover of a set   32
open set   29

ordered field   20
ordered n-tuple   13
ordered pairs   13
oscillate series   73
oscillation of a function in a point   116, 198
oscillation of a function on a nonempty set   198

P

partial sums   73
partition   121
period of a function   48
periodic function   48
point of inflection   151
pointwise limit of a sequence   211
positive part of a number   46
postulate   3
power of continuum   40
power set   41
proper maximum   155
proper subset   11
p-series   75

R

Raabe-Duhamel's test   79
radius of convergence   238
range   14
ratio test   59, 76
real sequence   53
real-valued function   45
relative (local) extremum   135
relative (local) maximum   135
relative (local) minimum   135
restriction of a function   14
Riemann integrable   167
Riemann integral   167
Riemann's integral test for sequences   191
Riemann sum   167

Riemann-Lebesgue theorem   199
Riemann-Stieltjes integral   210
right-hand limit   91
right-continuous   99
Rolle's theorem   131
Root test   77

    S

Schlömilch-Roche's form of the remainder   142
semicontinuous in a point   106
separable metric space   253
sequence bounded above   54
sequence bounded below   54
series Abel's summable   240
set   11
set of the first/second category   116
Sign Preserving Property   101
Snell's law   146
Squeeze theorem   55
strictly increasing/decreasing function   48
strictly increasing/decreasing sequence   63
strictly monotone function   48
subadditive function   106
subcover of a set   32
subset   11
superadditive function   106
suprema   32
supremum   21
surjective function   14

    T

Taylor's polynomial of degree n for a function in a point   143
Taylor's series of a function   243
Taylor's theorem   141
term-by-term differentiation   230
term-by-term integration   230
theorem   3

transcendental number   39
transfinite number   37

U

unbounded function   47
unbounded sequence   54
uncountable set   37
uniformly bounded   217
uniformly continuous   119
union of two sets   12
universal set   12
upper bound   21, 54
upper limit   32
upper Riemann sum   169
upper Riemann-Stieltjes sum   203
upper semicontinuous   106

W

Weierstrass M-test   230
Weierstrass's mean-value theorem   185

Z

zero of multiplicity n   148